SURFACTANT SCIENCE AND TECHNOLOGY

SURFACTANT SCIENCE AND TECHNOLOGY

THIRD EDITION

Drew Myers

WILEY-INTERSCIENCE
A JOHN WILEY & SONS, INC., PUBLICATION

Published by John Wiley & Sons, Inc., Hoboken, New Jersey.
Published simultaneously in Canada.

***Library of Congress Cataloging-in-Publication Data*:**

Myers, Drew, 1946-
Surfactant science and technology/Drew Myers. – 3rd ed.
p. cm.
Includes bibliographical references and index.
ISBN-13 978-0-471-68024-6 (cloth)
ISBN-10 0-471-68024-9 (cloth)
1. Surface chemistry. 2. Surface active agents. I. Title.
QD506.M94 2006
541′.33–dc22 2005007004

Printed in the United States of America

10 9 8 7 6 5 4 3 2 1

To friends gone, but not forgotten—

Johnny B.
Paul G.
Alan B.

Contents

Preface to the Third Edition

When a book reaches the third edition, it must be assumed that (1) the work has been useful to someone or (2) the publisher has lost its collective mind. As a simple matter of ego, I must assume that reason 1 is true in this case. For that reason, I have tried to maintain the same basic philosophy with regard to the style and content of the book, while endeavoring to incorporate new material where indicated. A good deal of the information presented is "old" in the sense that it represents work done many years ago by the virtual founders of the science of surface and colloid chemistry. In the mid-1950s a few names stood out as the "gurus" of the field—today the names are too numerous to mention, and the body of published literature is enormous. Surfactants and their applications continue to fill books and patents.

Important advances in the tools available for studying the activity of surfactants has significantly increased our understanding of what is happening at interfaces at the molecular level in both model and practical systems, although there is still a lot be learned. New knowledge obtained in the years since the publication of the second edition has added greatly to our understanding of the nature of the molecular interactions of surface-active materials and the consequences of their presence on system characteristics and performance. The basic concepts and principles, however, remain pretty much the same.

In this edition, some topics have been reduced or moved around and several new themes added. Two cases, those of phase transfer catalysis (PTC) and aerosols, are not directly related to surfactants, but their real or potential importance prompted me to include some introductory material related to them.

Without changing the fundamental philosophy and goals of the previous editions, this third edition was prepared with three major ideas in mind: (1) to maintain the basic content of the work, (2) to maintain the "readability" of the book for non-specialists, and (3) to improve the book's utility as a source of basic concepts concerning surfactants and their applications. A limited number of problems are provided at the end of each chapter (except Chapter 1) to illustrate some of the concepts discussed. In some cases, the problems provided may not have a unique solution, but are posed to stimulate imaginative solutions on the part of the reader. Some may also require some searching on the part of the problem solver to find missing pieces. While exact literature references are not provided, the Bibliography at the end of the book includes many of the better resources for more detailed information on each specific subject. It should serve as a useful guide to more detailed coverage for the interested reader.

I would like to thank my two "best friends," Adriana and Katrina, for their constant love and support, and the crew at ALPHA C.I.S.A.—Lucho, José, Guillermo, Lisandro, Gabriel, Soledad, Alberto, Carlos, Enrique, Rudi, and all the rest—for putting up with my presence and my absence. Gracias por haber soportado mi presencia y mi ausencia.

DREW MYERS

1 An Overview of Surfactant Science and Technology

Rapid evolution in the chemical-based nature of our modern society has made it increasingly difficult for scientists, engineers, regulators, and managers to remain abreast of the latest in the technologies impacting their work. The scientific and technical journals published worldwide number in the thousands, and this number increases yearly. Paralleling the proliferation of the scientific literature in general has been an apparent divergence into fields of "pure" science—studies in which the principal goal is a general advancement of human knowledge with no particular "practical" aim in mind—and "applied" science and technology, in which the research is driven by some anticipated application, quite often, but not always, profit-related. Few areas of chemistry have exhibited this growing dichotomy of purpose more than the study of surface and colloid science, especially as applied to surface activity and surface-active materials. Even the nomenclature used in discussing materials showing surface activity is widely varied, depending on the context of the discussion. It is not surprising, then, that the world of surface activity and surface-active agents, or surfactants, can appear complex and confusing to those not intimately involved in it on a day-to-day basis.

When one considers the impact of surface science in general, and emulsions, dispersions, foaming agents, wetting agents, and other related compounds in particular, in our day-to-day routines, the picture that develops reveals the great extent to which these areas of chemistry and chemical technology permeate our lives. From the fundamental aspects of biological membrane formation and function in living cells, which vividly illustrates the spontaneity and importance of colloidal phenomena, to the more "far out" problem of how liquids wet the walls of a rocket's fuel tank in a low-gravity environment, the physical chemistry of the interactions among various phases at interfaces lies at the root of much of our modern lifestyle.

Industrial concerns, whose very lifeblood may be intimately linked to application of the basic principles of interfacial interactions, often ignore the potential benefits of fundamental research in these areas in favor of an empirical trial-and-error approach, which may lead to a viable process but that possibly could be better understood and even significantly improved by the application of more fundamental science. In many cases the prevailing philosophy seems to be, to paraphrase an old adage, "A dollar in the hand is worth two in the laboratory." Unfortunately, such an approach often results in more dollars down the drain than many management-level decisionmakers care to admit. Academic researchers, on the other hand, are sometimes guilty of ignoring the potential practical aspects of their work in favor of experimental sophistication and the "Holy Grail" of the definitive theory or model. Neither philosophy alone truly satisfies the needs of our technological existence. Each approach makes its valuable contribution to the overall advancement of human knowledge; however, it sometimes appears that a great deal is lost in the communication gap between the two.

The science and the technology of surfactants have possibly suffered a double blow from the functional divergence of academic and applied research. Academic interest in surfactants, while increasing, has generally concentrated on highly purified, homogeneous materials [quite often limited to a few materials such as sodium dodecylsulfate (SDS), or cetyltrimethylammonium bromide (CTAB)] and elegant analytical techniques. While providing a wealth of useful information related to the particular system under investigation, the application of such information to more complex practical materials and processes is often less than obvious, and is sometimes misleading. The sad fact of life is that real surfactant systems are almost always composed of mixed chemical isomers, contaminants, and added materials that can dramatically alter the effects of a given surfactant on a system. High purity is necessary for the interpretation of delicate laboratory experiments, but requires the use of techniques that may be impractical at the industrial level.

In the results-oriented industrial environment, with some significant exceptions, surfactant research is often carried out on a "Make it work and don't worry about why!" basis. The industrially interesting materials are usually complex mixtures of homologs and structural isomers, or contain impurities resulting from chemical side reactions, unreacted starting materials, residual solvents or byproducts, and so on. Such "contamination" of the desired product is not only common, but commonly variable from batch to batch. For example, particularly significant surface property changes can be induced by the presence of such impurities as inorganic salts or long-chain alcohols remaining after processing. While the presence of such impurities and mixtures will often produce superior results in practice, analysis of the process may be difficult because of the unknown or variable nature of the surfactant composition. Considering the limitations imposed by each school of surfactant research, it is not surprising to find that a practical fusion of the two approaches can be difficult to achieve.

The different views of surfactant science and technology have spawned their own distinctive terminologies and literatures. While the academic or fundamental investigator may probe the properties of surface-active agents, surfactants, tensides, or amphiphiles, the industrial chemist may be concerned with the characteristics of soaps, detergents, emulsifiers, wetting agents, and similar compounds. The former group may publish their results primarily in the *Journal of Physical Chemistry*, *Colloids and Surfaces*, *Langmuir*, or the *Journal of Colloid and Interface Science*, the latter in the *Journal of the American Oil Chemists Society*, the *Journal of Dispersion Science and Technology*, or one of the other technologically specialized publications aimed at specific areas of application (foods, cosmetics, paints, etc.). All too often, the value of the results to each community can become lost in the sea of manuscripts and the philosophical and operational gulf that sometimes develops between the two, not to mention the almost impossible task of being abreast of all the information published in all the relevant literature.

Before beginning a discussion of specific aspects of the chemistry of surface-active materials and surfactant action, it may be useful to have some idea of the history of surfactants and how their synthesis and use have evolved through the years. Because of parallel developments in various areas of the world, the secrecy of industrial research efforts, and the effects of two world wars, the exact details of the evolution of surfactant science and technology may be subject to some controversy regarding the specific order and timing of specific developments. In any case, the major facts are (hopefully!) correct.

1.1. A BRIEF HISTORY OF SURFACTANT SCIENCE AND TECHNOLOGY

The pedigree of the synthetic surfactant industry is reasonably well documented, unlike that of the more ancient "natural" alkali soaps. However, it is not an easy task to pinpoint the exact time when the industry came into being. In a strictly chemical sense, a soap is a compound formed by the reaction of an essentially water-insoluble fatty acid with an alkali metal or organic base to produce a carboxylic acid salt with enhanced water solubility, sufficient to produce useful surface activity. Since the soaps require some form of chemical modification to be useful as surfactants, they could be considered to be synthetic; however, custom dictates that they not be classified in the same category as the materials prepared by more "elegant" synthetic routes.

The alkali metal soaps have been used for at least 2300 years. Their use as articles of trade by the Phoenicians as early as 600 B.C. has been documented. They were also used by the Romans, although it is generally felt that their manufacture was learned from the Celts or some Mediterranean culture. Early soap producers used animal fats and ashes of wood and other plants containing potassium carbonate to produce the neutralized salt. As the mixture of fat, ashes, and water was boiled, the fat was saponified to the free fatty acids, which were subsequently neutralized.

The first well-documented synthetic (nonsoap) materials employed specifically for their surface-active properties were the sulfated oils. Sulfonated castor oil, produced by the action of sulfuric acid on the castor oil, was originally known as "turkey red oil." It was introduced in the late nineteenth century as a dyeing aid and is still used in the textile and leather industries today. The first surfactants for general application that have been traditionally classified as synthetic were developed in Germany during World War I in an attempt to overcome shortages of available animal and vegetable fats. Those materials were short-chain alkyl-naphthalene sulfonates prepared by the reaction of propyl or butyl alcohol with naphthalene followed by sulfonation. The products, which proved to be only marginally useful as detergents, showed good wetting characteristics and are still in use as such. They are still sold under various trade names in Europe and the United States.

In the late 1920s and early 1930s, the sulfation of long-chain alcohols became common and the resulting products were sold as the sodium salt. Also in the early 1930s, long-chain alkylaryl sulfonates with benzene as the aromatic group appeared in the United States. Both the alcohol sulfates and the alkylbenzene sulfonates were used as cleaning agents at that time, but they made little impact on the general surfactant or detergent markets. By the end of World War II alkylaryl sulfonates had almost entirely overwhelmed the alcohol sulfates for use as general cleaning agents, but the alcohol sulfates were beginning to emerge as preferred components in shampoos and other personal care formulations.

In common with other chemical developments during that time, progress in the area of surfactants and detergents was not limited to one family of materials. The explosion of new organic chemical processes and the ready availability of new raw materials led to the development of a wide variety of new surface-active compounds and manufacturing processes. In a particular country, the limiting factor was almost always the availability of raw materials from which to prepare the desired product and the economics of each process.

Concurrent with the advance of alkylaryl sulfonates as economically viable surfactants, activities in the United States and Germany led to the development of the taurine (2-aminoethane-1-sulfonic acid) derivatives and the alkane sulfates, respectively. In the United Kingdom, secondary olefin sulfates derived from petroleum fractions were produced in large quantities. Each of those raw materials had its own special advantages and disadvantages; but in evaluating their feasibility, the producer had to consider such factors as the availability and cost of raw materials, ease of manufacture, the economics of manufacture and distribution, and overall product stability. As a result of their ease of manufacture and versatility, the propylene tetramer (PT)–based alkylbenzene sulfonates (ABS) very quickly gained a strong position in the world market. After World War II, the propylene tetramer, primarily a branched C_9H_{19} alkyl, $C_9H_{19}—C_6H_4—SO_3^- \; Na^+$, coupled to benzene became a predominant material. Thus, ABS materials very rapidly displaced all other basic detergents and for the period 1950–1965 constituted more than half of all detergents used throughout the world.

ABS materials held almost undisputed reign as the major ingredient used in washing operations until the early 1960s, with essentially 100% of the alkylbenzene detergents belonging to the PT family. Around that time it was noted that sewage effluents were producing increasing amounts of foaming in rivers, streams, and lakes throughout the world. In addition, where water was being drawn from wells located close to household discharge points, the water tended to foam when coming out of the tap. Such occurrences were naturally upsetting to many groups and led to investigations into the sources of the foaming agents. Such an undesirable phenomenon was ultimately attributed to the failure of the ABS materials to be completely degraded by the bacterial and other processes naturally present in wastewater treatment plants and effluents. It was further determined that it was the branched alkyl (PT) chain that hindered attack by the microorganisms. Fatty acid sulfates, on the other hand, were found to degrade readily, and since all naturally occurring fatty acids from which fatty alcohols are produced are straight-chained, it seemed probable that a straight-chain alkylbenzene might prove more easily biodegradable.

Test methods for determining degradability were developed and showed that, in fact, linear alkylbenzene sulfonates (LABS) were significantly more biodegradable and hence ecologically more acceptable. In most of the industrialized world, detergent producers, voluntarily or by legislation, have switched from ABS to LABS as their basic detergent building block. By the 1980s, more than 75% of synthetic detergents were of the LABS family.

The change to LABS feedstocks gave some rather surprising results. It was found that detergency in many heavy-duty cleaning formulations using LABS was approximately 10% better than when ABS were used. Solutions of the neutralized acid had a lower cloud point (see glossary in Section 1.8), and pastes and slurries had a lower viscosity. The first two results were obviously advantageous, and a lower viscosity in slurries had an advantage when the product was processed into a powder. When the LABS product was to be sold as a liquid or paste detergent, however, the lower viscosity was seen as a detriment to sales appeal and had to be overcome.

Today, even though many of the application areas such as detergents and cleaning products are considered to be "mature" industries, the demands of ecology, population growth, fashion, raw-materials resources, and marketing appeal have caused the technology of surfactants and surfactant application to continue to grow at a healthy rate overall, with the usual ups and downs that accompany most industries.

While a large fraction of the business of surfactants is concerned with cleaning operations of one kind or another, the demands of other technological areas have added greatly to the enhanced role of surfactants in our modern existence. Not only are personal care products becoming an even greater economic force in terms of dollar value and total volume; applications as diverse as pharmaceuticals, petroleum recovery processes, high-tech applications, and medicine are placing more demands on our ability to understand and manipulate interfaces through the action of surface-active agents. As a result, more and more scientists and engineers with little or no knowledge of surface chemistry are being called on to make use of the unique properties of surfactants.

1.2. THE ECONOMIC IMPORTANCE OF SURFACTANTS

The applications of surfactants in science and industry are legion, ranging from primary production processes such as the recovery and purification of raw materials in the mining and petroleum industries, to enhancing the quality of finished products such as paints, cosmetics, pharmaceuticals, and foods. Figure 1.1 illustrates a few of the major, high-impact areas of application for surfactants and other amphiphilic

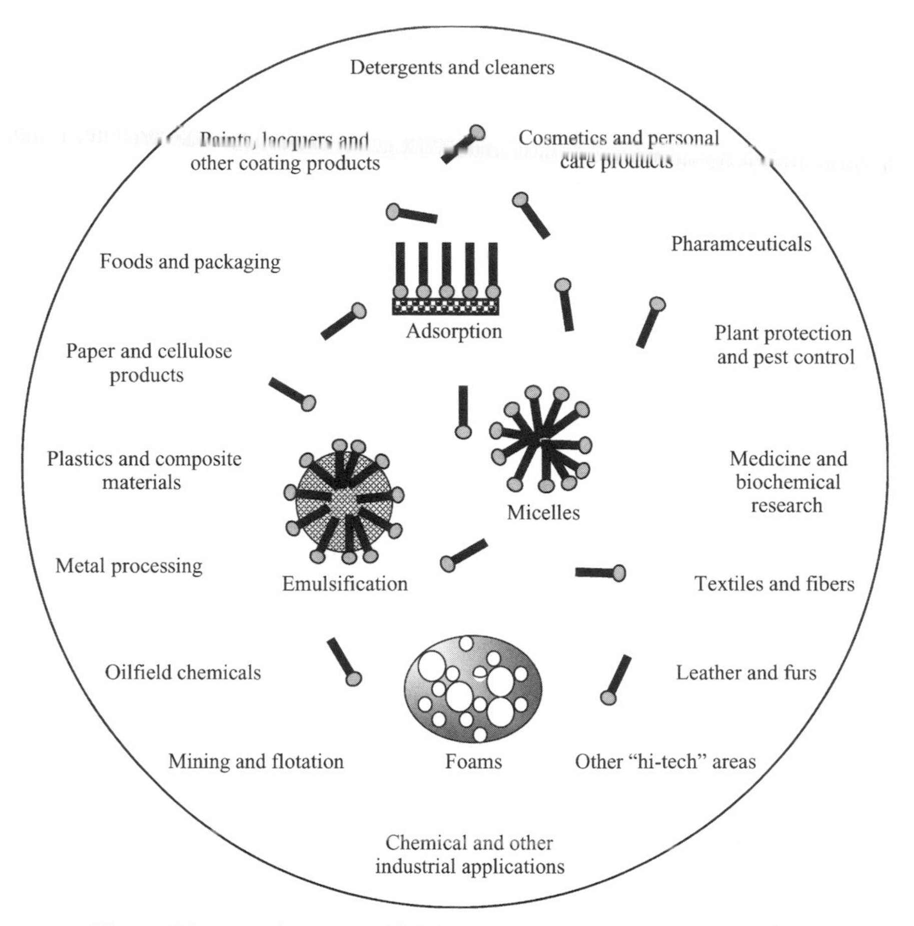

Figure 1.1. Some important, high-impact areas of surfactant applications.

materials. As the economic, ecological, and performance demands placed on product and process additives such as surfactants increase, it seems obvious that our need to understand the relationships between the chemical structures of those materials and their physical manifestations in particular circumstances becomes more important.

The properties and applications of surfactants are, as we shall see, determined by the balance between the lyophilic ("solvent-loving") and lyophobic ("solvent-hating") portions of the molecules. The desired properties will vary significantly for many of the applications noted in Figure 1.1. For that reason, such characteristics as solubility, surface tension reducing capability, critical micelle concentration (cmc), detergency power, wetting control, and foaming capacity may make a given surfactant perform well in some applications and less well in others. The "universal" surfactant that meets all the varied needs of surfactant applications has yet to emerge from the industrial or academic laboratory. The following chapters will

TABLE 1.1. Typical (But Not All) Characteristics for Surfactants that Must Be Evaluated for Various Applications

Application	Characteristics
Detergency	Low cmc, good salt and pH stability, biodegradability, desirable foaming properties
Emulsification	Proper HLB, environmental and biological (safety) for application
Lubrication	Chemical stability, adsorption at surfaces
Mineral flotation	Proper adsorption characteristics on the ore(s) of interest, low cost
Petroleum recovery	Proper wetting of oil-bearing formations, microemulsion formation and solubilization properties, ease of emulsion breaking after oil recovery
Pharmaceuticals	Biocompatibility, low toxicity, proper emulsifying properties

provide more detail on the molecular structural features that determine the various functional characteristics of surfactants. For now, suffice to say that each application will have specific requirements that will determine a specific surfactant's utility in a given system. Some of the fundamental characteristics that must be evaluated for a surfactant proposed for some specific applications are listed in Table 1.1.

The fast-paced, highly competitive nature of modern industrial developments often demands the fastest, most economical possible solution to a problem, consistent with the needs of the product. In the area of surfactant science and technology, it might often be the case that the fastest marginally acceptable solution could be replaced by a superior, possibly more economical, alternative if only the right minds and information could be brought together. Unfortunately, the world of surfactants and surface science historically has not received wide coverage in most academic training situations, and most workers have limited familiarity with the basic concepts and processes involved.

1.3. SOME TRADITIONAL AND NONTRADITIONAL APPLICATIONS OF SURFACTANTS

A comprehensive discussion of the myriad applications of surfactants in our daily activities is well beyond the scope of this work. Nevertheless, it is important to have a good concept of the impact, very often overlooked by those outside the field, of surfactants in our everyday lives—personally, professionally, socially, economically, and just about every other "-ly" word we can imagine. For that reason, the following notes will try to introduce some of those impact areas, without getting into too much detail for the moment.

1.3.1. Detergents and Cleaners

The primary traditional application for surfactants is their use as soaps and detergents for a wide variety of cleaning processes. As already noted, soaps have been used in personal hygiene for well over 2000 years with little change in the basic chemistry of their production and use. New products with pleasant colors, odors, and deodorant and antiperspirant activity have crept into the market since the early twentieth century or so, but in the end, soap is still soap.

On the other hand, the synthetic detergents used in cleaning our clothes, dishes, houses, and so on are relative newcomers. "Whiter than white" and "squeaky clean" commercials notwithstanding, the purpose of detergents is to remove unwanted dirt, oils, and other pollutants, while not doing irreparable damage to the substrate. In the past, due primarily to the shortcomings of available surfactants, such cleaning usually involved energy-intensive treatments—very hot water and significant mechanical agitation. Modern surfactant and detergent formulations have made it possible for us to attain the same or better results with much lower wash temperatures and less mechanical energy consumption. Improved surfactants and detergent formulations have also resulted in less water use and more efficient biological degradation processes that help protect our environment. Even with lower wash temperatures and lower energy consumption, extensive studies have shown that equivalent or improved hygiene is maintained. It is only in instances where particularly dangerous pathogenic agents are present, as in hospital laundries, for example, that additional germicidal additives become necessary to obtain efficient cleaning results.

More and more detergents and cleaners are being produced using feedstocks from "natural" or renewable sources, mainly vegetable oils and animal fats. The emotional or sociological impact of the "natural" label aside for the moment, that trend is important for several more practical standpoints—local availability, more constant prices (in general), relative ease of processing, and, of course, flexibility of production. The naturalness of the materials also helps out in terms of the ultimate biodegradation of the products, of course, since the building blocks fit naturally into the biological chain of life.

1.3.2. Cosmetics and Personal Care Products

Cosmetics and personal care products make up a vast multi-billion-dollar market worldwide, a market that continues to grow as a result of improved overall living standards in areas such as Asia and Latin America and continuing cultural driving forces in the already developed economies. Traditionally, such products have been made primarily from fats and oils, which often are perceived to have the advantage of occurring naturally in the human body and therefore present fewer problems in terms of toxicity, allergenicity, and so on. That perception is, of course, totally false, as shown by the large number of quite nasty allergens and toxins that come from the most "natural" of sources. Nonetheless, natural surfactants and other amphiphilic materials have been used in cosmetics since their "invention" in ancient Egypt (or before). The formulators of the day had no idea why certain things worked; they were interested only in the end results.

It is probably safe to say that few, if any, cosmetic products known to women (or men, for that matter) are formulated without at least a small amount of a surfactant or surface-active component. That includes not only the more or less obvious creams and emulsions but also such decorative products as lipstick; rouge; mascara; and hair dyes, tints, and rinses. An important aspect of such products is, of course, the interaction of the components of the cosmetic formulation with the human skin, membranes, and other tissues or organs with which it will come into contact during use. As mentioned above, merely because a product is "natural" or is derived from a natural source does not guarantee that it will not produce an adverse reaction in some, if not all, users. Just look at the poor peanut, a long-term staple for American kids and airline passengers for more than a century, banished from planes, schools, and other bastions of civilization because a few unlucky individuals have an allergic reaction to some component of that natural product. But that's another story.

The possible adverse effects of surfactants in cosmetics and personal care products must, of course, be studied in depth for obvious safety reasons as well as for questions of corporate liability and image. Unfortunately, our understanding of the chemical reactions or interactions among surfactants, biological membranes, and other components and structures is not sufficiently advanced to allow the formulator to say with sufficient certainty what reaction an individual will have when in contact with a surfactant. In the end, we unfortunately still need the rabbit's assistance.

1.3.3. Textiles and Fibers

Surfactants have historically played an important role in the textile-and-fibers industry. The dyeing of textiles is an obvious application of surfactants. The added surfactants serve to aid in the uniform dispersion of the dyes in the dying solution, the penetration of the dying solution into the fiber matrix, the proper deposition of the dyes on the fiber surface, and the proper "fixing" of the dye to that surface.

For natural fibers, the role of surfactants begins at the beginning—with the washing and preparation of the crude fiber in preparation for spinning. Once the crude material is ready for spinning, the use of surfactants as internal lubricants and static discharge agents allows the industry to produce yarns in extremely long and fine filaments that would be impossible to handle otherwise. Extremely fast modern spinning and weaving equipment requires that the fibers pass through the process without breaking or jamming, events that would produce very expensive production line stoppages. Sewing equipment that may work at more than 6000 stitches per minute requires that the fibers and needles pass in the night with a minimum of friction that could produce a significant amount of frictional heat and even burn the fibers. That interaction is controlled by the use of the proper surfactant and surfactant dosification.

Synthetic fibers also require surfactants at various steps in their evolution from monomeric organic chemicals to finished cloth. Depending on the type of polymer involved, the process may require surfactants beginning with the polymer synthesis, but certainly once the first extrusion and spinning processes begin. Even after the textile is "finished," it is common to apply a final treatment with a surface-active material to define the final characteristics of the product. In woven polyester rugs, for example, a final finish with an antistatic surfactant reduces or eliminates problems with static discharge (those shocking doorknobs in winter) and retards the adhesion of dirt to the fibers. The applications of fluorinated materials produces the stain repelling "Scotch Guard" effect by coating the fibers with a Teflonlike armor.

1.3.4. Leather and Furs

Surfactants are an important part of the manufacture of leather and furs, starting with the original untreated skin or hide and ending with the finished product. In leather tanning, for example, it is normal to treat the leather with a surfactant to produce a protective coating on the skin and hide fibers. This helps prevent the fibers from sticking together and keeps the fiber network flexible or supple while increasing the tensile strength of the finished leather product. Surfactants may also help the penetration of dyes and other components into the fiber network thereby improving the efficiency of various stages of the tanning process, saving time, energy, and materials while helping to guarantee a higher-quality, more uniform finished product.

The final surface finish of leather goods is now commonly applied in the form of lacquerlike polymer coatings that can be applied as emulsions and suspensions, using suitable surfactants, of course. Similar applications are found in the fur industry.

1.3.5. Paints, Lacquers, and Other Coating Products

It is probably not surprising to find that surfactants are required in many capacities in the production of paints and lacquers, and in related coating systems. In all paints that carry pigment loads, it is necessary to prepare a uniform dispersion that has reasonable stability to flocculation and coalescence. (See the glossary in Section 1.8 if those terms have slipped your mind.) In addition, the preparation of mineral pigments involves the process of grinding the solid material down to the desired particle size, which is an energy-intensive process. In general, it is found that a smaller, more uniform particle size results in a higher covering power for the same weight load of pigment, that is, a more efficient use of material and consequently a reduction in cost—always a nice effect in commerce.

The grinding process is helped by reducing the surface energy of the solid pigment, an effect achieved by the addition of surfactants. Since pigment solids are far from smooth surfaces at the molecular level, the raw material will have small cracks and holes that serve as initiation points for the rupture of the structure. In the presence of the proper surfactant, the molecules penetrate into the cracks and crevices, adsorb onto the solid surface, and significantly reduce the surface energy of newly exposed solid, facilitating the continued breaking of the large particles into smaller units. The adsorbed surfactant molecules also create a barrierlike coating that helps prevent the small particles from adhering or agglomerating. It is estimated that the use of surfactants in the grinding process can save up to 75% of the energy needed to achieve the same result without added surfactant.

Once the pigment is properly ground, it must be mixed into the basic liquid carrier and maintained stable or easily redispersible for an extended period of time, much against the natural driving force of thermodynamics. For the dispersion of the pigment in the final coating formulation, it may be necessary to add additional surfactant of the same or another class. In organic coating systems, the surfactant may in fact be a polymeric system that doubles as the final dried binder for the pigment. On the other hand, there are available low-molecular-weight surfactants specifically designed to act in organic solvents.

In aqueous or latex paints, the surfactant is important not only in the pigment grinding process but also in the preparation of the latex polymer itself. The chemistry of emulsion polymerization (i.e., latex formation) is a complex and interesting phenomenon and cannot be treated here. Very few emulsion polymers are produced without the addition of surfactants, and most of those so prepared are interesting laboratory novelties that never see the light of commercial exposure. In addition to surfactants for pigment grinding and dispersion and latex preparation, they are also important in the control of the wetting and leveling characteristics of the applied paint.

In painting applications that use lacquers such as the automobile industry, application and drying times are important. In such situations, wetting and leveling are also important. In powdered lacquers, the presence of the proper surfactants produces a net electrical charge on the surface of the particles, which allows them to be applied quickly and evenly by electrophoretic processes.

A potential drawback to rapid paint or lacquer application is that such speed can facilitate the introduction of air into the material resulting in foam formation at the time and point of application. If foam is produced, the drying bubbles on the painted surface will produce indentations and perhaps even bare spots that will significantly degrade the aesthetic and protective properties of the coating. To help prevent such foaming it is sometimes useful to add surfactants that also serve as antifoaming agents. Although it is common to relate surfactants with increased foam—as in beer, shaving cream, whipped toppings, and firefighting foams—we will see in Chapter 8 that surfactants can be either foam stabilizers or foam breakers, depending on the chemical structure and/or conditions of use.

1.3.6. Paper and Cellulose Products

Surfactants play several important roles in the papermaking industry. Several components of paper such as pigments for producing white or colored paper and sizing agents, often emulsion polymers that bind the cellulose fibers in the finished product and incorporate strength and dimensional stability, require surfactants in their preparation. In addition, the water-absorbing capacity of paper is often controlled by the addition of the proper surfactants.

Surfactants are also important in the process of recycling paper. A major step in the process is the removal of the ink and pigments present (deinking). That process is what is termed a *flotation process* (see Section 1.3.7), in which a surfactant is added to an aqueous slurry of old paper. The surfactant is chosen so that it will adsorb on the surfaces of pigment particle and ink droplets, causing them to become very hydrophobic. Air is then bubbled through the slurry. As the bubbles rise through the system, they become preferentially attached to the hydrophobic pigment and ink particles, acting like lifejackets and causing the particles to rise to the surface. At the surface they are skimmed off and separated from the cellulose slurry.

1.3.7. Mining and Ore Flotation

As just mentioned, the addition of the proper surfactant to a dispersion can produce a situation in which the solid particles, having a specific gravity much greater that that of water, can be made to float to the top and be easily (relatively speaking) separated from the aqueous phase. In the deinking mentioned above, there is no particular interest in being selective with respect to what is removed. It is essentially an "all out" proposition. In the mining industry the situation is quite different.

The flotation process has been important in mining for much longer than has deinking. In many instances, the desired mineral is present in small amounts that would be difficult or impossible to isolate and process while still "mixed" with the bulk of the mined rock. In that industry, therefore, it is necessary to have a more selective flotation process in which the desired mineral can be separated from the bulk of the ore in a continuous and relatively inexpensive process. Because different minerals tend to have slightly different surface properties, especially with regard to electrical charge characteristics, it is possible (with luck and perseverance) to design or formulate a surfactant system that will preferentially "float" a specific class of mineral while having little effect on other materials present. The selective surfactant or "collector" formulation allows the desired mineral to be skimmed from the top of the foaming slurry and thereby concentrated. The unwanted material can then be further processed or disposed of as slag.

While the theory of the adsorption of surfactants onto solid surfaces is highly developed and well understood in ideal systems, the reality of the universe is that in such complex multicomponent systems as mining ores, theory soon runs out of steam and success ultimately depends on hands-on laboratory and field trials, intuition, and art (or perhaps black magic).

Surfactants are also becoming more important in the coal mining industry. Aside from flotation processes, they are also employed as binders for the suppression of coal dust, and as dispersal aids and antifreezes for coal slurries that are pumped through pipelines.

1.3.8. Metal-Processing Industries

Surfactants are as important to the metal processing as to the mining industry. In order to perform as needed, metal surfaces must be cleaned and freed from deposits of oxides, oils, and other contaminants. Welding, painting, and other machining and surface treatments require a well-prepared surface. Even before that stage of fabrication, however, metals have a significant interaction with surfactants. High-speed metal rolling processes, for example, require the use of lubricating and cooling emulsions. With increased rolling speeds, heat production and buildup become significant problems that could lead to damage to equipment and a loss in the quality of the finished product. Properly formulated rolling oil emulsions containing surfactants reduce friction and the associated heat buildup, lessen the probability of rolling oils catching on fire, and help reduce the atomization of oil into the working environment and exhaust air.

In cutting and machining operations, cooling lubricants are required to carry away the heat produced by the cutting and drilling operations, thereby protecting the quality of the workpiece and prolonging the useful life of drillbits, and cutting surfaces. The components of cutting emulsions are critical, not only in terms of their direct action in metal processing but also because of worker and environmental exposure. The emulsions must be able to resist working temperatures in excess of 80°C, they must have significant antibacterial properties since they are routinely used for extended periods open to the atmosphere, and their components must meet rigid toxicological, dermatological, and environmental requirements because of the degree of operator exposure during their use.

1.3.9. Plant Protection and Pest Control

Surfactants are critical components in agricultural formulations for the control of weeds, insects, and other pests in agricultural operations. The roles of surfactants are varied, ranging from their obvious use as emulsifiers in spray preparations to their role as wetting and penetration aids and, in some cases, as active pest control agents. Surfactants also improve application efficiency by facilitating the transport of the active components into the plant through pores and membrane walls. Foam formation during application can also be a problem since the presence of foam will, in most cases, significantly reduce the effectiveness of the applied material.

In some applications, the choice of surfactant for a given active component can be critical. Since many pest control chemicals carry electrical charges, it is vital to use a surfactant that is electrically compatible with that ingredient. If the active material is positively charged, the addition of an anionic surfactant can, and probably will, result in the formation of a poorly soluble salt that will precipitate out directly before being applied, or the salt will be significantly less active, resulting in an unacceptable loss of cost-effectiveness.

1.3.10. Foods and Food Packaging

There are at least two important aspects to the role of surfactants in food-related industries. One aspect is related to food handling and packaging and the other, to the quality and characteristics of the food itself. Modern food-packaging processes rely on high-speed, high-throughput operations that can put great demands on processing machinery. Polymer packaging, for example, must be able to pass through various manufacturing and preparation stages before reaching the filling stage, many of which require the incorporation or use of surfactant containing formulations. Bottles and similar containers must be cleaned prior to filling, processes that usually require some type of detergent. The detergent, however, must have special characteristics that usually include little or no foam formation. Low-foaming detergents and cleaners are also important for the cleaning of process tanks, piping, pumps, flanges, and "dead" spaces in the process flow cycle. The presence of foam will often restrict the access of cleaning and disinfecting agents to difficult areas, reducing their effectiveness at cleaning the entire system and leading to the formation of dangerous bacterial breeding grounds.

In the food products themselves, the presence of surfactants may be critical for obtaining the desired product characteristics. Obvious examples would be in the preparation of foods such as whipped toppings, foam or sponge cakes, bread, mayonnaise and salad dressings, and ice cream and sherbets. Perhaps less obvious are the surfactants used in candies, chocolates, beverages, margarines, soups and sauces, coffee whiteners, and many, many more.

With a few important exceptions, the surfactants used in food preparations are identical or closely related to surfactants naturally present in animal and vegetable systems. Prime examples are the mono- and diglycerides derived from fats and oils, phospholipids such as lecithin and modified lecithins, reaction products of natural fatty acids or glycerides with natural lactic and fruit acids, reaction products of sugars or polyols with fatty acids, and a limited number of ethoxylated fatty acid and sugar (primarily sorbitol) derivatives.

1.3.11. The Chemical Industry

While surfactants are an obvious product of the chemical industry, they are also an integral part of the proper functioning of that industry. The important role of surfactants in the emulsion or latex polymer industry has already been mentioned. They are also important in other processes. The use of surfactants and surfactant micelles as catalytic centers has been studied for many years, and while few major industrial processes use the procedure, it remains an interesting approach to solving difficult process problems. A newer catalytic system known as *phase transfer catalysis* (PTC) uses amphiphilic molecules to transport reactants from one medium in which a reaction is slow or nonexistent into a contacting medium where the rate of reaction is orders of magnitude higher. Once reaction of a molecule is complete, the catalytic surfactant molecule returns to the nonreactive phase to bring over a new candidate for reaction. More information on PTC reactions is given in Chapter 6.

1.3.12. Oilfield Chemicals and Petroleum Production

The use of surfactants in the mining industry have already been mentioned. It is in the area of crude oil recovery, however, that surfactants possibly stand to make their greatest impact in terms of natural resource exploitation. As the primary extraction of crude oil continues at its hectic pace, the boom days of easy access and extraction have begun to come to an end and engineers now talk of secondary and tertiary oil recovery technology. As the crude oil becomes less accessible, more problems arises with regard to viscosity, pressures, temperatures, physical entrapment, and the like. While primary crude recovery presents its technological challenges, secondary and tertiary recovery processes can make them seem almost trivial.

Processes such as steam flooding involve injecting high-pressure steam at about 340°C into the oil bearing rock formations. The steam heats the crude oil, reducing its viscosity and applying pressure to force the material through the rock matrix toward recovery wells. Unfortunately, the same changes in the physical characteristics of the crude oil that make it more mobile in the formation also render it more susceptible to capillary phenomena that can cause the oil mass to break up within the pores of the rocks and leave inaccessible pockets of oil droplets. In such processes, surfactants are used to alter the wetting characteristics of the oil–rock–steam interfaces to improve the chances of successful recovery. Those surfactants must be stable under the conditions of use such as high temperatures and pressures and extremes of pH.

Although the use of surfactants for secondary and tertiary oil recovery is beneficial, it may also cause problems at later stages of oil processing. In some cases, especially where the extracted crude is recovered in the presence of a great deal of water, the presence of surfactants produces emulsions or microemulsions that must be broken and the water separated before further processing can occur. Naturally present surface-active materials in the crude plus any added surfactants can produce surprisingly stable emulsion systems. The petroleum engineer, therefore, may find herself confronted by a situation in which surfactants are necessary for efficient extraction, but their presence produces difficult problems in subsequent steps.

1.3.13. Plastics and Composite Materials

The importance of surfactants in the preparation of polymer systems such as emulsion or latex polymers and polymers for textile manufacture have already been mentioned. They are also important in bulk polymer processes where they serve as lubricants in processing machinery, mold release agents, and antistatic agents, and surface modifiers, and in various other important roles.

Surfactants can also play an important role in the preparation of composite materials. In general, when different types of polymers or polymers and inorganic materials (fillers) are mixed together, thermodynamics raises strong objections to the mixture and tries to bring about phase separation. In many processes, that tendency to separate can be retarded, if not completely overcome, by the addition of surfactants that modify the phase interfaces sufficiently to maintain peace and harmony among normally incompatible materials and allow the fabrication of useful composites.

1.3.14. Pharmaceuticals

The pharmaceuticals industry is an important user of surfactants for several reasons. They are important as formulation aids for the delivery of active ingredients in the form of solutions, emulsions, dispersions, gel capsules, or tablets. They are important in terms of aiding in the passage of active ingredients across the various membranes that must be traversed in order for the active ingredient to reach its point of action. They are also important in the preparation of timed-release medications and transdermal dosification. And in some cases, surfactants are the active ingredient. Surfactants for the pharmaceuticals industry must, of course, meet very rigid regulatory standards of toxicity, allergenicity, collateral effects and so on.

1.3.15. Medicine and Biochemical Research

Living tissues and cells (we and everything we know included) exist because of the physicochemical phenomena related to surface activity—in a sense, natural surfactants could be considered essential molecular building blocks for life. They are essential for the formation of cell membranes, for the movement of nutrients and other important components through those membranes, for the suspension and transport of materials in the blood and other fluids, for respiration and the transfer of gases between the atmosphere (the lungs, in our case) and the blood, and for many other important biological processes. It should not be surprising, then, that surfactants are finding an important place in research into how our bodies work and processes related to medical and biochemical investigations. Their roles in cosmetics and pharmaceuticals have already been mentioned, but their importance in obtaining a better understanding of life processes continues to grow. It is very probable that the years ahead will bring some surprising biochemical results based on surfactants and surface activity.

1.3.16. Other "Hi-Tech" Areas

Other industrial and technological areas that use surfactants include electronic microcircuit manufacture, new display and printing technologies, magnetic and optical storage media, and many more. New technologies, some seemingly far removed from classical surfactant-related technology, may begin to see the benefits or even necessity of using surfactants of some kind in order to achieve practicality. Such areas include the preparation of superconducting materials, nanotechnology related to nanofibers and buckey balls, molecular "motors," and a myriad of other exotic sounding areas. The unique character of surface-active materials make them natural candidates for investigation when interfacial phenomena, spontaneous molecular aggregation, specific adsorption, or similar ideas seem to offer a handle on a new idea. Surfactants may seem to be brutish bulk chemical commodities in their well-known, everyday applications, but the potential subtlety of their actions makes them prime candidates as special actors on the stage of technological progress.

New products are continually being developed to meet changing consumer and industrial demands, for new classical applications, and for new, unimagined uses. Surfactants are beginning to become more widely recognized as potentially useful tools in environmental protection and energy-related areas. They are being tried, with some success, in contaminated soil remediation, pollution control systems, less polluting paper-processing and recycling technologies, and for the coating of ultra-thin films. They are also being tried in such non-obvious technologies as the sintering of superconducting ceramics and in medical applications such as artificial blood for emergency or special-needs transfusions. The applications of surfactants mentioned briefly above constitute the bulk of the standard uses of surfactants in our world today. They represent an important direct and indirect driving force in our technological world.

1.4. SURFACTANT CONSUMPTION

The U.S. and world synthetic surfactant industries expanded rapidly in both volume and dollar value following World War II. Before the war, the great majority of cleaning and laundering applications relied for their basic raw materials on fatty acid soaps derived from natural fats and oils such as tallow and coconut oil. During and following the war, as already mentioned, the chemical industry developed new and efficient processes for the production of petroleum-derived detergent feedstocks based on the tetramer of propylene and benzene. In addition, economic and cultural changes such as increased use of synthetic fibers and automatic washing machines, increased washing frequency, population increases, and, of course, mass marketing through television and other media, all worked to increase the impact of non-fatty-acid-based surfactants. The percentage of U.S. surfactant consumption represented by the fatty acid soaps and synthetic detergents changed rapidly between 1940 and 1970. In 1945, synthetics represented only about 4% of the total domestic market. By 1953, the fraction had risen to over half of the total, and by 1970 the synthetic surfactant share had risen to over 80% of total soap and detergent production. The trend has leveled off since that time, with the fraction of total worldwide surfactant consumption as soap remaining in the range of 20–22%.

Beginning in the last years of the twentieth century, the surfactant industry began to undergo something of an upheaval in almost every area as a result of new formulations of surfactant-containing products brought about by changes in consumer demands, in local economies, raw-materials pricing, and changes in government regulatory practices.

In general, an improved economic situation leads to an increased demand for surfactants and surfactant-containing products. The reverse is also true, of course. Social and political forces have brought about demands for environmentally friendly "green" products that are milder for the end user and less potentially damaging to the environment. Cheaper products are also desirable, of course, from a marketing standpoint, which is made difficult by the ever-increasing price of crude oil and other surfactant feedstocks. Other reasons for price increases include the costs of regulatory compliance, insurance, and indirect environment-related expenses. Technical obsolescence is also a constant problem for any chemical-based industry.

In the United States, roughly one-half of the surfactants produced are used in personal care and detergent applications. For that reason, the industry is heavily influenced by consumer demands, fashion trends, and government oversight. The "green" movement is also exercising continually increasing pressure, especially in the areas of laundry and cleaning products, which constitute a significant chemical load in problems of water purity and wastewater treatment.

There is an increasing demand for mild, nontoxic, biodegradable products made from renewable or "natural" raw materials. Energy questions are becoming more important in relation to the production and use of surfactant containing products. Consumers are demanding products that function well at lower temperatures, as well as multifunctional products that allow them to save money and reduce the amounts of chemicals added to wastewater.

Federal, state, and local government regulatory requirements in areas of toxicology and environmental impact are beginning to influence industrial and consumer consumption. Government-imposed restrictions on the liberation of volatile organic compounds (VOCs) are affecting formulations in products ranging from cosmetics and toiletries, many of which use alcohols in their formulations, to paints and adhesives that carry along various classes of organic solvents and plasticizers. VOC regulations impact product performance and such functional characteristics as drying time, physical durability, and the final visual characteristics of coating products. The only way to reduce the VOC loads of such products is to increase the ability of new formulations to function as primarily water-based systems. Reformulations to achieve reduced VOC emissions require different surfactants or combinations of surfactants. All of these pressures pose a significant challenge to surfactant chemists and formulators.

Although the industry continues to place its major emphasis on the synthetic surfactants, demand for the traditional soap products remains relatively strong. In 2000, approximately 1.5 million metric tons of soaps were used in the three highly industrialized regions of the world—the United States, western Europe, and Japan. Much higher relative levels of use occur in the less industrialized nations in Africa, Asia, and Latin America. In many cases, those areas do not possess the sophisticated manufacturing capabilities or raw-materials availability for large-scale production of the synthetic precursors to the newer surfactants. In addition, there may be political and social reasons for high levels of soap usage—namely, the greater availability of natural fats and oils as a result of significant stocks of vegetable- or animal-derived materials. Even in the industrialized areas, however, soap demand is substantial. Because of the nature of the products, their specific applications, and the availability of the necessary raw materials, soaps will probably maintain a significant market share in the surfactant industry for the foreseeable future.

The approximate breakdown of surfactant consumption by class is shown in Figure 1.2. Six major surfactant types accounted for 60% of the total consumption. The "big six" are soaps, linear alkylbenzene sulfonates (LABS), alcohol ethoxylates (AE), alkylphenol ethoxylates (APE), alcohol ether sulfates (AES), and alcohol sulfates (AS).

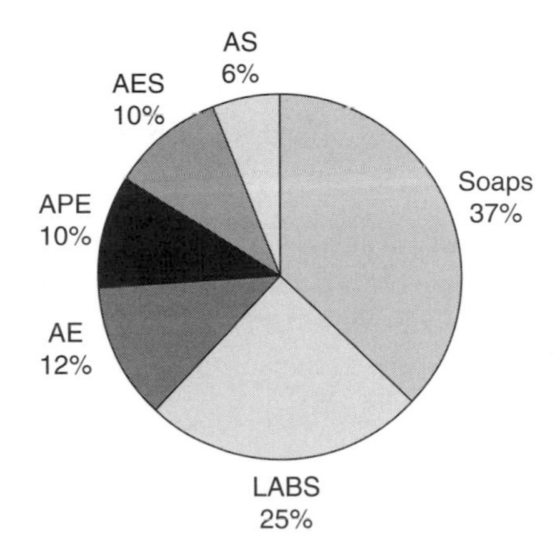

Figure 1.2. Surfactant consumption by type in the major industrialized areas for 2000.

The linear alkylbenzene sulfonates (LABS) family is probably the world's most important surfactant family, taking into consideration their wide applicability, cost-effectiveness, and overall consumption levels. If raw-materials prices and availability (i.e., normal paraffins and benzene) remain stable, there is little reason to expect the situation to change in the near future. If feedstock prices increase significantly, however, alcohol sulfates and related materials derived from fat and vegetable sources may become attractive alternatives.

There do exist some concerns and questions about the overall long-term ecological impact of LABS. Of particular importance are the following:

1. LABS are not easily biodegradable under anaerobic conditions.
2. Limited data are available on what happens when dissolved LABS enters a waterway and what its effects on adsorption and sedimentation will be.
3. When treated sewage sludge is transferred to the soil, what effects do residual LABS have on adsorption and soil wetting, and what is their final fate?
4. What is the true, ultimate biodegradability of LABS in terms of residues, metabolites, and other materials?

In terms of raw-materials availability, soaps are very desirable products. As already noted, soap is especially important in less industrialized countries because the sources are readily renewable, relatively speaking, and usually locally grown. In addition, the necessary production facilities and technology are relatively simple and inexpensive. In many modern applications, however, soaps are neither efficient nor effective, and cannot really replace the synthetic surfactants. While the use of "natural" soaps seems to have a high emotional rating among environmental groups due to their long tradition of use and "organic" sources, their inferior performance characteristics in many common situations require the use of much larger quantities of synergistic additives (e.g., phosphates and other "builders") to achieve results approaching those obtained using smaller quantities of synthetic detergents. The net result is a much higher organic load on the ecosystem—a very important factor in terms of sewage treatment and environmental impact.

Alcohol ethoxylate (AE) surfactants, representing about 12% of consumption, have shown better-than-average growth more recently relative to other surfactants. They exhibit several important advantages, including good detergency at low washing temperatures, low foaming characteristics, good detergency in phosphate-free formulations, good performance with synthetic fibers, and good performance in low-temperature industrial processes. Because they can be made from both petroleum and renewable raw materials, AE surfactants have a stable position in that respect.

Alkylphenol ethoxylate (APE) surfactants make up approximately 10% of overall consumption. While effective in many industrial applications, they face a number of environmental challenges that could greatly reduce their use in the future. Of major importance are questions concerning their relatively slow rate of biodegradation and the possible toxicity of degradation intermediates, especially phenols and other aromatic species. In the United States and western Europe, many detergent manufacturers have voluntarily discontinued their use in household products.

The alcohol ether sulfates (AES) represent approximately 9% of industrialized surfactant consumption. Because of their perceived "mildness," they are used primarily in personal care products. They have a strong position in terms of raw materials since they can be made from either petroleum or renewable (i.e., agriculturally derived) raw materials. One possible disadvantage of AES surfactants is the possible presence of dioxane derivatives as a byproduct of the ethoxylation process. Although modern processes have been shown to effectively eliminate the presence of such contaminants, emotional factors and lack of good information must always be considered, especially where consumer products are concerned.

The alcohol sulfates (AS) surfactants constitute approximately 6% of surfactant consumption. They have the advantage of being efficiently derived from renewable sources and can function as partial replacements for LABS in some applications. Their current major applications are in personal care products and emulsion polymerization processes. Because their biodegradability is essentially the same as that of soaps, AS surfactants seem to have a reasonably friendly reception on environmental grounds.

1.5. THE ECONOMIC AND TECHNOLOGICAL FUTURE

The wide variety of lyophobic ("hydrophobic" in aqueous systems) and lyophilic (or "hydrophilic" in water) groups available as a result of advances in synthetic technology and the development of new raw-materials resources provides an extremely broad menu from which the surfactant shopper can select a material for a particular need. By carefully analyzing the overall composition and characteristics of a given system, the investigator or formulator can choose from one of the available classes of surfactants based on charge type (i.e., the ionic properties of the surface-active species), solubility, adsorption behavior, or any of the other variations related to the chemical structure of the molecule and its interactions with other system components. From the trends in production and use, it is clear that surfactants, although they may seem to constitute a "mature" class of industrial chemicals, have a lot of room for additional growth.

Some classes of surfactants, in particular nonionic materials, may be especially favored for above-average growth in consumption. Their advantages in performance at lower temperatures, low-foaming characteristics, and relative stability at high temperatures and under harsh chemical conditions are definite pluses in many technological applications. Possible disadvantages may be in their dependence on petrochemical feedstocks, the potential security risks involved in the preparation of their oxide precursors, and lingering questions about the presence of very small amounts of reaction by products that are perceived to be particularly dangerous (peroxides, dioxins, etc.).

Because of their special characteristics, soaps will continue to be important surfactant products. Although increased industrialization in the third world will undoubtedly lead to greater use of synthetic alternatives, population growth alone can be expected to maintain the current levels of soap consumption worldwide.

While the "big six" surfactants will almost certainly continue to dominate the surfactant market, there will always arise the need for new and improved surfactant products. A few potentially fruitful areas of research include

1. Multifunctional surfactants (e.g., detergent and fabric softener in a single structure)
2. More ecologically acceptable chemical structures
3. New surfactants based on renewable raw materials
4. Surfactants with good chemical and thermal stability
5. Highly biocompatible surfactants
6. Polymeric materials that show good surfactant activity and produce viscosity enhancement
7. Materials that promise energy savings in terms of their manufacture or functionality at lower temperatures

These represent just a few ideas related to surfactant use and possible future growth potential. For a "mature" industry, surfactants remain an interesting area for research and development.

1.6. SURFACTANTS IN THE ENVIRONMENT

The use of surfactants throughout the world is increasing at a rate in excess of the population growth because of generally improved living conditions and processed material availability in the less industrially developed third world countries. Hand in hand with increased surfactant use go the problems of surfactant disposal. As the more developed nations have learned by painful and expensive experience, the ability of an ecosystem to absorb and degrade waste products such as surfactants can significantly affect the potential usefulness of a given material.

Of particular importance are the effects of surfactants on groundwater and waste treatment operations. Although it could be technologically possible to physically or chemically remove almost all residual surfactants completely from effluent streams, the economic costs would undoubtedly be totally unacceptable. The preferred way to address the problem is to allow nature to take its course and solve the problem by biodegradation mechanisms.

Biodegradation may be defined as the removal or destruction of chemical compounds through the biological action of living organisms. Such degradation in surfactants may be divided into two stages: (1) primary degradation, leading to modification of the chemical structure of the material sufficient to eliminate any surface active properties and (2) ultimate degradation, in which the material is essentially completely removed from the environment as carbon dioxide, water, inorganic salts, or other materials that are the normal waste byproducts of biological activity. Years of research indicate that it is at the first stage of primary degradation that the chemical structure of a surfactant molecule most heavily impacts biodegradability.

Some of the earliest reports on the biodegradability of synthetic surfactants were made in England, where it was observed that linear secondary alkyl sulfates (LAS) were biodegradable, while the alkylbenzene sulfonates (ABS) in use were much more resistant to biological action. It was soon found that the distinction between the LAS and ABS surfactants was not nearly as clear as first suggested. Specifically, it was determined that the biodegradability of a particular ABS sample depended to a large degree on the source, and therefore the chemical structure, of the sample. Early producers of ABS surfactants in England used either petroleum-derived kerosene or tetrapropylene as their basic raw material, without great consideration for the structural differences between the two. As a result, great variability was found in the assay of materials for determination of biodegradability. In fact, the materials derived from propylene showed little degradation while the nominally identical materials based on the kerosene were much more acceptable. The difference, of course, lay in the degree of branching in the respective alkyl chains.

In 1955 and 1956 it was suggested that the resistance of tetrapropylene-derived ABS surfactants to biodegradation was a result of the highly branched structure of the alkyl group relative to that of the kerosene-derived materials and the LAS materials. As a result of extensive research on the best available model surfactant compounds and analogs, it was proposed that the nature of the hydrophobic group on the surfactant determined its relative susceptibility to biological action, and that the nature and mode of attachment of the hydrophile were of minor significance. Research using an increasingly diverse range of molecular types has continued to support those early conclusions.

Although the chemical basis of surfactant biodegradation continues to be studied in some detail, leading to more specific generalizations concerning the relationship between chemical structure and biological susceptibility, the following general rules have developed, which seem to cover most surfactant types:

1. The chemical structure of the hydrophobic group is the primary factor controlling biodegradability; high degrees of branching, especially at the alkyl terminus, inhibit biodegradation.
2. The nature of the hydrophilic group has a minor effect on biodegradability.
3. The greater the distance between the hydrophilic group and the terminus of the hydrophobe, the greater is the rate of primary degradation.

1.7. PETROCHEMICAL VERSUS "RENEWABLE" OLEOCHEMICAL-BASED SURFACTANTS

As will be shown in Chapter 2, all surfactants have the same basic structure: a hydrophilic (water-loving) "head" and a hydrophobic (water-hating) "tail," which is almost always a long chain of carbon atoms. The tails, which are hydrophobic, interact with nonaqueous phases or surfaces (or themselves) while the heads try to improve the relationship of the system with the aqueous phase. One might think of the surfactant as the arbiter in the conflict between water and the nonaqueous world.

Presently, about 50% of the surfactants used in the surfactant industry are derived from petrochemical raw materials, and the other 50% are derived from oleochemical raw materials. The most important surfactants used in consumer detergents are anionic and nonionic materials. The alcohols used are linear or essentially linear, which results in a more rapid and complete biodegradation of both oleochemical- and petrochemical-derived detergent surfactants.

The surfactants currently available for industrial applications can be separated into two groups: those that have a "natural" or renewable origin derived from oil seed crops, animal fats, or trees, and those derived from petroleum distillates. There has been a great deal of debate on the pros and cons of these two types of sourcing. Renewable surfactant feedstocks are often perceived as being better for the environment and should therefore be the first choice for environmentally "friendly" products. But is that "analysis" of the situation scientific fact or spiritually pleasing fiction? Are renewable chemicals necessarily better for the environment because they are derived from plant and animal fats and oils? As with most scientific, political, and social questions, there is no easy answer.

The popular perception that "natural" products are always better for the environment than are "synthetics" has led to the suggestion that petrochemical surfactants should be replaced with surfactants based on renewable oilseed or animal-fat-derived materials because the change would improve the environmental profile or impact of surfactant containing products. While there may be good arguments for switching based on perceived long-term raw-materials availability and the renewable nature of the beast, a total substitution is not possible or possibly even desirable for many reasons.

In most applications, both renewable and petrochemical-based surfactants are available to product formulators. The flexibility of using both types of surfactants gives formulators the option of creating products that maximize the value of surfactant-based products by optimizing their functionality under a variety of conditions while keeping the cost to the consumer as low as possible.

A significant amount of research goes into the formulation of surfactant-based consumer products. In addition to the obvious concerns about performance and safety to the consumer and the environment, it is important to understand how different ingredients interact under various conditions and the stability of the product under extreme conditions of shipping and storage. Even before reaching the use phase, it is necessary to know what modifications may need to be made in the manufacturing process, the impact of those process modifications on costs, and the environmental impact of the production process.

Detergents, for example, are formulations that include surfactants, enzymes, and builders or additives that serve to "soften" the water to enhance the functionality of the surfactant component. Formulators generally have access to a broad range of surfactant structures, giving them a great deal of flexibility with which to achieve optimum detergent performance under a broad range of circumstances. As will be seen in later chapters, the hard reality of surfactant science is that seemingly small differences in the chemical nature of a surfactant molecule, usually related to the hydrophobic portion of the molecule—that is, its source—may significantly affect its performance in its final application. The question of renewable versus petrochemical feedstocks, then, becomes very complex for the following reasons (among others):

1. The functional characteristics of surfactants in a wide range of consumer products such as low-temperature and low-foaming detergents would be difficult to duplicate with renewable surfactant feedstocks alone.
2. Data from biodegradation, removal by sewage treatment, toxicity, and similar studies indicate that there is little or no measurable difference between surfactants based on petrochemical and renewable raw materials in terms of their direct impact on the environment.
3. Replacement of petrochemical-based surfactants by ones based on natural materials would not lead to any significant reductions in water or air emissions, nor would it reduce energy consumption across the use cycle of the surfactants.
4. Improved functionality of new detergent formulations at cooler wash temperatures will result in energy savings during use. This will have a positive impact for the environment, including reduced air emissions, conservation of petroleum stocks, and reduced waste.

For an in-depth discussion of the complex relationship between the chemical structure of surface-active materials and biodegradability, the interested reader is referred to work of Swisher (1986) cited in the Bibliography [at the end of this book, in the listings for this chapter (Chapter 1)].

1.8. A SURFACTANT GLOSSARY

As indicated above, the world of surfactants and their applications has become one in which the exact meaning of words and phrases is sometimes muddled by the growth of two basic schools of investigators—the industrial scientists and the academicians. Although it is more common to place a glossary at the end of a book or other publication, it seems more efficient in this case to see the meaning of the terms before encountering them in their normal context than to be flipping to the back each time a question comes up. The short list of some of the more common terms encountered in the practice of the art (as, in many cases, "art" is the best word for it) and science of surface activity and surfactant applications may be useful to help clarify some of the confusion that can arise on the part of the nonspecialist. Although these definitions may differ slightly from those found in other references, they are practical and meaningful for the understanding of the concepts and phenomena under discussion.

Aerosol. A dispersion of fine solid or liquid particles or droplets in a gaseous continuous phase; terms such as mist, fog, and smoke may be used for specific situations.

Amphiphilic. Refers to a molecular structure that contains distinct components that are on one hand soluble and on the other insoluble (or of limited solubility) in a given solvent environment.

Amphoteric surfactants. Surfactants that can be either cationic or anionic depending on the pH or other solution conditions, including those that are zwitterionic—possessing permanent charges of each type.

Anionic surfactants. Surfactants that carry a negative charge on the surface-active portion of the molcculc.

Bicontinuous phases. Surfactant aggregate structures related to liquid crystalline phases or mesophases that exhibit bicontinuous (two interwoven continuous phases) behavior. The most common is the cubic bicontinuous structure, often referred to today as "cubosome," although other structures are possible.

Biodegradability. A measure of the ability of a surfactant to be degraded to simpler molecular fragments by the action of biological processes, especially by the bacterial processes present in wastewater treatment plants, the soil, and general surface water systems.

Cationic surfactants. Surfactants carrying a positive charge on the surface-active portion of the molecule.

Cloud point. For nonionic surfactants—the temperature (or temperature range) at which the surfactant begins to lose water solubility and a cloudy dispersion results; the surfactant may also cease to perform some or all of its normal functions as a surfactant.

Coalescence. The irreversible union of two or more drops (emulsions) or particles (dispersions) to produce a larger unit of lower interfacial area.

Colloid. A two-phase system consisting of one substance (the dispersed phase) finely divided and distributed evenly (relatively speaking) throughout a second phase (the dispersion medium or continuous phase).

Contact angle. The angle formed between a solid surface and the tangent to a liquid drop on that surface at the line of contact between the liquid, the solid, and the surrounding phase (usually vapor or air), measured through the liquid drop.

Counterion. The (generally) non-surface-active portion of an ionic surfactant necessary for maintaining electrical neutrality.

Critical aggregation concentration (cac). A surfactant concentration at which micelle formation begins for a surfactant in the presence of polymer. The cac is an extensive characteristic of the specific surfactant–polymer system.

Critical micelle concentration (cmc). A concentration characteristic of a given surfactant at which certain solution properties change dramatically, indicating the formation of surfactant aggregates or micelles.

Detergency. The process of removing unwanted material from the surface of a solid by various physicochemical and mechanical means related to surfactant action.

Dispersion. The distribution of finely divided solid particles in a liquid phase to produce a system of very high solid–liquid interfacial area.

Dispersion forces. Weak quantum-mechanical interatomic or intermolecular forces common to all materials; generally attractive for materials in the ground state, although they can have a net repulsive effect in some solid–liquid systems.

Emulsifying agents (emulsifiers). Surfactants or other materials added in small quantities to a mixture of two immiscible liquids for the purpose of aiding in the formation and stabilization of an emulsion.

Emulsion. A colloidal suspension of one liquid in another. (A more specific functional definition is given in Chapter 9.)

Fatty acids. A general term for the group of saturated and unsaturated monobasic aliphatic carboxylic acids with hydrocarbon chains ranging from 6 to 22 carbons. The name derives from the original source of such materials, namely, animal and vegetable fats and oils.

Fatty alcohols. Primary alcohols with carbon numbers in the range of C_6–C_{22} historically derived from natural fats and oils, directly or by reduction of the corresponding fatty acids, but more recently obtainable from petroleum sources.

Flocculation. The (often) reversible aggregation of drops or particles in which interfacial forces allow the close approach or touching of individual units, but where the separate identity of each unit is maintained.

Foam booster. An additive that increases the amount or persistence of foam produced by a surfactant system.

Foam inhibitor. An additive designed to retard or prevent the formation of foam in a surfactant solution, usually employed at low concentrations.

Head group (surfactant). A term referring to the portion of a surfactant molecule that imparts solubility to the molecule. Generally used in the context of water solubility.

Hydrogen bonding. Interaction between molecules or portions of a molecule resulting from the Lewis acid or base properties of the molecular units. Most commonly applied to water or hydroxyl containing systems (e.g., alcohols) in the sense of Brønsted–Lowry acid–base theory, but also found in molecules having hydrogen bound to nitrogen (amines and amides).

Hydrophile–lipophile balance (HLB). An essentially empirical method for quantifying or estimating the potential surface activity of a surfactant based on its molecular constitution—used primarily in emulsion technology.

Hydrophilic ("water-loving"). A descriptive term indicating a tendency on the part of a species to interact strongly with water, sometimes equated with "lipophobic," defined below.

Hydrophobic ("water-hating"). The opposite of hydrophilic, having little energetically favorable interaction with water—generally indicating the same characteristics as lipophilic, except that some hydrophobic materials (e.g., perfluoro organics) can also be lipophobic.

Interface. The boundary between two immiscible phases. The phases may be solids, liquids, or vapors, although there cannot, in principle, be an interface between two vapor phases. Mathematically, the interface may be described as an infinitely thin line or plane separating the bulk phases at which there will be a sharp transition in properties from those of one phase to those of the other, although in fact it will consist of a region of at least one molecular thickness, but often extending over longer distances.

Interfacial tension. The property of a liquid–liquid interface exhibiting the characteristics of a thin elastic membrane acting along the interface in such a way as to reduce the total interfacial area by an apparent contraction process—thermodynamically, the interfacial excess free energy resulting from an imbalance of forces acting on molecules of each phase at or near the interface (see **Surface tension**).

Lipophilic ("fat-loving"). A general term used to describe materials that have a high affinity for fatty or organic solvents; essentially the opposite of hydrophilic.

Lipophobic ("fat-hating"). The opposite of lipophilic; that is, materials preferring to be in more polar or aqueous media; the major exceptions are the fluorocarbon materials, which may be both lipophobic and hydrophobic.

London forces. Forces arising from the mutual perturbation of the electron clouds of neighboring atoms or molecules; generally weak ($\approx$8 kJ/mol), decreasing approximately as the inverse sixth power of the distance between the interacting units.

Lyophilic ("solvent loving"). A general term applied to a specific solute–solvent system, indicating the solubility relationship between the two. A highly water-soluble material such as acetone would be termed *lyophilic* in an aqueous context.

Lyophobic ("solvent hating"). The opposite of lyophilic. A hydrocarbon, for example, would be lyophobic in relation to water. If the solvent in question were changed to octane, the hydrocarbon would then become lyophilic.

Micelles. Aggregated units composed of a number of molecules of a surface-active material, formed as a result of the thermodynamics of the interactions between the solvent (usually water) and lyophobic (or hydrophobic) portions of the molecule.

Nonionic surfactants. Surfactants that carry no electrical charge, as their water solubility is derived from the presence of polar functionalities capable of significant hydrogen bonding interaction with water (e.g., polyoxyethylenes, sugars, polyglycidols).

Oleochemicals. Products derived from vegetable oils and similar raw materials.

Soap. The name applied to the alkali salts of natural fatty acids, historically the product of the saponification of natural fats and oils.

Solubilization. The process of making a normally insoluble material soluble in a given medium. In the following chapters, the term is applied in two ways: the "solubilization" of a hydrocarbon chain in water by chemical modification—the addition of a head group—and the micellar solubilization of an oil phase in water or vice versa.

Spreading coefficient. A quantitative predictor of the ability or propensity of a given liquid to spread over the surface of a second liquid or solid on the basis of the surface tensions or surface energies of the two bulk phases and their respective interfacial tension.

Surface-active agent. The descriptive generic term for materials that preferentially adsorb at interfaces as a result of the presence of both lyophilic and lyophobic structural units, the adsorption generally resulting in the alteration of the surface or interfacial properties of the system.

Surface tension. The property of a liquid evidenced by the apparent presence of a thin elastic membrane along the interface between the liquid and a vapor phase, resulting in a contraction of the interface and reduction of the total interfacial area. Thermodynamically, the surface excess free energy per unit area of interface resulting from an imbalance in the cohesive forces acting on liquid molecules at the surface.

Surfactant. The widely used contraction for "surface-active agent."

Surfactant tail. In surfactant science, generally used in reference to the hydrophobic portion of the surfactant molecule.

2 The Organic Chemistry of Surfactants

In order to understand the relationship between the surface activity of a given material and its chemical structure, it is useful to have a handle on the chemistry of the individual molecular components that produce the observed phenomena. The following discussion introduces the basic chemical principles involved in common surfactants, ranging from basic raw materials and sources to the chemical group combinations that result in observed surface activity.

The chemical compositions and synthetic pathways leading to the formation of surface-active molecules are limited primarily by the creativity and ingenuity of the synthetic chemist and the production engineer. Therefore, it is practically impossible to discuss all potential chemical classes of surfactants, including their preparations and subtle variations. However, the majority of surfactants of academic and technological interest can be grouped into a limited number of basic chemical types and synthetic processes.

The chemical reactions that produce most surfactants are rather simple and easy to understand for anyone surviving the first year of organic chemistry. The challenge to the producer lies in the implementation of those reactions on a scale of thousands of kilograms, reproducibly, with high yield and high purity (or at least known levels and types of impurity), and at the lowest cost possible. With few exceptions, there will always be a necessity to balance the best surfactant activity in a given application with the cost of the material that can be borne by the value added to the final product. The challenge to the ultimate user is to understand the chemical, physical, and biological requirements that a candidate material must meet.

Before discussing specific details of surfactant types and possible synthetic pathways, it may be useful to introduce some of the many reactions that can produce surfactant activity in an organic molecule. In that way, the reader can begin to see some of the basic simplicity of surfactant science underlying the imposing variety of structural possibilities.

The chemical structures having suitable solubility properties for surfactant activity vary with the nature of the solvent system to be employed and the conditions of use. In "standard" surfactant terminology, the "head" refers to the solubilizing

---

Surfactant Science and Technology, Third Edition by Drew Myers

group—the lyophilic or hydrophilic group, in aqueous systems—and the "tail" refers to the lyophobic or hydrophobic group in water:

$$\underbrace{CH_3(CH_2)_n}_{\text{"tail"}}\underbrace{CH_2{-}S}_{\text{"head"}}$$

In water, the hydrophobic group may be, for example, a hydrocarbon, fluorocarbon, short polymeric chain, or siloxane chain of sufficient size to produce the desired solubility characteristics when bound to a suitable hydrophilic group. In aqueous systems, the hydrophilic group (the "head") will be ionic or highly polar, so that it can act as a solubilizing functionality. In a nonpolar solvent such as hexane the same groups will, in theory, function in the opposite sense. As the temperature, pressure, or solvent environment of a surfactant (e.g., cosolvent addition, pH changes, or the addition of electrolytes in aqueous systems) varies, significant alterations in the solution and interfacial properties of the surfactant may occur. As a result, modifications in the chemical structure of the surfactant may be needed to maintain a desired degree of surface activity. It cannot be overemphasized that a given surfactant effect will be intimately tied to the specific solvent environment in use. Any change in that environment may significantly alter the effectiveness of a surfactant and require major structural changes to retain the desired surface effects. Therefore, for surface activity in a given system, the prospective surfactant molecule must possess a chemical structure that is amphiphilic in the desired solvent under the proposed conditions of use. But how can one determine the best chemical structure for use in a given system?

For some time, a goal of surfactant–related research has been to devise a quantitative way to relate the chemical structure of surface–active molecules directly to their physicochemical activity in use. One of the earliest attempts to correlate surface activity and chemical structure came from the cosmetics industry and is known as the "hydrophile–lipophile balance" (HLB) system. Described more fully in Chapter 9, the HLB system relates the molecular composition of a surfactant (as mol% of hydrophile) to its surfactant properties. Although it is not a quantitative panacea for designing surfactant molecules, it continues to be an important tool in the practical arsenal of surfactant formulation technology. In more recent years, attempts have been made to use more theoretically "satisfying" tools such as the concepts of cohesive energy densities δ (also called *solubility* or *Hildebrand parameters*) and molecular geometry to correlate such relationships as surfactant chemical structure, the nature of the solvent, and surfactant activity on a more fundamental atomic and molecular level. Those and other schemes for predicting surfactant activity based on the specifics of molecular architecture will be addressed in more detail in subsequent chapters.

2.1. BASIC SURFACTANT BUILDING BLOCKS

One way to approach the concept of building a useful surfactant molecule is to look at the process much as a child building a boat with Leggo building blocks. The first step is to picture the desired structure. The second is to gather together the various types of blocks that the structure involves, and the third is to put together those varied blocks in the proper order and location to give the desired result. Modern organic chemistry has provided the synthetic chemist with a wide array of pieces. It is the imagination and skill of the chemist and engineer that must be used to attain the desired result. That image is, of course, oversimplified because there are many factors beyond a simple chemical structure that determine the potential utility of a given surfactant molecule such as toxicity, biocompatibility, environmental "friendliness," consumer acceptance, and energy needs. Nevertheless, the possibilities are seemingly limitless.

2.1.1. Basic Surfactant Classifications

In aqueous systems, by far the most important of surfactant applications in volume and economic impact, the hydrophobic group is generally a long-chain hydrocarbon group, although there are examples using fluorinated or oxygenated hydrocarbon or siloxane chains. The hydrophile or head will be an ionic or highly polar group that can impart some water solubility to the molecule. The most useful chemical classification of surface-active agents is based on the nature of the hydrophile, with subgroups based on the nature of the hydrophobe or tail. The four basic classes of surfactants are defined as follows:

1. *Anionic*—the hydrophile is a negatively charged group such as carboxyl ($RCOO^- M^+$), sulfonate ($RSO_3^- M^+$), sulfate ($ROSO_3^- M^+$), or phosphate ($ROPO_3^- M^+$).
2. *Cationic*—the hydrophile bears a positive charge, as for example, the quaternary ammonium halides ($R_4N^+ X^-$), and the four R -groups may or may not be all the same (they seldom are), but will usually be of the same general family.
3. *Nonionic*—the hydrophile has no charge, but derives its water solubility from highly polar groups such as polyoxyethylene (POE or $R{-}OCH_2CH_2O{-}$) or R–polyol groups including sugars.
4. *Amphoteric* (and zwitterionic)—the molecule contains, or can potentially contain, both a negative charge and a positive charge, such as the sulfobetaines $RN^+(CH_3)_2CH_2CH_2SO_3^-$.

In general, the nature of the hydrophobic groups may be significantly more varied than for the hydrophile. Quite often they are long-chain hydrocarbon groups; however, they may include such varied structures as

1. Long, straight-chain alkyl groups ($n = C_8$–C_{22} with terminal substitution of the head group)

$$CH_3(CH_2)_n{-}S$$

2. Branched-chain alkyl groups (C_8–C_{22}, internal substitution)

$$CH_3(CH_2)_nC(CH_3)H(CH_2)_mCH_2\text{—}S$$

3. Unsaturated alkenyl chains such as those derived from vegetable oils

$$CH_3(CH_2)_nCH{=}CH(CH_2)_m\text{—}S$$

4. Alkylbenzenes (C_8–$C_{15}C_6H_4$ with various substitution patterns)

$$C_9H_{19}(C_6H_4)\text{—}S$$

5. Alkylnaphthalenes (alkyl R usually C_3 or greater)

$$R_n\text{—}C_{10}H_{(7-n)}\text{—}S$$

6. Fluoroalkyl groups ($n > 4$, partially or completely fluorinated)

$$CF_3(CF_2)_n\text{—}S$$

7. Polydimethylsiloxanes

$$CH_3\text{—}(OSi[CH_3]_2O)_n\text{—}S$$

8. Polyoxypropylene glycol derivatives

$$CH_3CH(OH)\text{—}CH_2\text{—}O(\text{—}CH(CH_3)CH_2O)_n\text{—}S$$

9. Biosurfactants
10. Derivatives of natural and synthetic polymers

With such a wide variety of structures available, it is not surprising that the selection of a suitable surfactant for a given application can become a significant problem in terms of making the best choice of material for a given application.

2.1.2. Making Choice

The chemical structure of a surfactant is not the only determining factor in choosing between potential surfactant candidates for a given application. Economic, energetic, ecological, regulatory, and aesthetic considerations, in addition to questions of chemical functionality, are becoming more and more important in surfactant structure selection. Since most surfactants are used in formulations that include other ingredients, the relative role of the surfactant must be evaluated along with its physicochemical characteristics.

If the cost of the surfactant is significant compared to that of other components of a system, the least expensive material producing the desired effect will usually be preferred, all other things being equal. Economics, however, cannot be the only factor, since the final performance of the system will be of crucial importance. To make a rational selection of a surfactant, without resorting to an expensive and time-consuming trial-and-error approach, the formulator must have some knowledge of

1. The surface and interfacial phenomena that must be controlled in the specific application
2. The relationships among the structural properties of the available surfactants and their effects on the pertinent interfacial phenomena to be controlled
3. The characteristic chemical and physical properties of the available surfactant choices
4. Any special chemical or biological compatibility requirements of the system
5. Any regulatory limitations on the use a given class of materials (toxicity, allergenic reactions, ecological impact, etc.)
6. Public acceptance—"natural" versus "synthetic"

The following chapters will attempt to provide a basic foundation for making logical surfactant choices—or at least provide a good starting point and grounds for a good "educated guess."

2.2. THE GENERIC ANATOMY OF SURFACTANTS

Assuming an "aquocentric" point of view for the moment, surfactants, whether synthetic or of the "natural" fatty acid soaps family, are amphiphilic materials that tend to exhibit some solubility in water as well as some affinity for non-aqueous environments—they are nature's original bipolar chemical personalities, and we simply would not exist without them. (Perhaps the image of the universal arbiter would be more appropriate.) Such an ambivalent character occurs in materials that include two chemically distinct molecular groups or functionalities. For an aqueous system, as already noted, the functionality that would be readily soluble in water is termed the *hydrophile*; the other functionality, the *hydrophobe*, would, under normal circumstances, be essentially insoluble in water. It is the "push me, pull you" conflict within the particular molecular structure that produces the unique and amazingly useful family of chemical beasts that we know as surfactants.

Chemically speaking, the hydrophilic group is usually—not always—added synthetically to a hydrophobic material in order to produce a compound with some water solubility. The effectiveness of a given molecular structure as a surfactant will depend critically on the molecular balance between the hydrophile and the hydrophobe. Attaining the balance necessary to produce the desired result lies at the heart of surfactant science and technology, as does understanding the fundamental

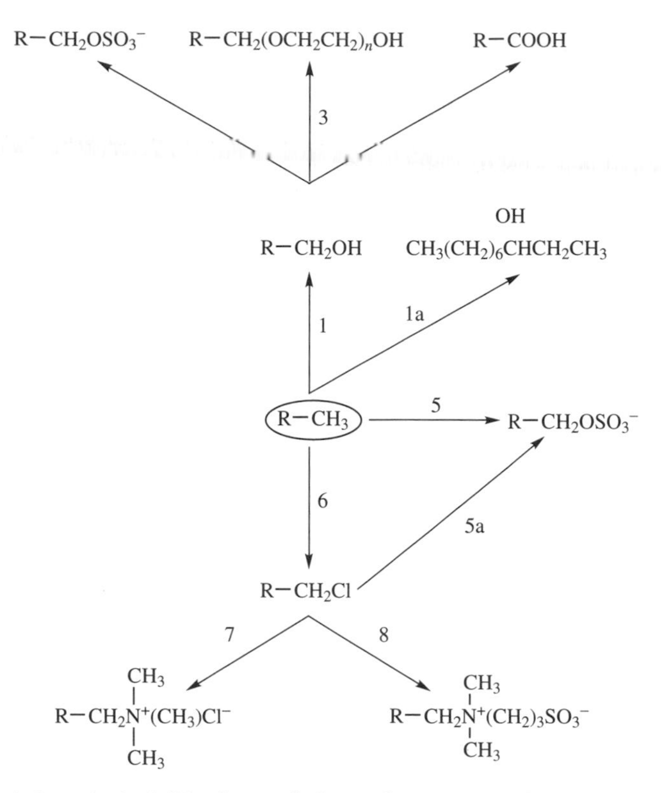

Figure 2.1. A hypothetical "family tree" for surfactant types that can be derived from dodecane through simple chemical reactions. In the scheme, $R = CH_3(CH_2)_{10}$–.

chemical principles leading to the observed phenomena. Several empirical schemes have been proposed for quantifying the critical balance between the two parts of surfactant molecules, and relating that balance to the activity of the material in a given application. Some of those ideas will be covered in more detail in later chapters. For now, we will see what some relatively minor structural changes can do to the character of the simple hydrocarbon molecule, *n*−dodecane (Figure 2.1).

2.2.1. The Many Faces of Dodecane

As the root of our surfactant family tree, we will consider the compound *n*−dodecane:

$$CH_3(CH_2)_{10}CH_3$$

This material is a hydrocarbon with essentially no solubility in water and is just about as hydrophobic as one could want, in a practical sense. The hydrocarbon chain is, by nature, somewhat unreactive chemically. In order to modify the molecule, it is necessary to open a "door" that will give chemical access to more of the tools of the synthetic trade. If one of the terminal hydrogen atoms on the chain is exchanged for an —OH group (path 1, Figure 2.1), the new material, 1-dodecanol, still has relatively low solubility in water, but it increases substantially relative to the parent hydrocarbon:

$$CH_3(CH_2)_{10}CH_2OH \quad \text{or} \quad n\text{-}C_{12}H_{25}OH$$

While such a modification seems pretty simple on paper, the actual process can be pretty messy, involving a rather random process of halogenation followed by substitution to form the alcohol. For purposes of the current discussion, however, we will pretend to live in a simpler world and assume that only the desired product is produced. If the alcohol functionality is placed internally on the hydrocarbon chain, as in 3-dodecanol (path 1a), the resulting material will be similar to the primary alcohol but will have slightly different solubility characteristics:

$$CH_3(CH_2)_2CH(OH)(CH_2)_5CH_3$$

Solubility differences among chemical isomers will generally be evident in other functional modifications. They may also be evident in changes in chemical reactivity and substitution patterns in subsequent chemical reactions. The effects of the position of substitution on surfactant properties can be quite significant and will be discussed in more detail later. Once formed, the alcohol can be sulfated (path 2) to produce dodecane sulfuric acid ester, a strongly acidic compound with good water solubility. When the sulfuric acid ester is neutralized with alkali, certain alkaline-earth metals, or organic amines, the material becomes highly soluble in water and an excellent surfactant. It is, in fact, probably the most extensively studied and best understood surfactant known to science—sodium dodecylsulfate (SDS):

$$n\text{-}C_{12}H_{25}OH + SO_3 \longrightarrow \underset{n\text{-dodecylsulfuric acid ester}}{n\text{-}C_{12}H_{25}\text{—}OSO_3H} + NaOH \longrightarrow \underset{\text{SDS}}{n\text{-}C_{12}H_{25}\text{—}OSO_3^-Na^+}$$

If the original n-dodecanol is treated with ethylene oxide (OE) and base catalyst (path 3), the material obtained is a docedylpolyoxyethylene (POE) polyether.

$$n\text{-}C_{12}H_{25}OH + n\text{OE} \longrightarrow n\text{-}C_{12}H_{25}\text{—}(OCH_2CH_2)_nOH$$

Such molecules can have widely varying solubility characteristics, depending on the value of n, the number of OE groups added to the molecule. Because of the nature of the reaction, the value of n will always be an average with a relatively large distribution of values. If $n = 10$, the material will be soluble in water and will show good surfactant properties. If n is as small as 5, its water solubility will decrease significantly, as will its usefulness as a surfactant. If n is taken to 20 or higher, high water solubility is attained, but most of the good surfactant qualities will be lost. For n less than 5, the material will have little significant water solubility.

If the original n-dodecanol is oxidized to dodecanoic acid (path 4), the resultant acid has very limited water solubility; however, when neutralized with alkali it becomes water-soluble, a classic soap. Although the dodecanoic acid in this example is hypothetically derived from a petroleum source, it will be chemically indistinguishable from the same molecule derived from a "natural" source such as vegetable oils or animal fats. It is a "fatty acid," its original source notwithstanding:

$$n\text{-}C_{12}H_{25}OH + [O] \longrightarrow n\text{-}C_{11}H_{23}COOH + NaOH \longrightarrow n\text{-}C_{11}H_{23}COO^{-}Na^{+}$$

The alkali carboxylate will be a reasonably good surfactant for many applications, but if the hydrocarbon chain length were increased to 16 or 18 carbons, many of the surfactant properties would be even better, illustrating the importance of obtaining the proper balance between the hydrophilic and hydrophobic portions of the molecule.

A major drawback to the use of the classic alkali soaps has always been their great sensitivity to their aqueous environment. They are often poorly soluble in cold water and are always sensitive to the presence of polyvalent metal ions in solution. The main components of so-called hard water are calcium, magnesium, and other divalent and trivalent cations. In the presence of such materials, the carboxylate soaps form salts of low water solubility that precipitate to produce scummy deposits, commonly encountered as the "bathtub ring." Their solubility in water is simply too low for the system to attain a sufficiently high concentration to produce useful results. In nonaqueous solvents, on the other hand, the polyvalent salts of carboxylate soaps exhibit significantly enhanced solubility and perform admirably in many surfactant functions.

Carboxylate soaps are also strongly affected by changes in solution pH and temperature. Their solubility in water increases significantly with increases in temperature, as does their usefulness as cleaners; the reverse is obviously the case. The sensitivity of the carboxylate soaps to the presence of commonly encountered ions, their sensitivity to pH changes, and their decreased solubility in cold water were the driving forces for the development of synthetic surfactants that would not be so adversely affected by the common circumstances of hard water and cool washing temperatures.

Continuing with the example of dodecane–based surfactant materials, the parent hydrocarbon can be sulfated to yield dodecane sulfonic acid (path 5), which closely resembles the sulfuric acid ester discussed previously and has similar water solubility:

$$n\text{-}C_{12}H_{25}OH \longrightarrow n\text{-}C_{12}H_{25}\text{—}SO_3H \longrightarrow n\text{-}C_{12}H_{25}\text{—}SO_3^{-}Na^{+}$$

When neutralized with the proper base, the resulting material is an excellent surfactant. It should be noted that while the sulfonic acid is chemically related to the ester, their solution and surfactant properties are not identical, so their potential applications may be different as well. The hydrocarbon sulfonates are generally more chemically stable than the sulfate esters, but the economics of their preparation is an impediment to their widespread use.

If the original dodecane molecule were terminally chlorinated (path 6) and reacted with trimethylamine (path 7), the resulting compound would be dodecyltrimethylammonium chloride, a water-soluble compound exhibiting some surfactant properties, but not generally as useful as the anionic analogs:

$$n\text{-}C_{12}H_{26} + Cl_2 \rightarrow n\text{-}C_{12}H_{25}Cl + N(CH_3)_3 \rightarrow n\text{-}C_{12}H_{25}N(CH_3)_3^+Cl^-$$

The utility of such compounds is limited not so much by their surface activity as by their interaction with various oppositely charged components found in practical systems (see Chapters 9 and 10).

Up to this point we have covered three of the four general classes of surfactants defined so far: anionic, nonionic, and cationic. To produce an example of the fourth class, an amphoteric or zwitterionic surfactant, it is only necessary to react the dodecylchloride prepared as described above with a difunctional material such as *N,N*-dimethyl-3-aminopropane-1-sulfonic acid (path 8):

$$C_{12}H_{25}Cl + (CH_3)_2NCH_2CH_2CH_2SO_3H \rightarrow C_{12}H_{25}N^+CH_2CH_2CH_2SO_3H\ Cl^-$$
$$+ MOH \rightarrow C_{12}H_{25}N^+CH_2CH_2CH_2SO_3^- + M^+ + Cl^-$$

The result is just one of several possible chemical types that possess the amphoteric or zwitterionic character of this class of materials. Under acidic conditions, the molecule carries a net positive charge. Under basic conditions, the acid is neutralized and the molecule carries both a positive charge and negative charge. In this context, we are talking about the electrical nature of the surface-active portion of the molecule and not any associated, but non-surface-active ions such as Cl^- or M^+.

The number of chemical modifications of the dodecane or similar simple hydrocarbon molecules that can lead to materials with good surfactant characteristics is impressive. When hydrocarbons containing aromatic groups, unsatruration, branching, heteroatom substitution, polymers, or other interesting functionalities are considered, the synthetic possibilities seem almost unlimited. Only imagination, time, and money seem to limit our indulgence in creative molecular architecture.

In each example discussed above, an aqueous "solubilizing group" has been added to the basic hydrophobe to produce materials with varying amounts of useful surfactant characteristics. When one considers the wide variety of hydrophobic groups that can be coupled with the relatively simple hydrophiles discussed so far and add in more complex and novel structures, the number of combinations becomes impressive. When viewed in that light, the existence of thousands of distinct surfactant structures doesn't seem surprising.

Using the evolution of dodecane-based surfactant structures as a jumping-off point, the discussion will now turn to more specific examples of surfactant building blocks. As noted, the chemical possibilities for surfactant synthesis seem almost limitless. The following discussions, on the other hand, will be limited by space and time, and are designed as guides rather than a comprehensive compilation.

2.2.2. Surfactant-Solubilizing Groups

The solubilizing groups of modern surfactants fall into two general categories: those that ionize in aqueous solution (or highly polar solvents) and those that do not. Obviously, the definition of what part of a molecule is the solubilizing group depends on the solvent system being employed. For example, in water the solubility will be determined by the presence of a highly polar or ionic group, while in organic systems the solubilizing functionality will be the organic portion of the molecule. It is important, therefore, to define the complete system under consideration before discussing surfactant types. As noted before, because the majority of surfactant work is concerned with aqueous environments, the terminology employed will generally be that applicable to such systems. Generality will be implied, however.

The functionality of ionizing hydrophiles derives from a strongly acidic or basic character, which leads to the formation of highly ionizing salts on neutralization with appropriate bases or acids. In this context, the carboxylic acid group, while seldom considered as such in acid–base theory, is classified as a strong acid. The nonionizing, or nonionic hydrophilic, groups, on the other hand, have functionalities or groups that are individually rather weak hydrophiles (alcohols, ethers, esters, etc.) but have an additive effect so that increasing their number in a molecule increases the magnitude of their solubilizing effect.

The most common hydrophilic groups encountered in surfactants today are illustrated in Table 2.1, where, as already noted, R designates some suitable

TABLE 2.1. The Most Commonly Encountered Hydrophilic Groups in Commercially Available Surfactants

General Class Name	General Solubilizing Structure
Sulfonate	$R—SO_3^- M^+$
Sulfate	$R–OSO_3^- M^+$
Carboxylate	$R–COO^- M^+$
Phosphate	$R–OPO_3^- M^+$
Ammonium	$R–N^+R'_xH_y X^-$ ($x = 1–3$, $y = 3–1$)
Quaternary ammonium	$R–N^+R'_3 X^-$
Betaines	$R–N^+(CH_3)_2CH_2COO^-$
Sulfobetaines	$R–N^+(CH_3)_2CH_2CH_2SO_3^-$
Polyoxyethylene (POE)	$R–OCH_2CH_2(OCH_2CH_2)_nOH$
Polyoxyethylene sulfates	$R–OCH_2CH_2(OCH_2CH_2)_nOSO_3^- M^+$
Polyols	$R–OCH_2–CH(OH)–CH_2OH$
Sucrose esters	$R–O–C_6H_7O(OH)_3–O–C_6H_7(OH)_4$
Polyglycidyl esters	$R–(OCH_2CH[CH_2OH]CH_2)_n–\cdots–OCH_2CH[CH_2OH]CH_2OH$

hydrophobic group that imparts surface activity, M^+ is an inorganic or organic cation, and X^- is an anion (halide, acetate, etc.). The list is in no way complete, but the great majority of surfactants available commercially fall into one of those classes. It is possible, and sometimes even advantageous, to combine two or more of the functionalities described above to produce materials with properties superior to those of a monofunctional material. Prime examples would be the alcohol ether sulfates in which a POE nonionic material is terminally sulfated, $R(OCH_2CH_2)_n OSO_3^- M^+$, and, of course, the zwitterionic and amphoteric materials noted, which often exhibit the advantages of both ionic and nonionic surfactants while having fewer of their potential drawbacks. The "hybrid" classes of surfactants, while not yet composing a major fraction of total surfactant consumption, can be particularly useful because of their flexibility and, especially in personal care items such as shampoos, because of the low level of eye and skin irritation that they are often found to produce.

Building on the basic hydrophilic functionalities discussed above, we now turn our attention to some of the specific structural subgroups derived from the more common hydrophobic groups.

2.2.3. Common Surfactant Hydrophobic Groups

By far the most common hydrophobic group used in surfactants is the hydrocarbon radical having a range of 8–22 carbon atoms. Commercially there are two main sources for such materials that are both inexpensive (relatively speaking) and available in sufficient quantity to be economically feasible: "natural" or biological sources such as agriculture and the petroleum industry (which is, of course, ultimately biological). Figure 2.2 indicates some evolutionary pathways from raw materials to final surfactant product. There are, of course, alternative routes to the same materials, as well as other surfactant types that require more elaborate synthetic schemes. Those shown, however, constitute the bulk of the synthetic materials used today.

Many, if not most, surfactant starting materials are not chemically pure materials. In fact, for economic and technical reasons, most surfactant feedstocks are mixtures of isomers whose designations reflect some average value of the hydrocarbon chain length included rather than a "true" chemical composition. In some cases isomeric composition may be indicated in the surfactant name or description, while in others the user is left somewhat in the dark. The term "sodium dodecylsulfate," for example, implies a composition containing only C_{12} carbon chains. The material referred to as "sodium lauryl sulfate," on the other hand, is nominally a C_{12} -surfactant, but will contain some longer- and shorter-chain homologues. Each source of raw materials may have its own local geographic or economic advantage, so that nominally identical surfactants may exhibit slight differences in surfactant activity due to the subtle influences of raw-materials variations. Such considerations may not be important for most applications, but should be kept in mind in critical situations.

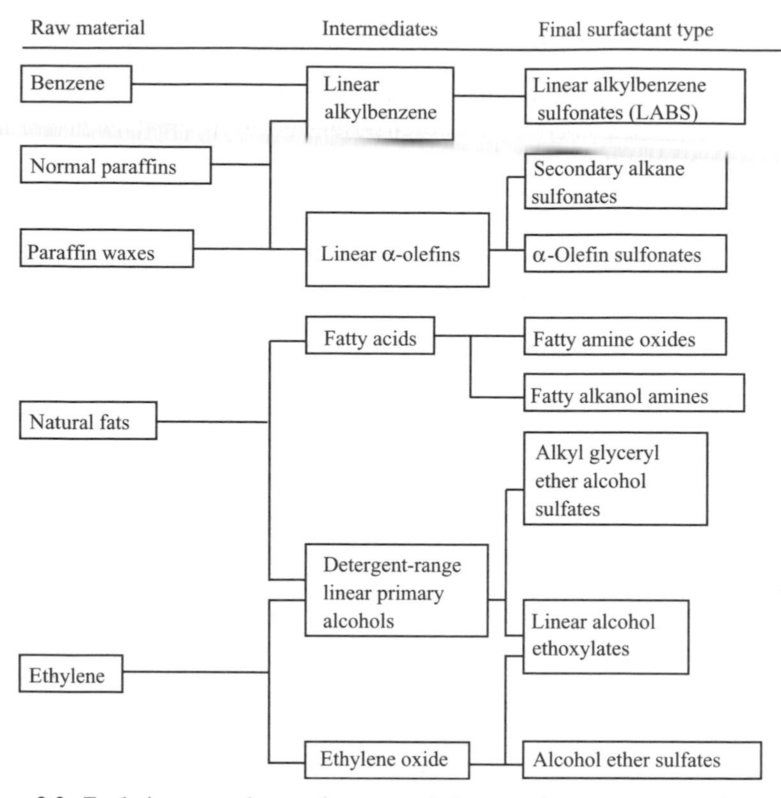

Figure 2.2. Evolutionary pathways for some of the more important types of commercial surfactants.

2.2.3.1. The Natural Fatty Acids

One of the major sources of raw materials for the commercial production of surfactants is also the oldest source—agriculture. Fats and oils (oleochemicals), products of nature's ingenuity and human labor, are triglyceride esters of fatty acids, which can be readily hydrolyzed to the free fatty acids and glycerol. Naturally occurring plant or animal fatty acids usually contain an even number of carbon atoms arranged in a straight chain (no branching), so that groups symbolized by an R in abbreviated nomenclature will contain an odd number that is one less than that of the corresponding acid. The carbons are linked together in a straight chain with a wide range of chain lengths; those with 16 and 18 carbons are the most common. The chains may be saturated, in which case the R group has the formula C_nH_{2n+1}, or they may have one or more double bonds along the chain. Hydroxyl groups along the chain are uncommon, but not unknown, especially in soaps made using castor oil (recinoleic acid). Other substitutions are rare.

Commercially, the largest surfactant outlet for fatty acids is conversion to soap by neutralization with alkali. In a strict sense, this may be considered to be a synthetic process, and soap therefore a synthetic surfactant. However, common usage reserves the term "synthetic" for the more modern products of chemical technology, primarily petroleum-derived, that have been developed in the twentieth century and generally show important improvements over the older soap technology. The chemical processes required for the production of modern surfactants and detergents are also usually much more complicated than the simple neutralization involved in soap manufacture. In this context, the term "simple" is relative, as is well known to anyone who has ever tried to prepare homemade soap.

2.2.3.2. *Saturated Hydrocarbons or Paraffins*

The hydrophobic groups derived from petroleum are principally hydrocarbons, originating from the paraffinic or higher-boiling fractions of crude oil distillates. The chain lengths most suitable for surfactant hydrophobes, C_{10}–C_{20}, occur in the crude oil cuts boiling somewhat higher than gasoline, namely, kerosene and above. The main components of kerosene are saturated hydrocarbons ranging from $C_{10}H_{22}$ to $C_{15}H_{32}$, ordinarily containing 10–25% of straight-chain homologs. There may be significant amounts of branched-chain isomers present, in addition to quantities of saturated cyclic derivatives, alkyl benzenes, and naphthalenes, and minor amounts of other polycyclic aromatics.

The paraffins have the disadvantage of being relatively chemically unreactive so that direct conversion to surfactants is difficult. As discussed above, substitution of one or more hydrogen atoms with halogen offers a pathway to some surfactant systems, but manufacturing complications can be an impediment. It is usually necessary to synthesize the surfactant by way of some more reactive intermediate structures, commonly olefins, alkyl benzenes, or alcohols. Such compounds contain reactive sites that are more easily linked to the required solubilizing groups.

2.2.3.3. *Olefins*

Olefins with the desired chain length are prepared by building up molecules from smaller olefins (oligomerization), by breaking down (cracking) larger molecules, or by direct chemical modification of paraffins of the desired chain length. An important historic example of surfactant-grade olefin production by the oligomerization process is the preparation of tetrapropylene, $C_{12}H_{24}$:

$$CH_3{-}CH{=}CH_2 \rightarrow \text{(mixed isomers of } C_{12}H_{24} \text{ olefins and higher and lower homologs)}$$

This may be prepared by the oligomerization of propylene, a byproduct of refinery operations, under the influence of a phosphoric acid catalyst. The reaction conditions are drastic, and extensive random reorganization of the product molecules occurs with substantial formation of intermediate isomers in the C_{10}–C_{14} range. The final product is composed of a variety of highly branched isomers and homologs, with the double bond usually situated internally in the molecule.

A second type of built-up olefin is that obtained by the polymerization of ethylene using a Ziegler–Natta catalyst. Such materials are predominantly linear with even carbon numbers, although branched isomers are present in small amounts. The ethylene raw material historically has been more expensive than propylene. The catalyst is also more expensive and the reaction conditions more sensitive and critical than those for propylene oligomerization. The formula is:

$$CH_2{=}CH_2 + \text{Z–N catalyst} \rightarrow \text{predominantly } CH_2{=}CH{-}C_9H_{19} \text{ to } C_{19}H_{39}$$

Production of surfactant or detergent-class olefins from higher-molecular-weight precursors is accomplished by the cracking process, which uses high temperatures to split high-molecular-weight paraffins into smaller units. A catalyst may also be employed in the process. Basically, the reaction involves the splitting of a paraffin into two smaller molecules, a paraffin and an olefin. In practice, a wide range of products is obtained because the original molecules may split at any spot along the chain and the resulting products may themselves be cracked further. Each olefin molecule that undergoes such secondary cracking produces two more olefins so that the paraffin content becomes progressively smaller. The olefins produced are predominantly α-olefins, with the double bond located at the terminus of the molecule, as indicated above. If the original cracking stock is linear, the product olefins will be predominantly linear; if branched or cyclic structures were originally present, such structures will also appear in the product.

The third route to detergent olefins is from paraffins of the same chain length. In principle, it is necessary only to remove two hydrogen atoms from an adjacent pair of carbons along the chain to produce the desired olefin, but the difficulties of dehydrogenation are such that a two-step process of chlorination and dehydrochlorination has been developed. In either process the reaction easily proceeds past the desired stage to give polychlorinated paraffins and polyolefins, all undesirable byproducts.

2.2.3.4. Alkyl Benzenes

Alkyl benzenes first made their appearance as hydrophobic groups in the late 1940s as a result of new industrial processing capabilities related to the chemistry and chemical engineering of aromatic alkylation reactions. They were prepared by the more or less random chlorination of kerosene and the subsequent Friedel–Crafts alkylation of benzene. Their real dominance in the field began in the early 1950s with the appearance of tetrapropylene-based alkyl benzenes, a functionally better and cheaper product obtained in a one-step process involving addition of benzene across the double bond of the olefin. A variety of Friedel–Crafts catalysts may be employed including aluminum chloride or hydrogen fluoride.

The alkyl group of the alkyl benzene might be expected to have a carbon backbone identical to that of the olefin from which it was derived. This is true in most, but not all, cases because the alkylation process may cause rearrangement of the carbons in the chain. Furthermore, the reaction of the benzene ring with the double bond of the olefin involves a number of intermediate steps during which isomerization may occur, so that the benzene may finally link on at some position other that of the original double bond. Thus each of the many species that make up the olefin feedstock may give rise to several isomeric alkyl benzenes, and the resulting material will be an even more complex mixture than the original. Gas chromatographic analysis of typical products may show more than 100 at least partially resolved components.

Propylene-derived alkyl benzenes were pretty much phased out in the 1960s because of their exceptional (and undesirable) biological stability and were replaced by the linear alkyl benzenes prepared from linear olefins or intermediate chlorinated paraffins. These products, like the tetrapropylenes, have a chain length that may range from C_8 to C_{16}, variously distributed according to the specific preparation procedure and the properties desired in the final surfactant. The nominal carbon values for most commercial products are C_8 and C_9 alkyl groups.

The alkylation of benzene with a linear olefin commonly results in a mixture of all possible linear secondary alkyl benzenes with that total carbon number, even though the original olefin may have been a pure compound. During the alkylation reaction a rearrangement occurs so that the ring bonds at any of the carbon atoms along the chain except the ends. Primary alkyl benzenes are not formed in this process. At equilibrium, the same isomers will be formed regardless of the original location of the double bond. Indeed, any of the five dodecylbenzene isomers can be converted into a mixture of all five by treatment with $AlCl_3$, the catalyst often used in the alkylation. A similar mixture is obtained using the chloroparaffin process. Since there is not much difference in the reactivity of the 26 hydrogen atoms in dodecane, all possible monochloro isomers are present in the chlorododecane and all possible phenyl isomers are present in the resulting dodecylbenzene mixture, including a small amount of the primary isomer.

2.2.3.5. Alcohols

Long-chain alcohols have been used as a source of surfactant hydrophobes since the earliest days of synthetic detergent manufacture. Linear alcohols have been used since the beginning; branched alcohols are a more recent addition to the chemical arsenal.

The classical route to a linear alcohol is by the reduction of the carboxyl group of a fatty acid. Actually, an ester of the carboxylic acid is usually employed, since the carboxyl group itself reacts rather sluggishly. Lauryl and tallow alcohols are two of the most commonly used substrates for surfactant synthesis. The first is derived from lauric acid, predominantly a C_{12} acid, but also usually containing some amounts of lower and higher homologs. The tallow alcohols average around C_{18}. Partially or completely hydrogenated fatty acids from oilseeds usually have carbon chains in the C_{16}–C_{18} range. Depending on the reduction process used, they may contain some unsaturated alcohols derived from the unsaturated acids in the original fatty acids:

$$CH_3(CH_2)_nCOOR + H_2 \rightarrow CH_3(CH_2)_nCH_2OH$$

Since the early 1960s linear primary alcohols have been available from petroleum sources, namely, ethylene. The process for their preparation is similar to the Ziegler process for linear olefins, except that the last step is an oxidative one yielding the alcohol directly instead of an olefin. As oligomers of ethylene, the Ziegler-derived alcohols are produced in even-numbered chain lengths. The average chain length and distribution of homologs can be controlled somewhat by the reaction conditions, and completely by subsequent distillation. The –OH groups are at the end of the chain (terminal), so that they are identical to the alcohols derived from the natural fatty acids. However, the two products may differ slightly because of variations in the amounts and distribution of minor products and impurities.

Branched-chain alcohols were used extensively for surfactant manufacture prior to the changeover to the more readily biodegradable linear products. They were usually derived from polypropylenes by the "oxo" process, which involves catalytic addition of carbon monoxide and hydrogen to the double bond in a sequence of reactions. Thus the tetrapropylene derivative is nominally a C_{13} alcohol, as highly branched as the original raw material.

If a linear α-olefin is used in the oxo process, addition may occur at either end of the double bond to give a mixture of linear primary and methyl branched secondary alcohols. Substitution further down the chain occurs only in small amounts, and with the proper choice of reaction conditions the proportion of linear primary alcohol may reach 80% or more.

The development of linear paraffin feedstocks for the production of linear alkyl sulfates (LAS) also made many secondary alcohols feasible as surfactant hydrophobes. Here the –OH group may be introduced by reaction of the paraffin with oxygen, or by chlorination and subsequent hydrolysis. In either case all possible isomers are formed and the OH group is found on any of the carbons along the chain.

2.2.3.6. Alkyl Phenols

The alkyl phenol hydrophobes are produced by addition of phenol to the double bond of an olefin. The alkyl group may be linked to the ring either ortho, meta, or para to the hydroxyl group, and the position can have a significant impact on the characteristics of the resulting surfactant. The earlier commercial products were derived from branched olefins such as octylphenols from diisobutylene and nonylphenols from tripropylene. More recently, linear alkylphenols have become available with the development of linear olefins for LAS.

2.2.3.7. Polyoxypropylenes

The polyoxypropylenes, oligomers of propylene oxide, can be cited as an example of nonhydrocarbon hydrophobes. A complete line of surfactants known commercially as the "pluronics" have been developed commercially. These are block copolymers of propylene oxide and ethylene oxide. By careful control of the relative amounts of each component incorporated into the polymer, it is possible to exercise a subtle control over the solubility and surfactant character of the product. The character of other hydrophobes such as alcohols and alkyl phenols may also be modified by addition of propylene oxide.

2.2.3.8. Fluorocarbons

An important modern addition to the spectrum of available hydrophobic functionalities for the design of surfactants is that in which fluorine is substituted for hydrogen on the carbon chain. Substitution may be complete to produce the "perfluoro" materials (R_f) such as sodium heptadecafluorooctane sulfonate

$$C_8F_{17}SO_3^-Na^+$$

or they may possess a terminal hydrogen (HR_f) with the general structure

$$HC_nF_{2n}{-}S$$

where S may be any of the solubilizing groups discussed. Hydrocarbon groups may also be encountered in association with linking functionalities such as

$$R_f{-}N(CH_3)CH_2CH_2{-}S$$

The most commonly encountered commercial fluoro surfactants are prepared by the electrolytic fluorination of the corresponding alkyl carboxylic acid fluoride or sulfonyl fluoride

$$CH_3(CH_2)_nCOF + E^\circ/HF \rightarrow CF_3(CF_2)_nCOF$$

or

$$CH_3(CH_2)_nSO_2F + E^\circ/HF \rightarrow CF_3(CF_2)_nSO_2F$$

The fluorinated products may then be hydrolyzed to the corresponding acid and neutralized, or functionalized by reaction of the reactive acid fluoride to introduce linking groups such as the sodium 2-sulfonamidoethane sulfonate

$$C_8F_{17}SO_2NHCH_2CH_2SO_3^-Na^+$$

Although normal straight-chain hydrocarbon starting materials may be used in the electrolytic process, some branching will always result, commonly yielding materials with up to 30% branched isomers.

Perfluorinated alcohols cannot be prepared directly from the electrolytic process since they require the reduction of the corresponding carboxylic acid to yield structures of the nature

$$CF_3(CF_2)_nCH_2OH$$

The presence of the two hydrogens alpha to the hydroxyl is synthetically useful; however, because the perfluorinated alcohols cannot be effectively used in esterification reactions, the perfluorinated products, when obtained at all, are very unstable to hydrolysis.

Hydrogen-terminated materials can be produced by the oligomerization of tetrafluoroethylene. In such a reaction, the product will be a mixture of homologous alcohols with an average molecular weight similar to those obtained in the preparation of POE-containing materials. It is also possible to obtain perfluorinated materials in the oligomerization process.

As a class, fluorocarbon compounds in which all carbon–hydrogen bonds have been replaced by carbon–fluorine possess the lowest surface tensions and surface energies of any substances currently known to science. If surface activity is defined as the tendency of a substance to reduce the total free energy of a system by preferential adsorption at available surfaces and interfaces, or aggregation to form micelles, the fluorocarbon surfactants can also be considered to be the most surface-active materials around. In addition, the electronic nature of the carbon–fluorine bond is such as to make them the most chemically stable of surfactants, able to withstand temperatures and chemical environments that would quickly destroy conventional hydrocarbon-based materials. As a result of those characteristics, the fluorocarbons have found applications in many technological areas in which hydrocarbons are either much less effective or chemically incompatible. The different chemical and electrical properties of the fluorine atom relative to hydrogen have also resulted in fluorinated surfactants finding applications for the modification of the surface properties of coated materials, especially with respect to triboelectric charging phenomena. Unfortunately, the wide use of fluorocarbon surfactants is limited by their relatively high cost, their environmental persistence, and the persistence and possible detrimental health effects of intermediates in their production. In some situations, however, their superior surface activity is essential and outweighs other factors.

2.2.3.9. Silicone Surfactants

Another type of non-hydrocarbon-based hydrophobic group gaining importance in the field of surfactant technology is that in which siloxane oligomers are functionalized to produce water-soluble materials:

$$CH_3O{-}Si(CH_3)_2{-}[O{-}Si(CH_3)_2]_n{-}O{-}R$$

The resulting materials usually exhibit surface activities falling somewhere between those of the normal hydrocarbons and the fluorinated compounds just discussed. Because of the nature of the siloxane linkage, such surfactants do not always follow the usual rules of surfactant activity with regard to such things as micelle formation and solubilization. They are very effective at the liquid–air interface, even in many nonaqueous solvents, and find wide application as antifoaming agents (see Chapter 8). A great deal of technical information on siloxane-based surfactants is not available in the open literature, possibly because of difficulties faced in characterizing the siloxane group and because much of the work in the area is of a proprietary nature.

2.2.3.10. Miscellaneous Biological Structures

In more modern developments, a great deal of new research is now under way in relation to surfactants and amphiphilic materials produced by fermentations and other biological processes. As might be expected on the basis of "normal" natural products chemistry, biologically produced amphiphiles are usually complex molecules including amino acid or protein functionalities, lipids, sugars, and all of the other twists and turns that nature has devised throughout evolution. In fact, it is found that the exact structure of the amphiphile produced by a given organism will vary according to the conditions of the fermentation and the exact nature of the "food" provided to the relevant organism. Biological surfactant systems will be discussed in a bit more detail in later chapters.

When properly combined, the major chemical components of surfactants produce the various classes of amphiphilic materials introduced above. The discussion will now turn to a more detailed description of each of the four primary classes with regard to their preparation and some general applications. While the coverage is not comprehensive, it does illustrate the major points of interest applicable to each class. With all the possible chemical structures available for surfactant synthesis, it is necessary to have some logical system of classification to guide the user to the material best suited to immediate and future needs.

2.3. THE SYSTEMATIC CLASSIFICATION OF SURFACTANTS

Surfactants may be classified in several ways, depending on the needs and intentions of the people involved. One of the more common schemes relies on classification by application. Surfactants may be classified as emulsifiers, foaming agents, wetting agents, dispersants, and the like. For the user whose work is confined to one type of application, such a classification scheme has certain obvious advantages. It does not, however, tell much about the specific chemical nature of the surfactant itself, nor does it give much guidance as to other possible uses of a material.

Surfactants may also be generally classified according to some physical characteristic such as primarily water or oil solubility or stability in harsh environments. Alternatively, some specific aspect of the chemical structure of the materials in question may serve as the primary basis for classification; an example is the type of linking group between the hydrophile and the hydrophobe (e.g., oxygen, nitrogen, amide, sulfonamido). Perhaps the most useful scheme from a general point of view, however, is that based on the overall chemical structure of the materials in question. In such a classification system, it is possible to more easily correlate chemical structures with interfacial activity, and thereby develop some general rules of surfactant structure–performance relationships.

As already mentioned, the simplest classification procedure is that in which the primary type is determined by the nature of the solubilizing functionality. Within each primary classification by solubilizing group there will exist subgroups according to the nature of the lyophobic group. It is possible to construct a classification system as complex as one might like, breaking down the lyophobic groups by their finest structural details such as branching and unsaturation in the alkyl chain. Such extremes, however, can introduce unnecessary complications in any discussion of structure–performance relationships, especially since surfactant systems consisting of several isomers or homologs are the rule rather than the exception. All subsequent discussion will adhere to what seems to be becoming the most widely accepted classification system based on the general chemical type of the lyophobic and solubilizing groups.

When discussing the commercial aspects of surfactant technology, especially with regard to the raw-materials sources, it is common to refer to materials with respect to their original starting materials—petroleum-based "synthetics" or "natural" oleochemicals-based materials. While such a classification may be useful from economic or sociopolitical points of view (where "natural" is "better" than synthetic), the complex natures of such materials make it very difficult to illustrate the role of chemical structures in determining surfactant properties. While modern, extremely sensitive analytical procedures available today may be able to differentiate between nominally identical products derived from agricultural or petrochemical feedstocks, the functional impact of such differences is unlikely to be evident in the use phase of the surfactant lifecycle. Later discussions, therefore, will be couched more in terms of chemical structure than of source. It should always be kept in mind, however, that nominally identical surfactants derived from different raw materials might exhibit differences in activity as a result of different isomer distributions.

In the following discussions, the previously introduced scheme of four primary surfactant classes has been employed, based on the nature of the principal solubilizing group. Sub-classifications are developed for specific hydrophobic or hydrophilic groups. Among the various books and reviews on surface activity and surfactants, there is often some disagreement as to the proper classification of amphoteric or zwitterionic materials. Some authors prefer to class such materials as nonionic because of their net electrical neutrality; however, the presence of discrete charges on the molecules, their potential sensitivity to environment (pH, temperature, etc.) in some cases, and personal preference cause me to favor the use of a separate classification. In the final analysis, the materials themselves could care less what we may call them; it is the manifestation of their chemical nature in solution and at interfaces that concerns us. As long as we have a firm concept of the chemical structures involved and their effects on the activity of a surfactant, nomenclature can be safely relegated to a position of secondary importance.

2.4. ANIONIC SURFACTANTS

The largest class of surface-active materials in general use today fall in the anionic classification, constituting 70–75% of total worldwide surfactant consumption. The major subgroups of this class are the alkali carboxylates or soaps, sulfates, sulfonates, and to a lesser degree, phosphates. The variety of anionic materials available arises primarily from the many types of hydrophobic groups that can be modified

TABLE 2.2. Approximate Consumption of Various Surfactant Types in the Major Industrialized Areas in 2000 ($\times 10^3$ metric tons)

Surfactant Type	Amount	Percent of Total (%)
Anionics (total)	4284	65
Nonionics (total)	1845	28
Other (amphoterics, cationics, etc)	461	7
Total, all surfactants	6590	100

by the addition of the proper anionic species. The overwhelming predominance of the anionic class of surfactants in economic importance is illustrated in Table 2.2. In 2000, approximately 65% of the total weight of surfactants consumed in the more industrialized countries was of the anionic variety.

Soaps derived from animal and vegetable fats and oils were the only surfactants available to humankind for thousands of years. The historical and economic advantage of the fatty acid soaps has always been their ready availability from natural, renewable sources. The exact properties of a given carboxylate soap will depend on the source of the raw material. The tallow acids (derived from animal fats), for example, are generally composed of oleic acid (40–45%), palmitic acid (25–30%), and stearic acid (15–20%). The materials derived from coconut fatty acids are usually composed of 45–50% C_{12} acids, 16–20% C_{14}, 8–10% C_{16}, 5–6% oleic acid, and 10–15% of materials of less than 12 carbons. Additional materials are derived from the tall oils (50–70% fatty acids and 30–50% rosin acids). In almost all the carboxylic acid materials, the acid is neutralized to the sodium or potassium salt, although amine salts are popular for some specific applications. As already mentioned, the major disadvantages of the carboxylic acid soaps are that they are very sensitive to the presence of di- and trivalent cations, high salt concentrations of any type, low pH (which produces water-insoluble free fatty acids), and low temperatures. They also have surface adsorption characteristics that make them hard to rinse off, leaving a generally undesirable "soapy" feel on the skin or clothes that may produce itching when dry.

The development of synthetic surfactants began when the functional or perceived shortcomings of the classical soaps in many modern industrial processes requiring surfactant action became apparent. Those same modern processes, especially in the field of organic synthesis, made it possible and practical to produce better alternatives on an industrial scale. It was found, for example, that the sensitivity of soaps to changes in pH and the presence of polyvalent ions could be overcome by the addition of another solubilizing group to the molecule. From an early synthetic point of view, the most readily accessible functionalities were the sulfonic acids and sulfuric acid esters of long-chain alkyl groups. Beginning with the development of the so-called turkey red oil used in the dyeing industry in the latter part of the nineteenth century, surfactants based on either sulfate esters or sulfonic acid salts have dominated the field.

The rapid growth in the availability of alternate raw materials after World War II led to the development of well-characterized synthetic surface-active materials to meet the needs of the new technological world. Because of their utility and relative ease of manufacture, the sulfate and sulfonate surfactants continue to lead the way, although many classes of ionic and nonionic compounds have been investigated and developed in the hope of finding a "super surfactant" with the surface activity and economic attractiveness needed to dominate a large portion of the market for such materials. The literature on the development of new surfactants, both academic and industrial, represents a wealth of useful knowledge for the surfactant chemist or the user in need of a solution.

2.4.1. Sulfate Esters

The neutralized organic esters of sulfuric acid, closely related chemically to the sulfonic acid salts to be discussed later, exhibit a number of significant differences in the chemistry of their preparation, in their hydrolytic stability after preparation, and in their ultimate activity as surfactants. Some of those differences are related to the different natures of the chemical linkages between the hydrophobic tail of the sulfates (carbon–oxygen–sulfur) versus that of the sulfonates (carbon–sulfur). Such seemingly minor differences in chemical structure lead to differences in the polarizability of the head group, different degrees of ion binding in solution, and different degrees of hydration, all of which may alter the surfactant characteristics of the materials. Since the most widely studied surfactant, sodium dodecylsulfate (SDS) is a member of the sulfate ester family, we will, out of deference to the "king," discuss that family first.

As the name implies, the sulfate ester surfactants contain a sulfuric acid ester group, which acts as the solubilizing group. Usually encountered as the alkali or ammonium salts, this class of materials has the generic formula $ROSO_3^-M^+$, where R is one of the hydrophobic groups described earlier. While the best-known members of this class are the simple straight-chain aliphatic materials such as SDS, many more complex structures are known and have found wide application.

The synthesis of the sulfate esters usually involves either the esterification of an alcohol with sulfuric acid, sulfur trioxide, or chlorosulfonic acid, or the addition of sulfuric acid across a double bond:

$$ROH + H_2SO_4 \text{ (or } ClSO_3H) \rightarrow ROSO_3H + H_2O$$

or

$$RCH{=}CH_2 + H_2SO_4 \rightarrow RCH(OSO_4H)CH_3$$

While the reactions as written appear quite simple, it must be remembered that the substrates for such processes are often mixtures of isomers or even functionalities, so that the resultant product may be much more complex than indicated. When agriculturally derived feedstocks are employed, the material being sulfated may contain both saturated and unsaturated alcohol functionalities. In such cases, any double bonds present normally will not be attacked until all the alcohol has reacted. As the substrates become more complex, as in the case of hydroxy fatty acids and esters, the rather harsh conditions of the reaction process can lead to other side reactions.

The surfactant properties of sulfated materials is sensitive to the starting material composition and conditions of reaction. While it is possible to characterize the general physical properties of such materials, a precise interpretation of experimental measurements of surface activity may be difficult. In the following chapters that are concerned with such measurements and interpretations, the properties of complex mixtures generally will be inferred from those of "pure" analogous materials. It must always be kept in mind, however, that an industrial-grade product may exhibit some characteristics significantly different from those of a laboratory purified analog.

The sulfate ester surfactants have attained their great technical importance based on several factors, including (1) good water solubility and surface activity, as well as reasonable chemical stability; (2) a relatively simple synthetic pathway amenable to low-cost commercial production; and (3) readily available starting materials from a number of agricultural and petroleum sources.

2.4.1.1. *Fatty Alcohol Sulfates*

Prior to World War II, fatty alcohols became available from the catalytic hydrogenation of natural fatty acids derived from vegetable and animal byproducts. The alcohol was then sulfated by reaction with chlorosulfonic acid and neutralized with alkali. The first of the commercial alcohol sulfate surfactants came onto the market in Germany and slowly gained popularity in the first half of the twentieth century. More recently, alcohols have been produced by various catalytic processes using ethylene as the starting material. The Ziegler process, for example, yields even-numbered alcohols, essentially equivalent to those found in the natural fatty acids. Any desired chain length can be produced, and the alcohols can be blended in almost any proportion. Myristic acid (n-$C_{13}COOH$), a starting material for the preparation of myristyl alcohol, for example, is not found in great quantities in natural sources but can be prepared in unlimited amounts by this process.

Alcohols prepared by the "oxo" process are slightly branched and always contain a portion of secondary alcohols. Because they are secondary alcohols, they are not as easily sulfated as the primary isomers, and since they contain both even and odd chain lengths, the physical properties of the resulting surfactants may differ somewhat from those of the natural and Ziegler process alcohols.

In addition to more varied alcohol sources, product quality is improved by the use of gas-phase SO_3 as the sulfating agent instead of the more harsh chlorosulfonic and sulfuric acids. As a result, the surfactants generally have a lighter color and less sodium sulfate as a contaminant.

2.4.1.2. *Sulfated Fatty Acid Condensation Products*

In addition to the simple alkyl sulfate esters discussed above, there exists substantial patent literature related to more complex sulfate esters of condensation products. Such materials would include linking groups such as amides, ethers (and polyethers), esters, and amines.

Condensation products containing sulfate groups and condensation linkages have the general structure

$$RCO–X–R'OSO_3^- M^+$$

where X can be oxygen (ester), NH, or alkyl-substituted N (amide), and R′ is alkyl, alkylene, hydroxyalkyl, or alkoxyalkyl. Such compounds have been found to exhibit good wetting and emulsifying characteristics, and are often used in cosmetics and personal care products because of their low skin irritation properties relative to the simple alcohol sulfate esters. A number of such structures have been described in the patent literature and have gained significant commercial attention.

The major members of this class of sulfated surfactants are the sulfated monoglycerides and other polyols, and sulfated alkanolamides. The sulfated monoglycerides are generally prepared by the controlled hydrolysis and esterification of triglycerides in the presence of sulfuric acid or oleum. The process and the properties of the final products are, as might be expected, sensitive to conditions of temperature, reaction time, and reactant concentrations. Because of the natural availability of the starting materials for such processes, the sulfated monoglycerides have a great deal of commercial potential in the so-called developing countries of Latin America, Asia, and Africa, where triglycerides from plant and animal sources may be more readily available than the more expensive petroleum-based raw materials. Surfactants based on coconut oil have historically been popular and can be found in a number of commercial products.

Other sulfated esters of polyols have been suggested as surfactants and have, in fact, been extensively patented. Sulfated monoesters of ethylene glycol and pentaerythritol have received some attention in that regard. They have not, in general, found much commercial acceptance, probably because of the hydrolytic instability of the ester linkage.

The drawbacks of the alkanol esters can be overcome by the use of the alkanolamide linkage. Hydroxyalkylamides can be conveniently prepared by the reaction of hydroxyalkylamines with fatty acids or esters, and by the reaction of epoxides with fatty amides. Since the latter process can result in the formation of a number of products, it is rarely employed for the preparation of materials where a single derivative is desired. The resulting materials are generally more hydrolytically stable than the corresponding ester, and increased stability is obtained if the amide linkage is separated from the sulfate group by more than two carbons. Such materials have been found to have good detergent properties when combined with soaps and are sometimes used in toilet bars and shampoos because of their low irritation.

2.4.1.3. Sulfated Ethers

Nonionic surfactants of the polyoxyethylene type, which will be discussed in a following section, generally exhibit excellent surfactant properties. The materials have been found to have two primary disadvantages, however, in that they are seldom good foam producers (an advantage in some applications such as automatic washing machines) and under some conditions give cloudy solutions, which may lead to phase separation. Fatty alcohol sulfates, on the other hand, generally have good foaming properties, but their more common sodium salts rarely produce clear solutions except at low concentrations. To achieve clarity it is often necessary to use some cation other than sodium, which may increase costs or introduce other difficulties.

When a fatty alcohol is ethoxylated, the resulting ether still has a terminal –OH, which can subsequently be sulfated to give the alcohol ether sulfate (AES):

$$R(OCH_2CH_2)_nOSO_3^-M^+$$

This class of surfactant has been extensively investigated because it has the potential to combine the advantages of both the anionic and nonionic surfactant types. In general the ethoxylation of the fatty alcohol is not carried sufficiently far to produce a truly water-soluble nonionic surfactant; usually five or fewer molecules of ethylene oxide are added and the unsulfated material is still of limited solubility in water. The water-insoluble nonionic material, however, can then be sulfated with chlorosulfonic acid or SO_3 and neutralized, usually with sodium hydroxide, to yield the desired product. Other counterions may be employed by slight modifications of the reaction or by the use of alternative reaction schemes.

The sodium salts of the ether sulfates have relatively limited water solubility. They seldom perform as well as do many other anionic surfactants as wetting agents, but their foaming properties are considerably better. They have found extensive use in shampoo formulations, and, in combination with other anionic and nonionic surfactants, they are being used more for household dishwashing detergents.

Lightly ethoxylated alkylphenols can also be sulfated to produce surfactants with the general formula

$$R{-}C_6H_4{-}O(CH_2CH_2O)_nSO_3^-M^+$$

The sulfated alkylphenol ethoxylates have found use in toilet soap preparations, but their main success has been in the area of light-duty household detergent liquids. In the preparation of such materials, however, the basic alkyl polyether phenol starting material has two possible reaction pathways: attack at the terminal –OH (to produce the desired sulfated polyether) and sulfonation of the benzene ring. Since the product intended in this case is the sulfated ether, it is desirable that no ring sulfation occur. A complete exclusion of ring attack can be achieved by several techniques, including the use of sulfamic acid as the sulfating agent. Such a procedure produces the ammonium salt, which may in some instances be more useful than the sodium analog.

The figure given for the degree of ethoxylation of an alcohol, the n in the chemical formula, is only an average. The low values of generally used in the sulfated products mean that there may be significant quantities of unethoxylated alcohol or phenol in the mixture as well as some molecules with significantly more OE units, which will be sulfated in subsequent steps and can have a significant impact on the characteristics of the final product.

2.4.1.4. Sulfated Fats and Oils

A final class of sulfated alkyl surfactants is that of the sulfated fats and oils in which the sulfate esters are obtained by the treatment of a variety of hydroxylated or unsaturated natural fats and oils with sulfuric or chlorosulfonic acids. These materials represent the oldest types of commercial synthetic surfactants, dating back to the original turkey red oils. Because of the nature of the starting materials and preparation processes, the sulfated fats and oils are chemically heterogeneous materials whose properties are very sensitive to their history. In fact, the preparation of such materials may correctly be considered to be more art than science. They will contain not only sulfated glycerides similar to those discussed above but also sulfated carboxylic acids and hydroxycarboxylic acids produced by hydrolysis of the starting materials. With the increased availability of more chemically pure surfactant materials, the use of the sulfated fatty oils has decreased considerably. They do still have their uses, however, especially where purity is not a major concern and cost is important.

From a technological point of view, the aliphatic sulfate ester surfactants have not attained a wide range of applicability, even though they are heavily used in areas where their properties fit the need. With few exceptions they are used as detergents, wetting agents, or both, with good detergency favored by the longer hydrocarbon chains and good wetting by the shorter homologs. The more chemically pure materials, especially the *n*-alkyl sulfates such as SDS, have found application in a number of areas related to emulsion polymerization and, of course, represent probably the most intensively characterized family of surfactants known.

2.4.2. Sulfonic Acid Salts

Although chemically similar to the sulfate esters, the sulfonic acid salt surfactants can differ considerably in properties and chemical stability. As with the sulfates, there exist a wide variety of hydrophobic groups that control the properties and applications of these materials. Some of the major groups or sources are discussed briefly below.

Some sulfonic acids were no doubt produced in the early sulfuric acid processing of the "sulfated oils" in the late nineteenth century. The first commercially available sulfonate surfactants, however, were produced as a result of raw-materials shortages in Germany during World War I. Some short-chain sodium alkylnaphthalene sulfonates were developed and despite relatively poor detergent properties were found to be good wetting agents and are still used as such today. The sulfonates in this series have also found use as emulsifying agents and dispersants in several agricultural and photographic applications. The postwar expansion of the chemical industries of Great Britain, Germany, and the United States led to the development of a number of new synthetic sulfonate surfactants, which have since gained a significant share of the commercial market.

Materials that are related through the sulfonate group include the aliphatic paraffin sulfonates produced by the photochemical sulfonation of refinery hydrocarbons, petroleum sulfonates derived from selected petroleum distillate fractions, olefin sulfonates, *N*-acyl-*N*-alkyltaurines, sulfosuccinate esters and related compounds, alkylaryl sulfonates, and ligninsulfonates, which are a byproduct of the paper manufacturing process. While complete coverage of the class would be prohibitive, some of the most important types are described below.

2.4.2.1. Aliphatic Sulfonates

The simple aliphatic sulfonic acids and their salts are represented by the general formula $R{-}SO_3^-M^+$ where R is a straight- or branched-chain, saturated or unsaturated alkyl or cycloalkyl group of a sufficient size to impart surface activity and M^+ is hydrogen, an alkali metal ion, an organic cationic species, or, in rare cases, a polyvalent metal ion. The classic example of these materials would be sodium dodecylsulfonate, not to be confused with its sulfate ester cousin SDS. As with the sulfates, appreciable surface activity is not attained until the length of the alkyl chain reaches eight carbons, and water solubility becomes a problem around C_{18}. Since they are salts of strong acids, the members of this class of surfactants seldom show a sensitivity to low pH conditions and are considerably more stable to hydrolysis than are the analogous alkyl sulfates. While the monovalent salts of the sulfonic acids are quite water-soluble and exhibit good wetting and detergency properties, their calcium and magnesium salts are of limited solubility; as a result, such materials are not generally favored as detergents, especially for use in "hard water" situations.

Paraffin sulfonates, or secondary *n*-alkylsulfonates, are mostly a European product prepared by the sulfoxidation of paraffin hydrocarbons with sulfur dioxide and oxygen under ultraviolet irradiation. They are used in applications similar to those of the linear alkylbenzene sulfonates (LABS) discussed below. It has been suggested that the paraffin sulfonates have higher water solubility, lower viscosity, and better skin compatibility than do the LABS of comparable chain length, although any such direct comparison must be qualified because of the distinctly different chemical and isomeric contents of the two classes of materials.

Petroleum sulfonates are prepared by the reaction of selected petroleum refining fractions with sulfuric acid or oleum. The raw materials are a complex mixture of aliphatic and cycloaliphatic isomers and the resulting surfactants are equally complex. As a result, the exact properties of the products may not always be sufficiently consistent to allow their use in critical applications. They are, however, relatively inexpensive and have found significant use in tertiary oil recovery, as frothing agents in ore flotation, as oil-soluble corrosion inhibitors, as components in dry cleaning formulations, and as oil/water emulsifying agents in metal cutting oils. Although they are inexpensive, they usually have a dark color and may often contain a significant amount of unsulfonated hydrocarbon as a contaminant.

α-Olefin sulfonates are prepared by the reaction of sulfur trioxide with linear α-olefins. The resulting products are a mixture of alkene sulfonates and primarily 3- and 4-hydroxyalkane sulfonates. The α-olefin sulfonates are gaining acceptance for use in detergents and other consumer products because of their tendency to be less irritating to the skin and their greater biodegradability.

The alkyl sulfonate salts can be prepared in a number of ways and have been the subject of considerable patent activity. They have not, however, achieved the widespread utility of the corresponding sulfates, primarily because of the easier and more economical methods for the synthesis of the esters and their better detergent activity. Special members of this class, the perfluoroalkyl sulfonates, have gained a great deal of attention because of their enhanced and sometimes unique surfactant properties. Although such materials are expensive relative to the analogous hydrocarbons, their great stability in harsh conditions and their enhanced surface activity at very low concentrations make the extra cost palatable in specialized applications. These materials will be discussed more fully in the subsequent discussions of fluorinated surfactants in general.

2.4.2.2. Alkylaryl Sulfonates

The sulfonation of aromatic nuclei such as benzene and naphthalene is a time-honored and reasonably well-understood organic chemical process. Simple sulfonated aromatic groups alone are not able to impart sufficient surface activity to make them useful in surfactant applications, however. When the aromatic ring is substituted with one or more alkyl groups, which in some cases can be rather small, the surface-active character of the molecule is greatly enhanced. This class of materials has become exceedingly important as a major fraction of the commercially important anionic surfactants.

Particularly important members of this family are the alkylbenzene sulfonates (ABS) and the linear alkylbenzene sulfonates (LABS). The ABS family, as mentioned earlier, is based on highly branched tetrapropylene alkyl groups that resist biodegradation processes, something that is less of a problem with the linear alkyl analoges.

Although usually sold as the sodium salt, LABS materials can be very useful in nonaqueous systems if the nature of the cation is changed. The calcium salt, for example, can have significant solubility in some organic solvents, as do many of the amine salts, especially that of isopropyl amine. Since the materials are the salts of strong sulfonic acids, they are essentially completely ionized in water, as are the free sulfonic acids. They are not seriously affected by the presence of calcium or magnesium salts or high concentrations of other electrolytes, and they are also fairly resistant to hydrolysis in the presence of strong mineral acids or alkali. While the sodium salts are not soluble in organic solvents, the free acid may be neutralized with amine bases to produce materials with high organic solubility.

Most of the commercially available alkylbenzene materials contain a C_{12} alkyl chain where the points of attachment to the benzene ring are a random distribution along the chain. Materials with longer alkyl chain lengths (C_{13}–C_{15}) are also available and have the potential advantage of being somewhat more soluble in organic solvents. A disadvantage of the LABS is that they are somewhat resistant to biodegradation under anaerobic conditions, although they pose no problem under aerobic conditions.

Members of the LABS family of surfactants exhibit several characteristics that explain their predominance in many surfactant applications. They not only show outstanding surfactant properties in general but are also derived from raw materials that are relatively easy to obtain and less expensive than those from which the other main types of surfactant are prepared. The main source of the alkylbenzene sulfonates is the petroleum industry, and it will probably remain the most economical raw material for these materials for some time to come.

As outlined earlier, this class of materials is the product of the Friedel–Crafts alkylation of benzene with a long-chain alkyl group; the prime example is dodecylbenzene. This compound may be derived from an alkyl chloride containing an average of 12 carbon atoms. The source of the alkyl chain may be a particular cut of petroleum distillate or a synthetic hydrocarbon chain obtained by the controlled oligomerization of a low-molecular-weight alkene—generally ethylene or propylene.

Because of the comparatively high cost of chlorine and the simultaneous production of hydrogen chloride, the use of olefin feedstocks has become the preferred route to these materials. For that reason, the propylene tetramer was for many years the primary source of the alkyl group. When the change to linear alkylbenzenes occurred for reasons of improved biodegradability, straight-chain olefins were frequently used.

Although the aromatic nucleus in surfactants is usually benzene, it is also possible to use toluene, xylene, phenol, and naphthalene. The oldest of the sulfonate-based, synthetic surfactants, in fact, are the alkylnaphthalene sulfonates developed in Germany during World War I. Although representing a relatively small part of the total surfactant market, they have shown considerable staying power for use as wetting agents for agricultural applications and paint formulations, and have long been used for the preparation of emulsions of dye precursors in color photographic products. When the nucleus employed is naphthalene, the alkyl sidechain may be as small as propyl; however, in that case, single substitution does not produce a material with useful surfactant properties, although some surface activity does result. To obtain significant wetting and detergency effects, the naphthalene nucleus must contain two or more propyl groups. If the chain length of the alkyl group is increased, the surfactant properties improve correspondingly until, above C_{10}, solubility problems begin to arise.

As the molecular weight of the alkyl chain attached to the aromatic nucleus increases, it becomes important to limit the final product to a single-chain substitution. In general, as the alkyl chain increases in length, water solubility decreases, and surfactant activity reaches a maximum and then begins to decrease. In practice, for benzene derivatives, dodecylbenzene sulfonate has generally been found to exhibit the best overall balance of desirable surfactant characteristics.

In the science and technology of surfactants, the requirements placed on a given material or process can vary greatly in different parts of the world. Except in the case of specially prepared and purified materials for basic scientific investigations, reference to a material such as sodium dodecylbenzene sulfonate implies a material whose average molecular weight and structure are those of the $C_{12}H_{25}C_6H_4SO_3^-Na^+$. The industrial product, in fact, will never be a single homogeneous material. If the feedstock is the propylene tetramer, the material may be contaminated with small amounts of ethylene and butylene, which, during polymerization, will produce various amounts of the trimer and pentamer of each, along with the trimer and pentamer contaminants from the propylene itself.

Mixed oligomers of all the ingredients are also possible, so that the final product will contain a distribution of molecular weights and isomers. When the isomers are coupled to the benzene nucleus, more isomeric possibilities arise, leading to a product containing an enormous variety of structures. With such a wide variety of molecular species present in the basic feedstock for sulfonation, it becomes difficult to reproducibly manufacture materials with the identical component distribution and surfactant characteristics. It has become common, therefore, to distill the alkylbenzene prior to sulfonation and employ the cuts with molecular weights in the ranges of 233–245 and 257–265, depending on the manufacturer and the intended use for the final material.

The material prepared from the lower-molecular-weight fraction is commercially referred to as *dodecylbenzene* and the higher, as *tridecylbenzene*. The so-called tridecyl material is actually a mixture of mostly C_{12} and C_{14} isomers; the exact mixture depends on the manufacturer. The tridecylbenzene sulfonate, in general, shows better detergent properties and better foaming in soft water, whereas dodecylbenzene sulfonate has a lower cloud point and better viscosity behavior in liquid formulations.

The change to linear alkylbenzene brought certain changes in surfactant properties. Since the alkyl portion is linear in the primary feedstock, there are no significant problems with different branched isomers at that stage, although some homologs will undoubtedly be present. When reacted with benzene, the coupling between olefin or alkylchloride can occur at any of the positions along the alkyl chain except at the termini, so that a number of isomeric alkyl substitutions will result. Branching of the alkyl groups, however, will not occur. It has been shown that the surfactant properties of the product depend very strongly on the final position of the phenyl group along the alkyl chain. It is to be expected, then, that nominally equivalent ABS and LABS surfactants will exhibit different properties. The effects of substitution patterns will be covered in later chapters concerned more directly with the surface-active properties of surfactants from a structural and physicochemical standpoint.

2.4.2.3. α-Sulfocarboxylic Acids and Their Derivatives

The α-sulfocarboxylic acids are represented by the general formula

$$RCH(SO_3^-M^+)COO^-M^+$$

in which R is CH_3 or a longer alkyl chain and M is hydrogen or a normal surfactant cation. The lower-molecular-weight analogs such as those derived from propionic, butyric, and possibly slightly longer acids are not sufficiently hydrophobic to be surface-active. However, if the free carboxylic acid is esterified with a fatty alcohol of proper length, usually C_8–C_{18}, the resultant materials will generally perform well as surfactants. A typical example of such a material is sodium dodecylsulfoacetate:

$$C_{12}H_{25}O_2CCH_2SO_3^-Na^+$$

The esters of sulfoacetic, α-sulfopropionic, and α-sulfobutyric acids with long-chain alcohols are highly surface-active but have found little general application because of their relatively high cost. They have several specialty applications, especially in personal care products such as toothpastes, shampoos, and cosmetics. Some have also been approved for use in conjunction with mono- and diglycerides in food applications.

The α-sulfocarboxylic acids of higher molecular weight, especially lauric, palmitic, and stearic acids, are highly surface-active. It is generally found that the water solubility of the unesterified α-sulfo acids decreases as the length of the carbon chain increases, as might be expected. Less obvious are the observed effects of the state of neutralization of the carboxylic acid and the nature of the counterion on solubility. In the case of the free acid sulfonate, solubility is found to increase in the order Li $<$ Na $<$ K, while the solubility of the neutralized dialkali salt increases in the opposite order. The use of alkylammonium counterions is also found to increase the solubility of a specific-chain-length material.

The presence of the free acid or carboxylate salt combined with the sulfonate group has suggested the use of such materials as corrosion inhibitors when applied to metal surfaces, which can form strong salts or complexes with carboxylic acid groups. Related applications would be in ore flotation. The materials have also found some utility as viscosity reducers in liquid detergent formulations.

The presence of the free carboxyl group in these surface-active materials suggests the possibility of additional derivatization, such as esterification or amidation. Esterification of the carboxyl group has usually been found to increase the water solubility of a given acid, as long as short-chain alcohols are employed. Unlike many esters, those of the α-sulfocarboxylic acids are generally resistant to hydrolysis, although they are readily biodegradable under both aerobic and anaerobic conditions:

$$R_2OOCCHR_1SO_3^-M^+$$

The surface properties of a series of α-sulfocarboxylic acid esters, in which the chain length of the acid (R_1) was varied from C_2 to C_{18} and that of the alcohol (R_2) from C_1 to C_{16}, were found to vary considerably with the location of the sulfonate group along the chain. Esters of short-chain alcohols and long-chain acids exhibit critical micelle concentrations roughly equivalent to those of the analogous alkyl sulfate surfactant of the same hydrocarbon chain length. Such materials show good detergent properties, while materials with equal carbon numbers, but with longer alcohol and shorter acid groups, were superior wetting agents with larger cmc's (see Chapter 4). The α-sulfocarboxylic acid esters have found application as lime-soap dispersants, components in dry cleaning formulations, detergent powders, mineral flotation, and soap bars.

Esterification is not the only method of derivatizing the carboxyl functionality, and it is not surprising that amides and *N*-alkylamides of the α-sulfo acids have also received extensive attention in the surfactant patent literature. The α-sulfoalkanolamides have the general structure

$$R_2NHOCHR_1SO_3^-M^+$$

where R_1 can be H or an alkyl group. The simple amide materials generally have surface properties similar to those of the analogous sulfoacid, while the N-alkyl-substituted materials closely correspond to the sulfoesters in their solubility. Although they are not widely used as industrial surfactants, they are receiving some attention for use in personal care formulations.

2.4.2.4. Miscellaneous Sulfoester and Amide Surfactants

The sulfoesters and amides discussed in the preceding section are those in which the two functionalities were located on adjacent carbon atoms, a situation that imparts some special properties to the resultant surfactant molecules. A second class of related materials is that in which the two groups are more widely separated. General examples of such compounds would be

$$RCOX(CH_2)_nSO_3^-Na^+$$

and

$$RCOX{-}Ar{-}SO_3^-Na^+$$

in which R is the usual alkyl chain of 8–18 carbon atoms, where n = 2–4, X is O, NH, or *N*-alkyl, and Ar is an aromatic nucleus such as benzene or naphthalene.

The fatty acid esters and amides having the general structures shown above were some of the earliest and most commercially successful materials that adequately overcame the problems of hard water and acid sensitivity that plagued many surfactant applications. The sulfoethyl ester of oleic acid was the first of a line of such materials. As with many esters, however, those materials showed some tendency to hydrolyze in solution and have been generally replaced by the more stable amide derivatives. The presence of the amide nitrogen not only enhanced the hydrolytic stability of the sulfoacid materials, but it also provided a locus for the further customization of the surfactant molecule and, therefore, modification of its surface properties. By the judicious choice of variables such as R, X, and *n*, a wide range of wetting, foaming, detergency, and related properties could be designed into the surfactant. As many as 15 different sulfoesters and amides of this family are currently in wide commercial use.

The *N*-acyl-*n*-alkyltaurines are prepared by the condensation of taurine or *N*-alkyltaurine (usually methyl or ethyl) with a fatty acid:

$$R_1COOH + NR_2CH_2CH_2SO_3^-M^+ \rightarrow R_1CONR_2CH_2CH_2SO_3^-M^+$$

The resulting material will usually exhibit solubility, detergency, foaming, and other properties very similar to those of the corresponding fatty acid soap. They have the distinct advantage, however, of being much less sensitive to low-pH and high-electrolyte concentrations and to the presence of di- and trivalent ions. They are also stable to hydrolysis and show good skin compatibility. As a result, they are very popular for use in toilet soaps, bubble baths, and shampoos. They are also excellent as wetting agents.

Related to the sulfoacid esters and amides are materials derived from sulfonated polycarboxylic acids. Such polyacids include succinic, itaconic, phthalic, and isophthalic acids and the tricarballylates. The presence of several carboxylic acid groups in these materials opens the way for the preparation of a variety of multifunctional surfactant molecules.

The most commonly encountered members of this class of materials are the sulfosuccinate and sulfosuccinamate esters and diesters. Those materials, which have excellent wetting, emulsifying, dispersing, and foaming properties, are relatively cheap and easy to prepare. As a result, they have gained wide acceptance for many applications.

The mono- and diesters of sulfosuccinic acid, especially the sodium di-2-ethylhexyl ester, have found wide application as wetting agents and emulsifiers in paints, printing inks, textiles, agricultural emulsions, and photographic materials. They can be prepared to have good solubility in either water or organics, so that many desirable properties can be designed into the molecule. The synthesis is straightforward and will produce materials essentially free of inorganic salt contaminants, unlike many other members of the sulfonate class of surfactants. Their major disadvantage is that the ester group is susceptible to hydrolysis in either alkaline or acidic solutions.

The sulfosuccinate esters have the general structure

$$R_1O_2CCH_2CH(SO_3^-M^+)CO_2R_2$$

where $R_1 = R_2$ for most diesters and $R_2 = H$ for the monoester. They were first commercialized under the trade name "Aerosols." The diesters where R_1 is C_9 or less generally have good water solubility, although the di-*n*-octyl diester does have some solubility problems in water at room temperature. The branched-chain esters usually have higher water solubility than do the straight-chain analogs.

The higher alkyl analogs of these materials are somewhat unique in that they have good solubility in a wide variety of organic solvents and undergo apparent micelle formation in many nonaqueous systems. A possible disadvantage of the sulfosuccinate esters is that they do undergo significant hydrolysis under both acidic and basic conditions. The secondary alcohol esters are more stable in this regard than the primary analogs, especially under basic conditions.

Related to the sulfosuccinic diesters are the sulfosuccinamides, succinimides, and succinamates, which may show some advantage in stability but usually require more harsh conditions for formation. The sulfosuccinic diamide materials are prepared in a manner similar to that for the diesters from maleic anhydride or the free acid and a primary or secondary amine, followed by the addition of sodium bisulfite across the double bond to give materials of the general structure

$$R_1R_2NCOCH_2CH(SO_3^-M^+)CONR_1R_2$$

where R_1 is an alkyl group of suitable size to impart the desired surface activity and R_2 may be H or a second alkyl group.

The sulfosuccinamates can be obtained by the reaction of maleic anhydride with a primary or secondary amine to form the monoamide, followed by esterification and sulfonation to the desired product

$$R_1R_2NCOCH_2CH(SO_3^-M^+)COOR_3$$

in which the Rs have their usual meaning.

The sulfosuccinimdes may be of two varieties, one analogous to the previously mentioned succinic acid derivatives, where the sulfonate group is added across the double bond of maleic acid and subsequently reacted to produce the cyclic imide. The second variety requires a different synthetic approach, since the locations of their hydrophilic and hydrophobic groups are reversed from those of the succinic acid derivatives. In this case, the starting material is a 2-alkyl, alkenyl, or similar succinic anhydride, which is reacted with the appropriate sulfoalkylamine to produce the amide acid, followed by dehydration to the imide. Similar reaction schemes can be used to prepare di- and higher polyesters and amides of malonic acid, itaconic acid, and other polycarboxylic acids.

Unlike the sulfosuccinic acid derivatives, sulfopolyesters and amides, and imides of the phthalic acids and other sulfonated aromatic carboxylic acids can be prepared by a process such as the direct esterification or amidation of the corresponding acid.

Various potential surfactant structures can be prepared from the other polycarboxylic acids available to the synthetic chemist. Their properties as surfactants have received a great deal of attention because surface activity and solubility in water and organic liquids can be varied almost continuously by the proper choice of the esterifying alcohol. As mentioned earlier, the use of shorter-chain alcohols ($<C_9$) results in materials with substantial water solubility, with branching favoring higher solubility. As the alcohol chain passes through C_8, significant oil solubility can be achieved so that those compounds become useful as nonaqueous surfactant systems. Surfactant characteristics such as wetting power also vary with chain length, with longer chains usually resulting in a greater effectiveness as a wetting agent. Other surfactant properties of these materials will be discussed in later chapters.

Because the sulfosuccinates and sulfosuccinamates have shown excellent wetting, dispersing, emulsifying, and foaming properties, they have found a wide variety of applications in industry. Their applications range from use in lubricating oils, ink dispersions, mineral ore flotation, and oil emulsion breakers to stabilizers for emulsion polymerizations and emulsifiers, and coating aids in photographic products. Some members of this family have also found extensive use in shampoos, toilet bars, and other cosmetics because of their low skin irritation, and for ingestion as stool softeners.

2.4.2.5. Alkyl Glyceryl Ether Sulfonates

The alkyl glyceryl ether sulfonate class of surfactants, represented by the general formula

$$ROCH_2CH(OH)CH_2SO_3^-M^+$$

were first synthesized in Germany relatively early in the twentieth century. However, there are very few data published about their specific properties, other than the information available in the patent and trade literature. An early patent described the reaction of epichlorohydrin with a fatty alcohol to produce the chloroether, which was subsequently treated with sodium sulfite to yield the desired sulfonate salt:

$$ROH + C_3H_5ClO \rightarrow ROCH_2CH(OH)CH_2Cl$$

$$ROCH_2CH(OH)CH_2Cl + Na_2SO_3 \rightarrow ROCH_2CHOHCH_2SO_3^-Na^+$$

The specific details of the reaction pathways to these materials can be quite complex, not in the chemistry involved, but in the conditions necessary to obtain a commercially useful product and process. The major problems are control of the formation of di- and higher ethers from the epichlorohydrin and the removal of unreacted starting materials and byproducts from the final material.

Patent references to the alkyl glyceryl ether sulfonates describe them as efficient wetting, foaming, and dispersing agents, with improved water solubility and stability to acids and bases. The materials are, however, inherently more expensive than most other sulfonates and therefore have not received a great deal of interest in most high-volume, low-cost applications. They have found extensive use only in specialty areas where the added cost for the specific advantages of such materials can be tolerated.

2.4.2.6. Lignin Sulfonates

Lignin sulfonates are, technically, low-molecular-weight polymers containing a witch's brew of primary and secondary alcohols, phenols, and carboxylic acid functionalities. They are prepared by the sulfonation of lignin byproducts of pulp and paper manufacture, followed by neutralization to the sodium, calcium, or ammonium salt. The resulting materials are useful as dispersing agents for solids and oil/water emulsions, and as stabilizers for aqueous dispersions of dyes, pesticides, and cement. They are very inexpensive relative to other materials and are very low-foaming. They have the disadvantages of being highly colored, insoluble in organic solvents, and relatively ineffective at lowering surface tensions.

2.4.3. Carboxylate Soaps and Detergents

Their long and well-documented history notwithstanding, it is important to include some mention of the classical carboxylate soaps and related materials in this review of anionic surfactant classes. While a complete history of soap technology will not be presented, a few special carboxylic acid salts have been prepared that are claimed to be superior to the classical soaps in both surface activity and stability to hard water and pH variations, as well as providing a number of functional advantages in product formulations. Such materials often contain additional polar groups such as esters and amides, different fatty acid compositions, or non-fatty-acid components that impart greater solubility and stability to the neutralized soap product. This class of materials includes not only normal esters and amides of fatty acids, but also those of the amino acids. Because the starting materials for such materials are often mixtures, the final products are usually complex and may often be poorly characterized.

The natural fatty acid soaps derived from tallow (animal fat) continue to form an important group of surfactants even though their portion of the market has tended to decline relative to the synthetics. The salts of the coconut oil acids and some acids derived from oilseed sources, on the other hand, have found increased use in hard-water applications and liquid hand soaps, where high soap concentrations are desired. The improved performance of coconut oil soaps probably results from their higher content of the lower-molecular-weight C_{12} and C_{14} acids (60–70%) versus the tallow soaps (80–95% C_{14} or greater).

Other sources of carboxylate salt surfactants include those derived from (1) tall oil byproducts of wood and paper manufacture, (2) condensation products of fatty acids with sarcosine [2-(*N*-methylamino)ethanoic acid, $(CH_3)HNCH_2COOH$] and other amino acids and proteins, and (3) fatty acid esters of common hydroxy-acids such as lactic, tartaric, cirtic, and succinic acids.

The tall oil soaps are particularly attractive because they are very inexpensive. They are actually complex mixtures of fatty acids and rosin acids, which are difficult to characterize and control. The presence of the rosin acids in these materials generally imparts better water solubility as well as enhanced surfactant properties. Because of their complex nature and difficulties in obtaining sufficiently "clean" materials, the tall oil soaps have generally found use only in the most tolerant areas of heavy-duty industrial cleaning and within the processes from which they are derived.

Condensation products of fatty acids with amino acids to produce materials such as *N*-laurylsarcoside

$$C_{11}H_{23}CON(CH_3)CH_2COO^-Na^+$$

have been shown to possess several properties that make them useful in personal care products. They are often less sensitive to hard water and low pH than are the tallow soaps and, because of their structures, are nonirritating to skin and eyes. Many of the amino acid derivatives are nontoxic and have found uses in oral applications such as toothpastes and mouthwashes. Additional uses can be found in such diverse areas as antistatic agents and lubricants for food packaging polymers, textiles, petroleum recovery, metal processing fluids, carpet cleaners, and photographic emulsions. Acylated proteins derived from leather and bone also show good detergent properties and perform well as fabric softeners and "control" agents in shampoos.

Esters of fatty acids, primarily stearic and palmitic, and distilled monoglycerides with lactic, tartaric, citric, succinic, and related hydroxy acids have found wide application as emulsifiers and improvers in the food industry. Sodium or calcium salts of esters of stearic acid with lactic acid [sodium stearoil-2-lactylate (SSL)]

$$CH_3(CH_2)_{16}COOCH(CH_3)COCH(CH_3)COO^-Na^+$$

and diacetyltartaric acid esters of monoglycerides (DATEM)

$$C_{17}H_{35}COOCH_2CH(OH)CH_2OOCCH(OOCH_3)CH(OOCH_3)COOH$$

are widely used throughout the world in the production of bread and other bakery products. The "natural" character of such products derived from lactic acid (an α-hydroxy acid) has led to their occasional use in skin and hair care products. Because of the nature of the starting materials for such products, they are rather complex mixtures of compounds ranging from free fatty acid salts to low-molecular-weight polymers and cyclic isomers of the corresponding hydroxyl acids.

2.4.4. Phosphoric Acid Esters and Related Surfactants

The esters of phosphorus-based acids constitute a relatively uniform group of anionic surfactants that can fill a special role in the overall scheme of surfactant applications. The esters and diesters of phosphoric acid have the general formula

$$RO{-}PO_3^-M^+$$

where the R group is usually a long-chain alcohol or phenol. The materials may be obtained as the free acid (M = H) or sodium or amine salts. In fact, they are generally found as a mixture of mono- and dibasic phosphates. The surfactant properties of the alkyl phosphates vary. They have been reported to be somewhat superior to related sulfates and sulfonates in some applications because of their low foaming characteristics, good solubility in water and many organic solvents, and resistance to alkaline hydrolysis. However, their performance as detergents seems to be inferior in most cases. The phosphate esters have the general disadvantage of being more expensive than sulfonates and sulfates; but wider solubility and good activity in harsh environments have made them useful in dry cleaning formulations, highly alkaline cleaners, and various emulsion formulations.

Useful variations of the alkyl phosphate surfactants are those in which a polyoxyethylene chain is inserted between the alkyl and phosphate ester groups. Such materials are usually found to exhibit only slight anionic character, having properties more in common with the nonionic analogs. An advantage of the addition of the phosphate group to the nonionic is that the resultant material will quite often have better solubility in aqueous electrolyte solutions. Additional advantages can be gained from the fact that the pour point, the temperature at which the nonionic surfactant solidifies, can be significantly decreased by phosphorylation. These materials are somewhat limited by the fact that they hydrolyze in the presence of strong acids, although their stability to strong alkali, like that of the simple phosphates, is quite good. The wetting, emulsifying, and detergent properties of these surfactants are generally better than those of similar phosphate surfactants not containing the added solubilizing groups.

Lecithin, glyceryl esters containing two fatty acid residues and one phosphate ester (usually as phosphatidylcholine), constitute one of the oldest of the commercially important phosphoric acid derivative surfactants. However, since they contain nitrogen and are normally either nonionic or amphoteric, they are not discussed further in this section. Additional phosphorus-containing surfactants include derivatives of phosphonic and phosphinic acids.

Although not reprsenting a particularly large portion of the total surfactant market, on either a weight or a dollar basis, the phosphorus-containing surfactants have carved themselves a substantial niche in many industrially important applications. That importance is emphasized by the large number of commercially available materials. The organophosphorus surfactants exhibit a number of useful surface properties related to emulsification, wetting, detergency, solubilization, and other processes. They have also found a number of significant uses as a result of nonsurfactant-related characteristics such as antistatic and lubrication properties and corrosion inhibition, and as fuel additives. More recent ecological problems have led to a reduction in the use of phosphate-containing formulations for many applications, especially laundry detergents; however, their unique properties and general safety make them a useful tool in the overall repertoire of surfactant types available today.

The foregoing discussion of anionic surfactants only touches the tip of the iceberg in relation to the exceedingly wide variety of materials available. More comprehensive coverage of the topic is available in several volumes cited in the Bibliography list for this chapter. Leaving the pursuit of more detail to the interested reader, we now turn or attention to the positively charged cationic family of surfactants.

2.5. CATIONIC SURFACTANTS

Cationic surfactants first became important when the commercial potential of their bacteriostatic properties was recognized in 1938. Since then, the materials have been introduced into hundreds of commercial products, although their importance does not approach that of the anionic materials in sheer quantity or dollar value. Currently, cationic surfactants play an important role as antiseptic agents in cosmetics, as general fungicides and germicides, as fabric softeners and hair conditioners, and in a number of bulk chemical applications. Many new applications for cationic surfactants have been developed since World War II, so that these compounds can no longer be considered to be specialty chemicals; rather, they truly fall into the category of bulk industrial surfactant products.

Relative to the other major classes of surfactants, namely, anionics and nonionics, the cationics represent a relatively minor part of worldwide surfactant production, probably less than 10% of total production. However, as new uses and special requirements for surfactants evolve, their economic importance can be expected to continue to increase.

Commercial cationic surfactants, like their anionic and nonionic counterparts, are usually produced as a mixture of homologs, a point that must always be kept in mind when discussing physical properties and applications of such materials. As previously noted, slight variations in the chemical structure or composition of the hydrophobic group of surfactants may alter their surface-active properties, leading to the possibility of important errors in the interpretation of results and in performance expectations. That possibility holds true for all surfactant classes and will be emphasized repeatedly throughout this work. When the sources of hydrophobic groups for cationic surfactants are natural fatty acids such as coconut oil or tallow, there may be significant variations in both chain length and the degree of unsaturation in the alkyl chain. When the alkyl group is derived from a petrochemical source, the components may be found to vary in molecular weight, branching, the presence of cyclic isomers, and the location of ring substitution in aromatic derivatives.

Pure cationic surfactants such as cetyltrimethylammonium bromide (CTAB) have been used extensively for research into the fundamental physical chemistry of their surface activity. Such investigations have led to a vast improvement in our basic understanding of the principles of surfactant action. Because of the significant differences in purity and composition between commercial- and research-grade materials, however, care must be taken not to overlook the effects of such differences on the action of a given surfactant in a specific application.

Prior to the availability of straight-chain, petroleum-based surfactants, the sole sources of raw materials for cationic surfactants were vegetable oils and animal fats. All those materials could be considered to be derivatives of fatty amines of one, two, or three alkyl chains bonded directly or indirectly to a cationic nitrogen group. The most important classes of these cationics are the simple amine salts, quaternary ammonium compounds, and amine oxides:

$C_nH_{2n+1}NHR_2^+X^-$ (R = H or low-molecular-weight alkyl groups)

$C_nH_{2n+1}NR_3^+X^-$ (R = low-molecular-weight alkyl groups)

$C_nH_{2n+1}N^{\delta^+}R_2 \rightarrow O^{\delta-}$ (R = low-molecular-weight alkyl groups)

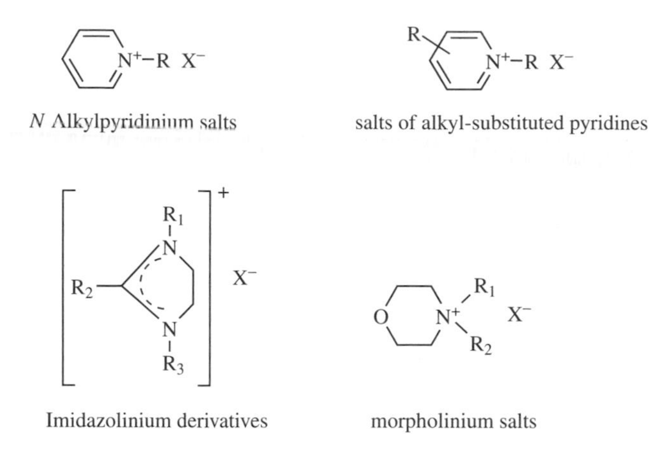

Figure 2.3. Typical structures of heterocyclic cationic surfactants ($X = Cl^-$, Br^-, CH_3COO^-, SO_4^{2-}, etc.).

There are two important categories of cationic surfactants that differ mainly in the nature of the nitrogen-containing group. The first consists of the alkyl nitrogen compounds such as simple ammonium salts containing at least one long-chain alkyl group and one or more amine hydrogen atoms, and quaternary ammoniun compounds in which all amine hydrogen have been replaced by organic substituents. The amine substituents may be either long-chain or short-chain alkyl, alkylaryl, or aryl groups. The counterion may be a halide, sulfate, acetate, or similar compound. The second category contains heterocyclic materials typified by the pyridinium, morpholinium, and imidazolinium derivatives as shown in Figure 2.3. Other cationic functionalities are, of course, possible, but are much less common than these two major groups.

In the pyridinium and other heterocyclic amine surfactants, the surfactant properties are derived primarily from the alkyl group used to quaternize the amine. As a variation to that approach, however, it is possible to attach a surfactant-length alkyl (or fluoroalkyl) group directly to the heterocyclic ring and quaternize the nitrogen with a short-chain alkyl halide. Many commercial cationic surfactants with the general structures $R\text{–}(CH_2)_5NH^+R'X^-$ and $R\text{–}(CH_2)_5N^+(R_2)X^-$, where R is the surfactant-length hydrophobic group and R′ is a short-chain alkyl or hydroxyalkyl chain, are available. As is the case with the anionic materials, the structures of such materials are limited mainly by the skill and imagination of the preparative organic chemist.

Some types of amphoteric surfactants (to be discussed below) in which the nitrogen is covalently bound to a group containing an anionic (e.g., $-CH_2CH_2SO_3^-$) or potentially anionic (e.g., –COOH) functionality are also classed as cationic in some publications; however, under the classification scheme employed in this work, such materials are covered in a separate category. The examples above represent the simplest types of cationic surfactants. Many modern examples contain much more complex linkages; however, the basic principles remain unchanged.

The economic importance of the cationic surfactants has increased significantly because of some of their unique properties. Most cationics are biologically active in that they kill or inhibit the growth of many microorganisms. They have also become extremely important to the textile industry as fabric softeners, waterproofing agents, and dye fixing agents. Because many important mineral ores and metals carry a net negative charge, the cationic surfactants are also useful in flotation processing, lubrication, and corrosion inhibition, and they are gaining importance as surface modifiers for the control of surface tribological properties, especially electrostatic charge control. Since the hydrophobic portions of the cationic surfactants are essentially the same as those found in the anionics, individual discussion of those groups is not repeated here.

2.6. NONIONIC SURFACTANTS

Although the two surfactant types discussed so far can be characterized by the existence of a net electrical charge on the molecule that imparts a required degree of solubility, such a situation is not a fundamental requirement for the existence of surface activity in water. In fact, there may be several very good reasons for having surfactant materials that are electrically neutral. Some of the most important advantages can include a significantly lower sensitivity to the presence of electrolytes in the system, a reduced effect of solution pH, and the synthetic flexibility of the ability to design the required degree of solubility into the molecule by the careful control of the size of the hydrophilic group.

An interesting characteristic of many nonionic surfactants, especially the polyoxyethylene (POE) family, is that they exhibit an inverse temperature–solubility relationship; that is, as the solution temperature is increased, their solubility in water decreases. The phenomenon is attributed to a disruption of specific interactions, in this case, hydrogen bonding, between the water and the POE units in the molecule. The temperature at which components of the POE surfactant begin to precipitate from solution is defined as the "cloud point." In general, the cloud point of a given family of surfactants (with the same hydrophobic group) will increase with the average number of OE groups.

2.6.1. Polyoxyethylene-Based Surfactants

Of all the nonionic surfactant classes available, which are few, but increasing in number, the polyoxyethylenes are easily the most numerous and the most important technically. These materials have the general formula

$$RX(CH_2CH_2O)_nH$$

where R is normally a typical surfactant hydrophobic group, but may also be a hydrophobic polyether such as polyoxypropylene, and X is O, N, or another functionality capable of linking the POE chain to the hydrophobe. In most cases, n, the average number of OE units in the hydrophilic group, must be greater than 5 or 6 to impart sufficient water solubility to make the materials useful. The "average" nature of n should be emphasized, since the reaction pathways leading to the product will always result in a distribution of molecular weights. Pure, homogeneous samples of materials with low n, usually less than ~10, can be prepared, but at great expense. Such materials have little technical significance beyond the information gained by studying the effects of stepwise addition OE groups.

The most common reaction pathway to the POE surfactants is the reaction of ethylene oxide with a hydrophobic material containing at least one reactive hydrogen. Alternative routes include the reaction of the hydrophobe with a preformed POE chain or the use of ethylene chlorohydrin instead of ethylene oxide. Neither of the latter reactions has achieved industrial importance for mass surfactant production, although the use of preformed glycols can be a useful route to homogeneous surfactant samples.

As mentioned above, the reacting hydrophobe must contain at least one active hydrogen, as in the case of alcohols, acids, amides, mercaptans, and related compounds. If primary amines are used, there exists the possibility of forming double-chain surfactants of the form

$$RN[(CH_2CH_2O)_nH]_2$$

In such materials, the values of n for each POE chain will be averages and may be quite different, so that the complexity of the resulting product will likely be greater than that of the single-chain analog.

Some of the specific effects of the variables (R, X, N, etc.) on the surfactant properties of the POE surfactants will be seen in more detail in the following chapters. In general, however, it is usually found that a number of the properties of materials with the same hydrophobic group and varying POE chain lengths will change in a regular, predictable manner. By way of illustration, the following patterns have been determined for n-dodecyl–POE surfactants:

1. Water solubility increases regularly as the number of OE groups is increased from 3 to 16.
2. The surface tension of aqueous solutions of the materials decreases regularly over the same composition range.
3. The interfacial tension between aqueous solutions and hydrocarbons reaches a maximum at around $n \sim 5$ and decreases from there.
4. Foaming power reaches a maximum at about $n = 5$ or 6 and remains relatively constant from that point.
5. Other technologically important characteristics of these materials such as detergency, wetting power, and dispersing ability are also found to vary regularly with OE content.

2.6.2. Derivatives of Polyglycerols and Other Polyols

Polyol surfactants, because of their general biocompatibility and their relatively low cost, in most cases, have found a number of important uses, including many that require strenuous testing and clearance by the U.S. Food and Drug Administration and other governmental agencies throughout the world. Their principal areas of impact include foods and food emulsions, pharmaceuticals, cosmetics, and agricultural applications of pesticides and herbicides. Additional uses include lubrication, cutting oils, detergents, dry cleaning fluids, and miscellaneous specialized applications.

Like the POE-based surfactants, compounds derived from the condensation products of hydrophobic groups containing an active hydrogen and glycerol or other polyols constitute an important family of nonionic surfactants. Represented by the general structure

$$RX(OCH_2CHR'CH_2O)_nH$$

where R and n have the usual significance, R′ may be H or OH, and X may be CH_2 (an ether), CO_2 (an ester), or other linkages. Some such materials, especially the polyglycerols, are more complex mixtures than those previously discussed because of the potential for branching in the oligomerization reaction. There are three possible isomers of diglycerol (Figure 2.4), and each subsequent addition of a glycerol unit increases the isomeric possibilities in the final product. A completely random distribution of isomers will not be obtained because of differences in the reactivity of the various hydroxyl groups. The mixture will, however, be so complex that accurate analysis is difficult.

There are two generally encountered routes to the synthesis of surfactant-grade polyglycerols. One involves the polymerization of glycidol to form the polyglycerol followed by esterification with a surfactant-class fatty acid. The alternative method is the reaction of glycerol with a hydrophobic group containing a reactive hydrogen in a way similar to the preparation of the POE surfactants. Sufficiently reactive materials include alcohols, amines, acids, phenols, mercaptans, and amides.

The products of the reactions above have been shown to have good wetting and dispersing properties, and they are good emulsifiers in many cases. They have the advantage that they are generally nontoxic and are employed in a number of useful food-related applications. These would include food emulsions, such as margarine,

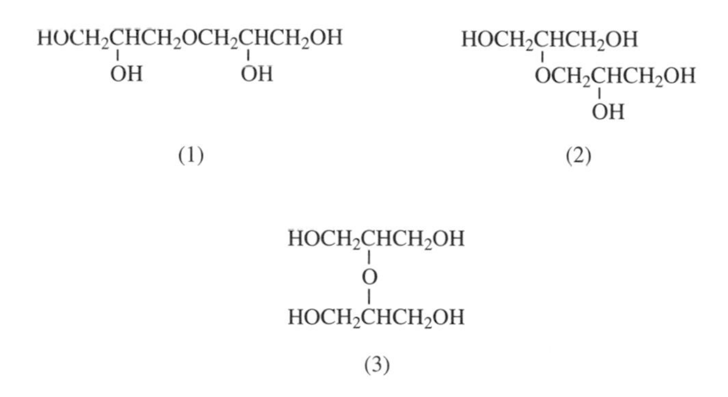

Figure 2.4. Potential isomers of polyglycidyl addition products.

and moisturizing additives for breads and other baked goods. A disadvantage of such materials is that the reaction process can result in the formation of a significant amount of nonsurfactant polyglycerol, which is difficult to remove from the mixture. In addition, there seems to be a practical limit of about 10 glycerol units, above which the polyglycerol becomes intractable and reaction with the hydrophobic group becomes difficult.

An interesting member of the polyol surfactants that is even more complex is the family of polyglycerol polyrecinoleates, which are prepared by the high-temperature, low-pressure oligomerization of castor oil or recinoleic acid with glycerol. Recinoleic acid is a hydroxy acid that on reaction produces an ester unit containing 3–5 recinoleate units esterified to a polyglycerol of up to 10 glycerol units. Such compounds have been found to have very useful effects in chocolate manufacture and in whipped dessert products.

Other classes of polyol surfactants have as their starting materials polyhydroxy compounds with 2–6 hydroxyl units per chain. Those hydrophilic groups include sugars, ethylene and propylene glycol, and other related materials. The surfactants are usually prepared by the esterification of fatty acids with the desired polyol, normally resulting in the formation of mixtures of mono-, di-, and often polyalkyl esters. The commercial products, therefore, may be complex mixtures of compounds whose final properties must be determined by the careful control of feedstock composition and reaction conditions. Relatively pure monoglyceride esters ($>90\%$ α-mono) may be obtained by molecular distillation. Distilled monoesters of propylene glycol are also important in the food industry.

Esters of fatty acids with sorbitan are frequently ethoxylated to various degrees. The esters and ethoxylated esters (the well-known Spans, Tweens, polysorbates, etc.) form a very important family of nonionic materials for use in many applications such as foods, cosmetics, and pharmaceuticals.

A number of surfactants based on the fatty acid esters of sucrose and other sugars are now becoming important because of the more "natural" character of the polyol versus POE and related materials. These "sugar detergents" are reported to have good detergent properties and to be nontoxic. In contrast to the POE family of surfactants, the sugar-based materials exhibit few solubility problems below 100°C and appear to be very effective in mixed micelle formation in combination with anionic surfactants such as sodium dodecylsulfate. As the ester, the sugar surfactants have some problems with regard to hydrolytic stability in acid media. Alternative linkages that have been suggested to overcome such problems include amides and urethanes.

A major drawback of the sucrose ester surfactants is the difficulty of synthesis due to the lack of mutual solubility between the sugar and the hydrophobic components, usually fatty acids. To obtain a commercially viable process, a good solvent for all components such as dimethylsulfoxide (DMSO) must be used. To make the final product meet FDA standards, the level of residual DMSO must be extremely low, requiring rigorous purification, implying high production costs. Nevertheless, the advantages of the family of products has led to a their commercial productions in limited quantities.

2.6.3. Block Copolymer Nonionic Surfactants

An interesting class of nonionic surfactants that has developed as a result of advances in block polymerization techniques is that of the polyalkylene oxide copolymers. Such materials exhibit many interesting and useful properties that have allowed them to carve out a special niche in the surfactant formulation world. Although they are relatively low-molecular-weight materials as polymers go, they are much larger than normal surfactant molecules and for that reason will be discussed in more detail in Chapter 7.

2.6.4. Miscellaneous Nonionic Surfactants

A number of technologically important nonionic surfactants do not fall into any of the main categories discussed so far. These include the alkanolamides derived from fatty acids, amine oxides, sulfoxides, and phosphine oxides, and others too numerous to cover adequately in a brief overview of surfactant technology. Such materials are reported frequently in the patent literature and are discussed in more detail in the works of Burnette and Schwartz et. al., cited in the Bibliography.

Perhaps the most commercially important family of surfactants falling in this category consists of the fatty alkanolamides given by the general formula

$$RCONR'(OH)R''$$

where the R group is a fatty acid or mixture of acids, while R′ can be a simple short-chain hydroxyalkyl (e.g., $-CH_2CH_2OH$) or a more complex ether, amine, or other such structure containing hydroxyl group. The R″ can be H, alkyl, or the same as R′ to yield a diol surfactant. The fatty alkanolamides are found in a wide variety of products, especially where high foaming and foam stability are perceived to be advantageous, either in the action of the formulation or for esthetic reasons. Prime examples of such applications would be shampoos and dishwashing liquids. Such materials have also been found to be generally hypoallergenic and readily biodegradable, resulting in their increased use in cosmetic and other formulations.

Amine oxides are prepared by the oxidation of tertiary amines with peroxides or peracids to yield materials that, while possessing a formal charge separation on the nitrogen and oxygen atoms, behave as nonelectrolytes:

$$RR'R''N^+{-}O^-$$

The R, R′, and R″ groups may be alkyl, aryl, or any of several other structures, and at least one will be a fatty acid residue. Most commonly, the R′ and R″ will be CH_3.

Additional classes of nonionic surfactants can be found in the patent and other specialized literature. In many cases the materials described are of little practical significance, but they should not be totally ignored, especially when very special or unique applications and properties are being considered.

2.7. AMPHOTERIC SURFACTANTS

The family of surfactants commonly referred to as "amphoterics" are surface-active materials that contain, or have the potential to form, both positive and negative functional groups under specified conditions. Their definition as a separate class of surfactants has historically been somewhat controversial, since they may be electrically neutral and their general properties under many conditions make them functionally similar to some nonionic surfactants. For purposes of discussions related to chemical structures, however, they have been separated into a distinct family. In the final analysis, a surfactant by any other name is still a surfactant, so that too much importance should not be placed on nomenclature.

Although amphoteric materials represent only a small portion of total worldwide surfactant production, their market position is increasing significantly because of the unique properties that such materials can impart to a formulation. It is particularly important that they often show considerable synergism when employed with other classes of surfactants. Their nature can make them especially useful in applications requiring biological contact as, for example, in "no tears" baby shampoos. In addition, for uses that might involve the presence of charged polymeric species, the "dual" nature of the materials reduces or eliminates the possibility of undesirable polymer–surfactant interactions (see Chapter 7).

Although there exists a rather large group of organic functionalities that hold the potential for producing amphoteric surfactants, in general four classes of materials are most often encountered: imidazoline derivatives, betaines and sulfobetaines, amino acid derivatives, and lecithin and related phosphatides. Charge-separated compounds such as amine oxides and sulfoxides, mentioned in Section 2.6, could easily be included in the amphoteric classification; however, the more intimate nature of their charge separation, as compared to the internal salts to be discussed here, seems to justify their inclusion in the nonionic category.

It may be noticed that the discussion of amphoteric surfactants of the first three classes allows for considerable overlap as far as chemical structures are concerned. Surfactants that can be considered to be amino acid derivatives may be prepared via the imidazoline intermediate or by some more direct route as described below. The betaines, on the other hand, are merely a special subclass of amino acid derivatives in which the amino group has been quaternized, again resembling a subclass of the imidazoline-derived materials. Clearly, an unequivocal distinction between the general types of amphoteric surfactants is impossible. A detailed description of the wide range of possibilities that exist in this class of materials is given in the works cited in the Bibliography. A brief description of some of the main groups of amphoteric surfactants will illustrate the general chemical nature of this family of materials.

2.7.1. Imidazoline Derivatives

Imidazoline-derived surfactants represent one of the most commercially important classes of amphoteric surfactants. Although mentioned in the patent and surfactant literature as early as 1940, they received little commercial attention until the early 1950s, when it became evident that they possessed a number of potentially useful properties, including a significantly greater mildness to skin and eyes than other surfactants. As a result, the imidazolines have developed a significant place in the cosmetics and personal care products market. The growing availability of these materials as a result of their importance in the cosmetics industry has led to their application in a number of other unrelated fields.

Most commercially important imidazoline-derived amphoteric surfactants can be described as fatty acid/aminoethylethanolamine condensates of the general structure

$$RCONHCH_2CH_2NR'R''$$

where R is the fatty acid residue and R′ and R″ can be any of several functionalities to be described below. The free tertiary amine can be further alkylated to produce a quaternary ammonium compound possessing a permanent positive charge. Most commercial imidazoline surfactants are prepared in a two-step process. First a fatty acid is condensed with a polyamine such as ethylenediamine or aminoethylethanolamine, with the loss of approximately 2 mol of water to yield a cyclic imidazoline. The second step is the alkylation of the imidazoline with an alkylating agent containing the anionic portion of the final molecule and ring opening to give the final product.

Although there are a wide variety of chemical structures possible, four main classes can be found in most commercially available materials. Those classes, as well as their predominant ionic forms and ionization characteristics, can be summarized as follows:

Class 1. Amine/carboxylic acids (not to be confused with the amino acid derivatives to be discussed in following sections) containing both free amine ($-NR_2$) and free acid ($-COOH$) functionalities. These materials will be cationic at low pH ($-N^+R_2H + -COOH$), isoelectric near neutral pH ($-N^+R_2H + -COO^-$), and anionic at high pH ($-NR_2 + -COO^-$).

Class 2. Quaternary ammonium/carboxylic acids. These materials contain a permanent cationic site ($-N^+R_3$) as well as the carboxyl group. At low pH they will, of course, be cationic. At slightly alkaline pH they will become isoelectric and remain so. They can never become anionic in the way that the class 1 materials can.

Class 3. Amine/sulfonic acids (or sulfate ester). Having the strong acid sulfonic acid group, which is highly ionized even at low pH, these materials will form internal salts and will be essentially isoelectric in very acidic media. As the pH is raised to the alkaline side, the protonated amine is neutralized to yield a net anionic species. These materials are essentially the opposite of the class 2 surfactants.

Class 4. Quaternary ammonium/sulfonic acids (or sulfate esters). Possessing both the permanent cationic charge of the quaternary ammonium group and the highly ionizing strong acid, these materials will be isoelectronic over most of the pH range, except at very low pH under conditions where ionization of the acid may be suppressed.

These classifications represent the most fundamental members of this family of surfactants. It should be noted that the betaine and sulfobetaine surfactants discussed in the following section are in fact special members of classes 2 and 4. A number of additional structural modifications commonly encountered in commercial products add to the diversity of the materials and properties that can be obtained.

Most commercial carboxylated imidazoline surfactants are actually mixtures of classes 1 and 2 listed above, while the sulfated materials are combinations of classes 3 and class 4. The carboxylated materials will usually have a buffering action in solution so that the native pH will be slightly alkaline. The class 3 and class 4 materials possess slightly less buffering capacity, but will lie just to the acidic side.

As mentioned above, the increased importance of the imidazoline-derived surfactants stems primarily from their mildness and low toxicity. The extent of their use in shampoos and body care products has followed closely the overall increase in the use of such products worldwide. Their amphoteric nature also makes them useful in a wide range of water types, ranging from hard to soft water and high to low pH. Such flexibility makes them useful in cleaning formulations that will see a variety of conditions.

The variable electronic characteristics of the amphoteric surfactants also make them useful in textile applications where antistatic and "softening" properties under various conditions and on different fabric types are advantageous. Treatment of various metal surfaces is also facilitated by the ambivalent nature of such materials, leading to their use in metal treating and finishing products.

2.7.2. Surface-Active Betaines and Sulfobetaines

Betaine is a naturally occurring material, first identified in the nineteenth century, having the chemical structure of trimethylaminoacetate:

$$(CH_3)_3N^+CH_2COO^-$$

The compound is an internal salt that, in its natural form, has most of the characteristics of a totally un-ionized material. When one (or more) methyl groups is replaced by a long-chain alkyl such as a fatty acid residue, materials with significant surface activity can result. Such materials are now commonly referred to as *betaine surfactants*.

As mentioned above, the betaine surfactants can be considered to be special members of the ring-opened, imidazoline surfactants. However, they do not exhibit many of the characteristics of other amphoteric surfactants, especially with regard to their solubility and electrical nature in alkaline solution. Even at high pH, the betaines do not acquire any significant anionic character, and they appear to maintain their good water solubility, even at the isoelectric pH. They are compatible with anionic surfactants at all pH and do not appear to have problems of complex formation. The carboxyl-containing betaines have been found to form external salts in very strong acids (e.g., hydrochlorides in HCl), while the sulfobetaines do not do so. Members of this class of surfactants are generally insensitive to the presence of electrolytes and usually perform well in hard water.

The carboxyl betaines have found a number of commercial applications, including use in textile processing as leveling and wetting agents, detergents, scrubbing compounds, antistatic agents, and fabric softeners. They have also found use as lime soap dispersants, in detergent formulations, dry cleaning fluids, and personal care products. The sulfobetaines seem to have gained much less attention commercially, although they have been found to be useful in a number of special areas such as the control of static charge in photographic films.

The use of the term "betaines" is, of course, not technically correct from a chemical nomenclature standpoint, even though *Chemical Abstracts* does maintain such a listing in its Chemical Subjects Index. The betaine surfactants may be found named according to the parent amino acid, from which they can, in principle, be derived. For example, the compound

$$C_{12}H_{25}(CH_3)_2N^+CH_2COO^-$$

could be named *dodecylbetaine*, *N-dodecyl-N,N-dimethylglycine*, or *dimethyldodecylammonioacetate or ethanoate.*

Probably the oldest and still most common route to the carboxybetaines is through the quaternization of long-chain alkyldimethyl tertiary amines with chloroacetic acid. A similar route can be employed to prepare the analogous sulfobetaines from 2-chloroethane sulfonic acid. Other synthetic pathways can be found in the review by Ernst and Miller.

2.7.3. Phosphatides and Related Amphoteric Surfactants

Phosphatidyl surfactants, of which the commercially popular materials referred to as "lecithins" are members, are generally composed of di(fatty acid), monophosphoric acid esters of glycerol combined with an amine containing radical such as ethanolamine or choline. Typically, their structure can be written as shown below, in which R and R' have their usual meanings. The location of the phosphate ester on the terminal carbon is designated an α-phosphatide, while location internally would be designated as β. Because of their natural sources, the phosphatidyl or lecithin surfactants are normally found as rather complex mixtures, especially with regard to the nature of R and R$'$. In general, the natural lecithins will have one saturated and one unsaturated R group, with the saturated group usually found at the α position.

The natural lecithins generally have limited solubility in water. Having good oil solubility, these materials have found extensive commercial application as nonaqueous emulsifiers, dispersants, and wetting agents in such diverse areas as marine paints, inks, foods, and cosmetics. More water-soluble materials can be prepared by the enzymatic removal of the β-fatty acid group. Such materials are commonly referred to as "lysolecithin."

The natural sources of lecithins range from egg yolks to many seed oils such as cottonseed, sunflower, and soybean; soybean is the most common because of its relatively lower cost and greater availability. It is, of course, possible to prepare synthetic phosphatidyl surfactants; however, a number of practical pitfalls make such approaches difficult at best. The preparation of an unsymmetrical diester of phosphoric acid presents inherent difficulties, as does the fact that the amine functionality can lead to the formation of salts of phosphatidic acid. The inherent reactivity of the ester functionalities also can lead to extensive ester interchange during preparation.

The phosphatide family concludes the general summary of the main classes of surfactants that are commonly encountered by the scientist or technician venturing into the field of surfactants and surface activity. In the discussions above, no attempt was made to provide a comprehensive review of all the surfactant classes and subclasses, and the seemingly infinite structural variations that can arise in the design and synthesis of surfactant molecules. Just a few of the general references cited in the Bibliography represent thousands of pages and several volumes of material on the subject. In addition, a scan of each volume of *Chemical Abstracts* reveals a number of new materials being disclosed in the patent and academic literature. Even the miracle of modern electronic communications and information transfer does not allow one to remain constantly abreast of the changing world of surfactant science and technology. The following chapters represent an attempt to summarize as clearly and succinctly as possible the physicochemical ramifications of chemical structure in surface activity and the application of such phenomena to modern technological needs.

The reader is again reminded that no attempt is made to provide an in-depth theoretical review of surfactant activity at interfaces or an encyclopedic listing of the various properties of every surfactant known to humankind. Suitable references are provided for the reader desiring more information on any given subject.

PROBLEMS

2.1. Using basic principles of organic synthesis, suggest a process for the synthesis of sodium dodecylsulfate [$CH_3(CH_2)_{11}OSO_3^- Na^+$] from lauric acid.

2.2. Detergents were originally developed to replace carboxylate soaps for use under conditions of "hard" water and low temperatures. Discuss the basis for the difference in surfactant activity of the two classes of materials based on solution theory.

2.3. Most higher-molecular-weight carboxylic acids ($>C_8$) do not have sufficient water solubility to be effective surfactants and must be neutralized with alkali to produce classical soaps. Strongly acidic materials such as alkylsulfonic acids (R–SO_3H) and sulfuric acid esters (R–OSO_3H), however, are usually surface-active as the free acid, although normally employed as the alkali salt. If R is taken as a C_{16} hydrocarbon chain, explain why or why not you would expect the following reaction product to be a good surfactant:

$$\text{R–SO}_3\text{H} + \text{R}_3\text{N} \rightarrow \text{R–SO}_3^{-}\ {}^{+}\text{NHR}_3$$

Explain your reasoning based on general concepts of solutions, solubility, and other parameters.

2.4. Why do PT-based detergents have a longer persistence in lakes, rivers, and underground water tables than do LABS materials?

2.5. Explain why amphoteric surfactants, while having discrete electrical charges, tend to behave more like nonionic than ionic surfactants.

2.6. Alkylphosphonic acids and their salts (R–PO_3H/M^+) and phosphoric acid esters of long-chain alcohols (R–OPO_3H/M^+) can make good surfactants, but in fact represent a small percentage of commercial products. Suggest some reasons for their lack of "popularity."

2.7. Other things being equal, rank the following materials in order of their desirability as raw-materials sources for detergent manufacture: coconut, soybean, cotton, palm nuts, beef fat, coal, crude oil, olives, and natural gas. Explain briefly the basis for your ranking.

2.8. Most commercial fluorocarbon surfactants are produced by the electrolytic substitution of fluorine for hydrogen on the carbon backbone of a carboxylic acid fluoride or sulfonic acid fluoride

$$\text{H}_3\text{C–(CH}_2)_n\text{–COF} + \text{HF/NaF} + \text{e}^- \rightarrow \text{F}_3\text{C–(CF}_2)_n\text{COF}$$
$$\text{H}_3\text{C–(CH}_2)_n\text{–SO}_2\text{F} + \text{HF/NaF} + \text{e}^- \rightarrow \text{F}_3\text{C–(CF}_2)_n\text{SO}_2\text{F}$$

folllowed by hydrolysis to the acid and neutralization. Other materials may be prepared by the oligomerization of tetrafluoroethylene to produce the alcohol followed by further reaction

$$\text{F}_2\text{C=CF}_2 + \text{OH}^- \rightarrow \text{HO(CF}_2\text{–CF}_2)_n\text{H}$$

Would you expect two surfactants, one prepared by each process having the same carbon chain length and hydrophilic group, to be equivalent in terms of surface activity? Explain.

2.9. Chemically speaking, what is a potential disadvantage of using a fat source rich in polyunsaturated acids as a raw material for soap manufacture?

3 Fluid Surfaces and Interfaces

For the chemist or chemical technologist, a surface or an interface may be described as the boundary between at least two immiscible phases. In any such system, the boundaries between the phases may be of primary importance in determining the characteristics and behavior of the system as a whole, although the bulk characteristics of each phase are, in theory, unaffected. The reality is that even very low levels of solubility among the phases can alter the bulk-phase characteristics to some extent. The viability of many scientific and commercial applications of multiphase systems depends on an ability to control and manipulate phase boundaries or interfacial interactions.

Geometrically, it is obvious that only a single surface can exist between two immiscible phases—ignoring for the moment such systems as emulsions, dispersions, and foams in which there are many interfaces, but all of one type. If three phases are present, only a single line can be common among the various elements (Figure 3.1). Geometric considerations simplify the concepts needed for understanding the fundamentals of surface activity. Once the physical and chemical concepts of a surface or interface are clear, it becomes easier to understand the role of surfactants in their modification.

In what may be referred to as the "real" world, five basic types of interfaces are encountered: (1) solid–vapor (S/V), (2) solid–liquid (S/L), (3) solid–solid (S/S), (4) liquid–vapor (L/V), and (5) liquid–liquid (L/L). Traditionally, interfaces involving one vapor and one condensed phase (e.g., S/V and L/V interfaces) are referred to as "surfaces." In the following discussions that tradition will generally be followed, although "interfaces" may be used where generality is implied. Most obvious effects of surface-active agents on the physical properties of a system are seen in systems where at least one phase is a liquid. The true importance of interfaces goes much deeper, however. As will hopefully become apparent in the following sections and chapters, the modification of an interface with less than a one-molecule-thick layer may bring about dramatic changes in the nature of mixed phase systems.

Although adhesion and lubrication, with at least one solid phase, involve significant interfacial interactions and often employ the actions of surfactants, they also involve other physical phenomena that may ultimately be of greater importance to the system. They also quite often involve nonequilibrium conditions that make

Surfactant Science and Technology, Third Edition by Drew Myers

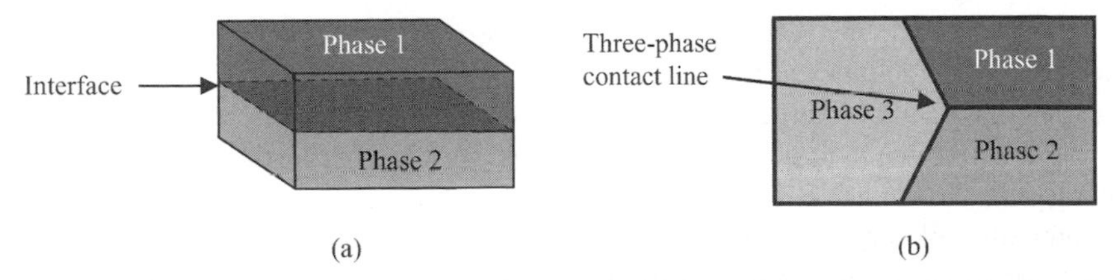

Figure 3.1. Geometric considerations for the definition of interfaces: (a) the interface between two phases, which will be a plane or curved section; (b) the contact line between three phases, the three-phase contact line, functioning as the interface between the phases.

them difficult or impossible to analyze. The major portion of the subsequent discussions, therefore, will be concerned with conditions and systems in which the surface activity is of primary importance, and usually under conditions at or near equilibrium. It must be recognized, however, that the modification of solid surfaces to control such characteristics as friction, water repellency, and static charging is becoming increasingly important and constitutes a broad area for potential surfactant application.

When one undertakes a discussion of surface and interfacial phenomena, several key points must be kept in focus:

1. The requirements of mathematics and a touch of "common sense" cause most of us to visualize an interface as a sharp boundary or plane having a thickness of at least one molecular diameter; the effect of the boundary in its simplest manifestation does not extend a great distance into either bulk phase. In more rigorous theoretical treatments, the interface may be assumed to be several molecules thick, and in some instances, the energetic consequences of the interface may extend several hundred nanometers into one or both phases.
2. Matter at an interface is usually found to have physical properties distinct from those of the bulk material and, as a result, a definite free energy is associated with each unit of interfacial area. In particular, atoms and molecules located at an interface will experience significantly different force fields from those in the bulk material because of different numbers and types of neighboring atoms and molecules.
3. Another characteristic of phase boundaries, especially those involving water, is the probable existence of an electrical potential across the interface. Although such charging phenomena are not always present, where they do exist they can be extremely important in determining the interfacial characteristics of the system. Electrostatic effects are most important in aqueous suspensions, dispersions, emulsions, foams, and aerosols in which one phase is very finely divided, creating a large interfacial area. The presence of electrical charges at interfaces often contributes significantly to the overall stability of a system. In addition, interfacial charges make possible many important industrial processes such as electrostatic or electrophoretic paint and rubber deposition.

The interfacial energy and electrical characteristics of a system are determined by the usual quantities such as temperature and pressure, and the chemical composition of the different phases. Charge characteristics may be altered significantly by the addition of ionic materials such as neutral electrolytes, by changes in the system pH, or by changes in the chemical composition of the aqueous phase, such as the addition of a water-miscible organic solvent that results in a change in the dielectric constant.

Interfacial energies will also be affected by other changes in phase compositions. Because of the relatively large distance between molecules, the nature of a contacting gas will normally have little effect on the surface tension of a contacting liquid phase; however, if specific chemical or physical interactions are possible, some effect may be seen, especially for high-energy surfaces such as metals and metal oxides. The solubility of the gas in the liquid phase may also produce a change in the surface tension.

The surface tension or surface energy of a material, especially a liquid or high-energy solid, may be very significantly altered by small changes in its bulk composition. At interfaces between two condensed phases (e.g., liquid–liquid or liquid–solid), compositional changes in either or both phases may greatly alter the interfacial energy. Generally, the addition of a solute will change the surface tension of a liquid if the nature of the solute is such that its presence at the surface will result in a lower net free energy for the system (positive adsorption). Such an effect is especially important in liquids that have a relatively high surface tension, such as water. The presence of solutes may either raise or lower the surface tension at the liquid–vapor interface, although the latter is normally observed. Interestingly, the presence of very small amounts of a surface-active organic molecule may lower the interfacial tension of water by 50% or more, while a fully saturated electrolyte solution may exhibit an increased surface tension of only a few millinewtons per meter (mN/m).

The liquid phase of most academic and industrial interest is water, which has a surface tension of 72–73 mN/m at 20°C. A 1% aqueous sodium hydroxide solution will have a slightly higher value of about 73 mN/m, while one of 10% will approach 78 mN/m. Relatively high concentrations of NaOH, or other electrolyte, are required to significantly increase the surface tension of water. The more normal result of dissolution of a material is to lower the surface tension of the liquid. In addition, the lowering effect is usually apparent at concentrations much lower than those required to raise the surface tension.

Conditions producing changes in surface tension at a liquid–vapor interface will usually also result in modification of the interfacial energy between the liquid and another liquid or a solid phase. In general, there will not be an obvious direct relationship between changes in the surface tension of a liquid phase and its other interfacial interactions under a given set of conditions. Nevertheless, the change in the surface tension of a liquid on addition of a solute can be a qualitative indicator of changes to be expected in its interactions with other phases. The following chapters discuss in more detail the chemical and physical principles responsible for observed changes in interfacial interactions with the addition of solutes to one or more phases of a system.

3.1. MOLECULES AT INTERFACES

Atoms or molecules at an interface will have a higher potential energy than those in the bulk of a material as a result of their "geography." Their location at the interface means that they will experience a net asymmetric force field due to interactions with neighboring units significantly different from units in the bulk (Figure 3.2). For two immiscible phases, surface or interfacial units will normally interact more strongly with identical units in the bulk rather than the "foreign" components in the adjacent phase. Because of the increased energy of units at the interface, thermodynamics demands that work be required to move them from the bulk phase to the surface. The minimum-energy rule for systems in equilibrium will therefore lead to surface conditions yielding minimum interfacial area, or minimum asymmetric interactions.

Surface-active materials are compounds that, because of their characteristic molecular structures, are natural fence-sitters. Their split personalities drive them to reduce unfavorable energetic interactions by moving to more comfortable interfacial neighborhoods (adsorption), by getting together with their "own kind" (aggregation), or by simply getting out of town (precipitation or phase separation). When present in relatively low concentrations, such materials will preferentially adsorb at available interfaces, replace the higher-energy bulk-phase molecules, and result in a net reduction in the free energy of the system as a whole.

Materials that possess chemical groups leading to surface activity are generally referred to as being "amphipathic" or "amphiphilic." When a material exhibiting the characteristics of surface activity is dissolved in a solvent (whether water or an organic liquid), the presence of the lyophobic group causes a distortion of the solvent liquid structure (and, in principle, that of a solid phase as well), increasing the overall free energy of the system. In an aqueous surfactant solution, for example, such a distortion of the water structure by the hydrophobic group increases the overall energy of the system and means that less work is required to transport a surfactant molecule to a surface or interface. The surfactant may therefore concentrate or preferentially adsorb at those locations, or it may undergo some other process to lower the energy of the system (e.g., aggregation or micellization). Since less work is required to bring surfactant molecules to the available interfaces, the

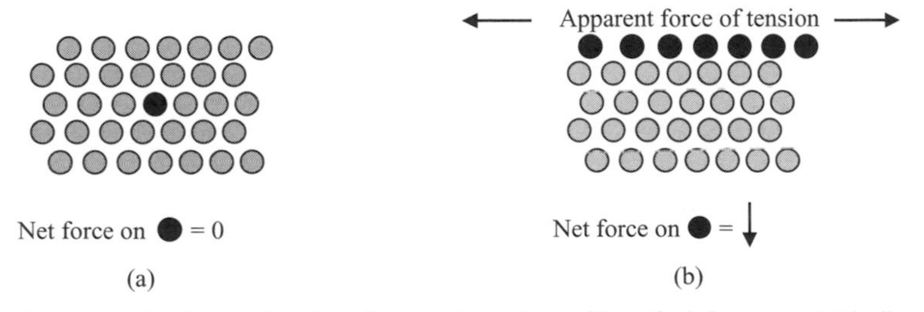

Figure 3.2. Atoms (molecules) at interfaces—the origin of interfacial energy: (a) bulk atoms or molecules; (b) surface atoms or molecules.

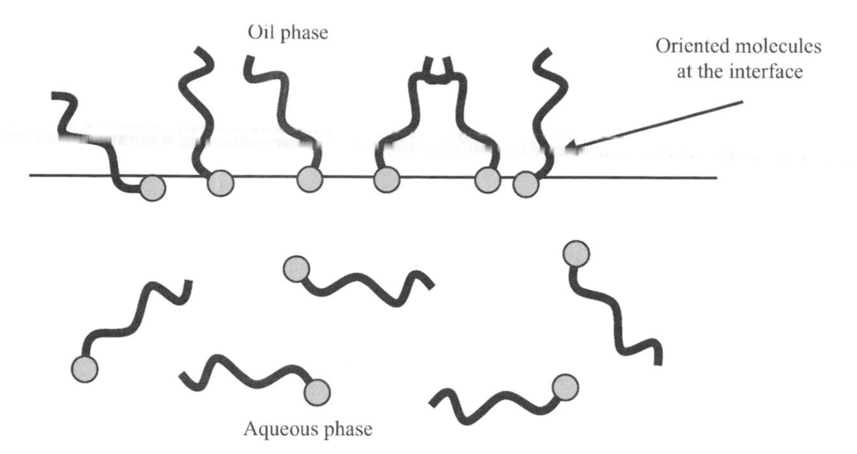

Figure 3.3. The preferential orientation of surfactant molecules at interfaces.

presence of the surfactant decreases the work required to increase the interfacial area. The work per unit area required to form that new interface is the surface free energy or surface tension of the system, σ, usually reported in units of millinewtons per meter (mN/m) or dynes per centimeter (dyn/cm) in non-SI units. For solid surfaces, to be discussed in Chapter 10, the convention is to use millijoules per square meter (mJ/m^2) or ergs per square centimeter ($ergs/cm^2$).

The presence of a lyophilic group on the surfactant molecule prevents or retards the complete expulsion of the solute molecules from the solvent as a separate phase, at least at low concentrations. As discussed in Chapter 5, high concentrations of surfactant can lead to a number of interesting and useful phenomena related to the varied aggregate structures that they can form.

The amphipathic structure of surfactant molecules not only results in their concentration at a liquid surface and consequent alteration of the surface tension but also causes orientation of the adsorbed molecules such that the lyophobic groups are directed away from the bulk solvent phase (Figure 3.3). The resulting controlled molecular orientation produces some of the most important macroscopic effects observed for surface-active materials, as will be discussed in subsequent chapters. For now, it is more important to understand the qualitative relationships between the nature of interfaces and the general chemical structures required for a molecule to exhibit significant surface activity.

3.2. INTERFACES AND ADSORPTION PHENOMENA

The region of space forming the boundary between two immiscible phases is generally referred to as the "interface" and represents a transition region in which the chemical and physical characteristics of one bulk phase undergo an abrupt (on a macroscopic scale) change to those of the adjacent one. On a microscopic scale, however, that change must occur over the distance of at least one, but more

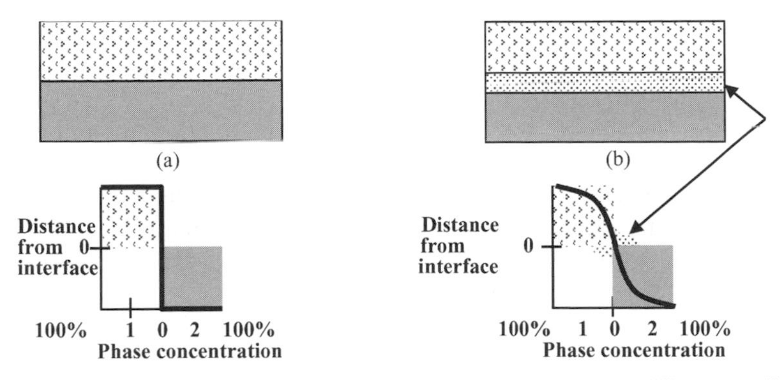

Figure 3.4. Dividing surfaces at fluid interfaces: (a) ideal interface; (b) diffuse ("realistic") interfacial region.

often several, molecular distances. To quantitatively treat the phenomena related to surface activity, especially adsorption phenomena, it is necessary to mathematically define the location of the dividing line or surface at which the change is assumed to occur.

For convenience, it is usually assumed that a dividing surface can be defined as shown in Figure 3.4a, where an ideal plane lies between phases 1 and 2. Such an ideal model is unrealistic, however, especially in the event of adsorption at the interface. Such an adsorbed interfacial film not only will possess a finite thickness related to the size of the adsorbed molecule but may also alter the nature of molecules of phases 1 and 2 located near the interface and result in an interfacial region in which the composition changes more or less continuously over a considerable distance (Figure 3.4b).

At this point it may be useful to reiterate that the general term "interface" refers to the above mentioned boundary between any two phases. In common usage, however, the term "surface" is used with reference to systems in which one phase is a gas, such as in the "surface tension" of a liquid. Throughout the subsequent discussions, the use of "interfacial" will imply applicability to multiple types of boundary region, while "surface" will apply only to S/V and L/V systems.

3.2.1. A Thermodynamic Picture of Adsorption

In the consideration of adsorption processes, two aspects must be addressed: (1) the effect of the adsorbed species on the final equilibrium interfacial energy of the system and (2) the kinetics of the adsorption process. For the most part, the discussions to follow are concerned only with equilibrium conditions, and dynamic processes are not addressed. For many surfactant applications, such a restriction will not result in significant limitations to the validity of the concepts involved. Obvious exceptions would be processes involving dynamic interfaces (i.e., coating operations), polymer adsorption, and other kinetically "slow" processes.

If the idealized concept of a mathematical dividing plane between phases is employed to represent the interface, the adsorption of a solution component can conveniently be pictured as the existence at the interface of a concentration of the adsorbed material, n_i^{s}, that differs from its concentration, n_i^{b}, in one or both of the bulk phases, where "s" denotes the surface phase and "b," the bulk. The amount of component i in the surface phase in excess of that amount would have been present had each phase extended up to the dividing plane S–S without changing composition is referred to as the "surface excess concentration" of i, specifically, Γ_i. Formally, it is given by

$$\Gamma_i = \frac{n_i^{\mathrm{s}}}{A} \tag{3.1}$$

where A is the interfacial area. In principle, Γ_i may be either positive or negative, and its value will be determined by the somewhat arbitrary choice of the location of the dividing surface S–S.

When interfacial adsorption occurs, the energy of the interface changes. To understand and predict the role of surfactant adsorption, it is necessary to know the amount of material adsorbed at the interface of interest. The Gibbs equation, which relates changes in the interfacial energy of a system to the degree of adsorption of a species at the interface and the compositions of the bulk phases, forms the basis for understanding the thermodynamics of the adsorption process. Under conditions of constant temperature and pressure, the basic equation is given as

$$\delta\sigma_i = -\Gamma_1\,\delta\mu_1 - \Gamma_2\,\delta\mu_2 - \Gamma_3\,\delta\mu_3 - \cdots \tag{3.2}$$

where σ_i is the interfacial energy, Γ_i is the surface excess of component i at the interface, and μ_i is its chemical potential in each bulk phase.

The change in the free energy G of a system may be given by

$$dG = -S\,dT + V\,dP + \sigma\,dA + \sum \mu_i\,dn_i \tag{3.3}$$

where G, S, P, V, and T have their usual thermodynamic meanings and A and n_i are as defined above. At equilibrium and under constant conditions of T, P, and n_i, Eq. (3.3) reduces to

$$dG = \sigma\,dA \tag{3.4}$$

If the surface excess of component i is allowed to vary by adsorption, then

$$\delta\sigma = \frac{-\sum n_i^{\mathrm{s}}}{A d\mu_i} = -\sum \Gamma_i\,\delta\mu_i \tag{3.5}$$

As pointed out above, the value of Γ_i is defined by the choice of the location of the dividing surface. To simplify the mathematics, it is convenient to define S–S so that the surface excess of one component, usually one of the bulk solvent phases, will be zero. For a simple two-component system (2 dissolved in 1, for example), Eq. (3.5) reduces to

$$\delta\sigma = -\sum \Gamma_2 \, \delta\mu_2 \tag{3.6}$$

The chemical potential of a species is related to the species activity by

$$\mu_2 = \mu_2^\circ + RT \ \ln a_2 \tag{3.7}$$

so that

$$d\mu_2 = RT \, d \ln a_2 \tag{3.8}$$

$$d\sigma = -\Gamma_2 RT \, d \ln a_2 \tag{3.9}$$

For dilute solutions where the activity coefficient of the solute is approximately 1, concentration c_2 can be substituted for a_2. The Gibbs equation is then written in its most commonly encountered form

$$\Gamma_2 = \frac{-\frac{1}{RT} d\sigma}{d \ln c_2} \tag{3.10}$$

In systems where the interfacial energy can be directly determined (e.g., in L/L and L/V systems), Eq. (3.10) can be used to determine the surface excess concentration of the adsorbed species and, in principle, to relate that quantity to the structure of the molecule. It therefore becomes a useful tool for characterizing a surfactant species at the molecular level and aids in the interpretation of surface phenomena on the basis of chemical composition and molecular structure. In systems where the interfacial energy cannot be measured directly (e.g., in systems involving a solid interface) but the surface concentration can, the equation allows one to calculate changes in the interfacial energy of the system that would otherwise be inaccessible.

The form of Eq. (3.10) is general and can be encountered in several alternative formulations depending on the systems and interfaces in question. It is particularly important to remember that in the case of adsorbed species that are ionized, the adsorption of both ionic species (surfactant and counterion) must be considered. For a more complete derivation and explanation of the consequences of such a situation, the reader is referred to one of the references listed in the Bibliography for this chapter.

Liquid interfaces are normally well defined and, in principle, easier to handle on a theoretical basis than are solid surfaces. As a result, interfaces involving liquid–liquid and liquid–gas phases have attracted a large portion of the attention devoted to the theoretical understanding of surfaces and interfaces. Such interfaces also constitute a large fraction of the practical applications of surfactants. Even considered their relative simplicity, liquid systems can involve a wide variety of interfacial phenomena. From a practical standpoint, the most important interfacial aspects of L/L and L/V systems are related to interfacial tension and the effects of adsorbed surfactants on that property. To have a concept of the role of surfactants at such interfaces, it is necessary to understand, in a qualitative way at least, the molecular forces involved.

3.2.2. Surface and Interfacial Tensions

The interfacial tension between a pure liquid and its vapor or between two immiscible or partially miscible liquids reflects the difference in the forces of attraction acting on molecules at the interface as a result of differences in the density or chemical compositions of the two phases. It has long been accepted that the existence of condensed phases of matter, especially the liquid state, is a result of van der Waals attractions between molecules. That is especially true for materials that do not possess any chemical structure that could lead to the action of forces of an electrostatic, dipolar, or other related specific character. For the sake of simplicity, consider a liquid whose molecules interact only through van der Waals or dispersion forces. In the bulk of the phase under consideration, all molecules will be surrounded by an essentially uniform force field, so that the net force acting on each will be zero. Molecules located at or near an interface, on the other hand, will experience a distorted field resulting in a net attraction for the surface molecules by the bulk. The unbalanced force of attraction acting on the surface molecules will cause the liquid to contract spontaneously to form, in the absence of gravity, a spherical drop.

To visualize the concept of the surface tension of a liquid, it is convenient to define it as a force acting tangentially to the surface at all points, the net result of which is the apparent formation of a surface "skin," which contracts to confine the liquid into a shape of minimum interfacial area. Such a definition, while facilitating the understanding of the results of the phenomenon, may be misleading in the sense that no actual tangential force is acting at the surface of a pure liquid—it only produces the appearance of such a force. A more thermodynamically appropriate definition of surface tension and surface free energy is the work required to increase the area of a surface reversibly and isothermally by a unit amount. The interface between two immiscible liquids can be viewed similarly, except that the presence of a second, more dense liquid phase will usually result in a less severe imbalance in the forces acting on the molecules at the interface and consequently a lower value for the interfacial tension.

Most commonly encountered room temperature liquids have surface tensions against air or their vapors that lie in the range of 10–80 mN/m. Liquid metals and other inorganic materials in the molten state will exhibit significantly higher values as a result of the much greater and more diverse interactions occurring in such systems. Water, the most important liquid that we will consider, lies at the upper scale of what are considered to be normal surface tensions with a value in the range of 72–73 mN/m at room temperature, while hydrocarbons reside at the lower end, falling in the lower to middle 20s. The interfacial tension between water

TABLE 3.1. Typical Surface and Interfacial Tensions of Liquids at 20°C

Liquid	Surface Tension (mN/m)	Interfacial Tension (vs. Water, mN/m)
Water	72.8	
n-Octane	21.8	50.8
Benzene	28.9	35.0
n-Octanol	27.5	8.5
Mercury	485	375

and a hydrocarbon liquid will fall somewhere between the surface tensions of the two phases. For reference purposes, some typical surface and interfacial tensions of liquids are listed in Table 3.1.

Modern treatments of van der Waals and related forces have made it possible to calculate with good accuracy the expected interfacial tensions of many systems that do not involve specific interactions such as dipoles and hydrogen bonding. While no further coverage of the more theoretical aspects of interfacial interactions is given here, a deeper understanding of the principles involved can aid greatly in the extension of the concepts covered to new systems and applications.

The concept of interfacial tensions given above is simplistic in the sense that it implies that the surface or interface is a static entity. In reality, there is a constant and, for liquids and gases, rapid interchange of molecules between the bulk and interfacial regions and between the liquid and vapor phases. If it is assumed that molecules leave the interfacial region at the same rate that they arrive, it is possible to estimate the exchange rate β of an individual molecule from the relationship

$$\beta = \alpha(2\pi mkT)^{1/2} p_0 \tag{3.11}$$

where α is the so-called sticking coefficient, p_o is the equilibrium vapor pressure of the liquid, m is the mass of the molecule, and k is Boltzmann's constant. Assuming α to lie in the range of 0.03–1.0, a water molecule at 25°C will have an average residence time of 3 μs or less at the air–water interface. The corresponding residence time for a mercury atom would be roughly 5 ms, while that for a tungsten atom would be 10^{37} s at room temperature.

With such molecular mobility, it is clear that the surface of a pure liquid offers little resistance to forces that may act to change its shape; that is, there will be very little viscous or elastic resistance to the deformation of the surface. A physical consequence of this fact is that a pure liquid will not support a foam for more than a small fraction of a second (see Chapter 8). A similar situation exists for the L/L interface. That fact, as we shall see in later chapters, has significant implications for many technological applications such as emulsions and foams, and it forms the basis for many of the most important applications of surfactants.

Because of the mobility of molecules at fluid interfaces, it is not surprising to find that temperature can have a significant effect on the interfacial tension of a system. As the temperature of a system is increased, the surface tension of almost all liquids will decrease. From Eq. (3.3), it is clear that a negative temperature coefficient for the surface free energy indicates that the surface excess entropy is positive. At temperatures near the critical temperature of the liquid, the cohesive forces acting between molecules in the liquid become very small and the surface tension approaches zero. While it is intuitively attractive to assume that molecules at a surface possess more degrees of freedom and are more disordered and, possibly, that the surface region has a lower density than does the bulk liquid, finding the proper choice for a model has made the calculation of surface configurational entropy difficult. A number of empirical equations that attempt to predict the temperature coefficient of surface tension have been proposed, with one of the most useful being

$$\sigma\left(\frac{Mx}{\rho}\right)^{2/3} = k_s(T - T_c - 6) \tag{3.12}$$

where M is the molar mass of the liquid, ρ its density, x the degree of association, T_c the critical temperature, and k_s a constant. There do exist a few exceptions to the observation of negative temperature coefficients, but such exceptions are found in molten metal and metal oxide melts, where the atomic and molecular interactions are much more complex.

3.2.3. The Effect of Surface Curvature

Because so many applications of surfactants involve surfaces and interfaces with high degrees of curvature, it is often important to understand the effect of curvature on interfacial properties. What is usually considered to be the most accurate procedure for the determination of the surface tension of liquids, the capillary rise method, depends on a knowledge of the relationship between surface curvature and the pressure drop across curved interfaces. Because of the existence of surface tension effects, there will develop a pressure differential across any curved surface, with the pressure greater on the concave side of the interface; that is, the pressure inside a bubble will always be greater than that in the continuous phase. The Young–Laplace equation

$$\Delta p = \sigma\left(\frac{1}{r_1} + \frac{1}{r_2}\right) \tag{3.13}$$

in which Δp is the drop in pressure across a curved interface, r_1 and r_2 are the principal radii of curvature, and σ is the surface (or interfacial) tension, relates the quantities of interest in this situation. For a spherical surface where $r_1 = r_2$, the equation reduces to

$$\Delta p = \frac{2\sigma}{r} \tag{3.14}$$

For a very small drop of liquid in which there is a large surface : volume ratio, the vapor pressure will be higher than that over a flat surface of equal area. The movement of liquid from a flat interface into a volume with a curved interface will require an input of energy into the system, since the surface free energy of the curved volume will increase. If the radius of a drop is increased by dr, the surface area will increase from $4\pi r^2$ to $4\pi(r + dr)^2$, or by a factor of $8\pi r\,dr$. The free-energy increase will be $\sigma \times 8\pi r\,dr$. If during the process δn moles of liquid are transferred from the flat phase with a vapor pressure of p_o to the drop with vapor pressure p_r, the free-energy increase will also be given by

$$\Delta G = \delta n\; RT\; \ln\frac{p_r}{p_o} \tag{3.15}$$

Equating the two relationships leads to what is known as the *Kelvin equation*:

$$RT\;\ln\frac{p_r}{p_o} = \frac{2M\sigma}{\rho r} = \frac{2V_m\sigma}{r} \tag{3.16}$$

In Eq. (3.16), ρ is the density, M the molar mass, and V_m the molar volume of the liquid. It can be shown that extremely small radii of curvature can lead to the development of significant pressure differences in drops. For a drop of water with a radius of 1 nm, the partial pressure ratio from Eq. (3.15) will be ~3. It is obvious, then, that the condensation of molecules in systems where the seed nuclei are exceedingly small will be retarded by a relatively high-energy barrier. Such a relationship helps to explain the ability of many liquid and vapor systems to become supersaturated. It is the input of energy induced by scratching, agitation, and other motion, or the provision of a seed site of sufficient size, that brings about the rapid condensation or crystallization of the system.

3.3. THE SURFACE TENSION OF SOLUTIONS

In the most general sense, the surface tension of a liquid refers to the equilibrium excess surface energy at the boundary between the liquid and its own vapor. In practice, the vapor phase will usually be a mixture of the vapor and other gases such as air. The difference, however, is not significant for most purposes. When the liquid phase is not a pure liquid but a homogeneous mixture (solution) of two or more components, it seems intuitively obvious that the surface tension of the system should be some mathematical average of that of the two pure components. The simplest such combination for a binary mixture would be an additive combination

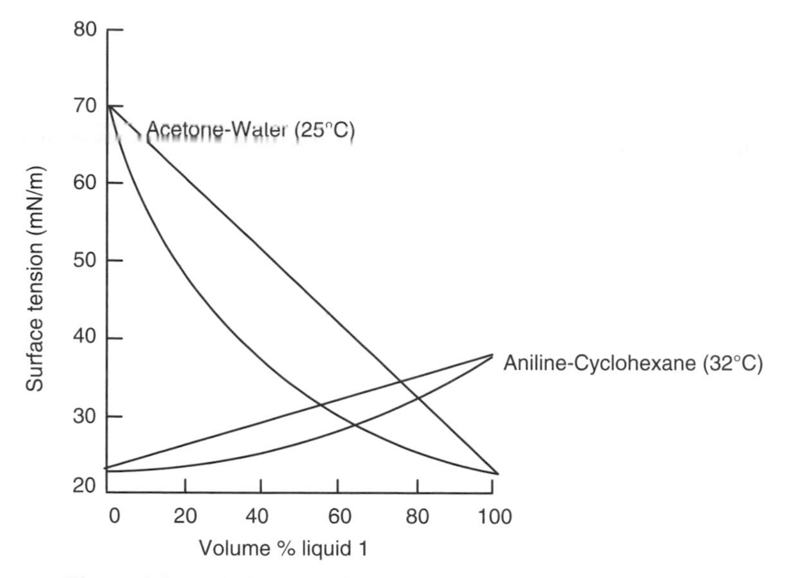

Figure 3.5. Variation in surface tension of various liquid mixtures.

related to the quantity of each component in the mixture, such as mole fraction. The relationship may be written as

$$\sigma_{mix} = \sigma_1 X + \sigma_2(1 - X) \tag{3.17}$$

where σ_{mix} is the surface tension of the solution, σ_1 and σ_2 are the surface tensions of the respective components, and X is the mole fraction of component 1 in the mixture. In ideal systems where the vapor pressure of the solution is a linear function of the composition, such relationships are found. Normally, however, there will be some positive or negative deviation from linearity; the latter is most commonly encountered. Some examples of the variation of the surface tension of mixtures with composition are shown in Figure 3.5.

Taking water as an example, when the second component of a mixture is a solute such as an inorganic electrolyte or other material that requires significant solvation, relationships between surface tension and composition can be expected to be quite varied, depending on the exact nature of the solute–solvent interaction. It is generally found, for example, that the addition of inorganic electrolyte to water results in an increase in the surface tension of the solution, although the effect is not dramatic and requires rather high salt concentrations to become significant (Figure 3.6). The relative effectiveness of cations at increasing the surface tension of water generally follows the Hofmeister series: $Li^+ > Na^+ > K^+$, and $F^- > Cl^- > Br^- > I^-$.

Unlike inorganic electrolytes, the presence of an organic material in aqueous solution will result in a decrease in the surface tension of the system. The extent of such lowering will depend on a number of factors, including the relative miscibility of the system (or the solubility of the organic solute) and the tendency of

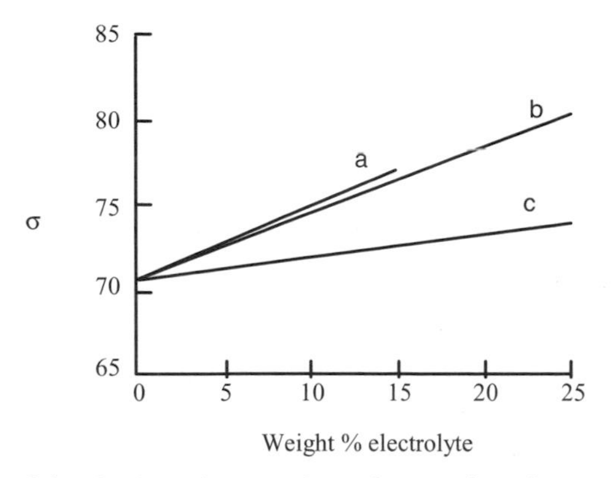

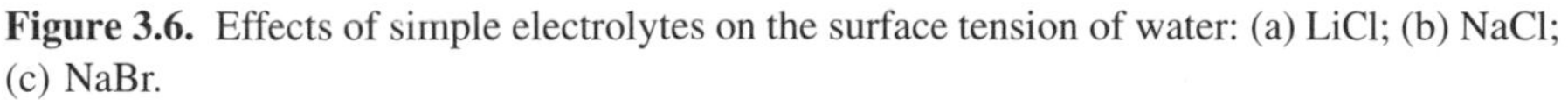
Figure 3.6. Effects of simple electrolytes on the surface tension of water: (a) LiCl; (b) NaCl; (c) NaBr.

the organic material to preferentially adsorb at the water–air interface. Fully miscible liquids such as ethanol or acetic acid result in slight, gradual decreases in the surface tension of their aqueous solutions, while longer-chain organics such as 1-butanol can produce more dramatic effects (Figure 3.7). When the organic solute has a limited solubility in water, the effect on surface tension becomes characteristic of surfactant solutions, where a minimum value of σ will be obtained as the solute concentration increases before surface saturation or some form of solute

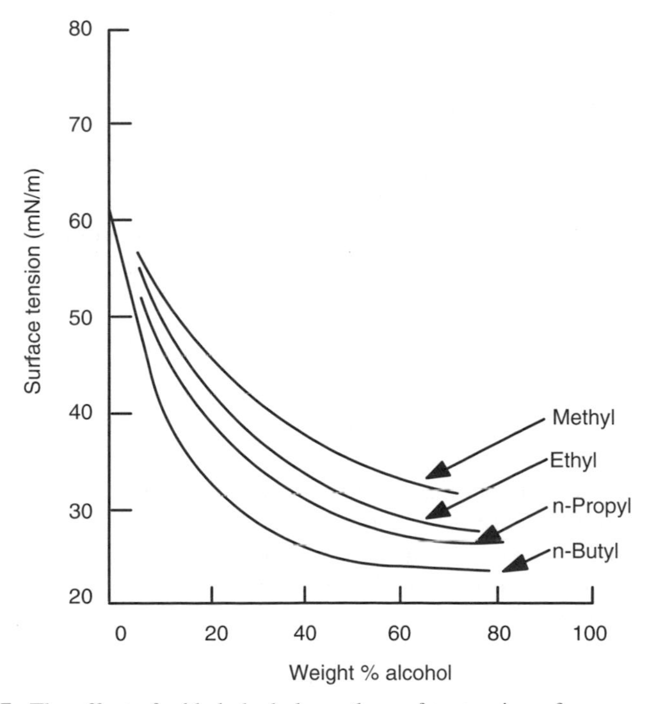

Figure 3.7. The effect of added alcohols on the surface tension of aqueous solutions.

behavior change (precipitation, micelle formation, etc.) prevents further change in the surface tension.

3.3.1. Surfactants and the Reduction of Surface Tension

As stated previously, the measurement of the surface tension of a surfactant solution is possibly one of the most common physical properties of such solutions used to characterize the properties of surfactants in general. Since the surface tension of a liquid is determined by the energy of the molecules in the interfacial region, the displacement of surface molecules by adsorbed solute will directly affect the measured value. It is the relationship between the chemical structure of an adsorbing molecule and the rate and extent of adsorption under given circumstances that differentiates the various surfactant types and determines their utility in applications where surface tension lowering is of importance.

In aqueous solutions, the interface between the liquid and vapor phases involves interactions between relatively densely packed, highly polar water molecules, and relatively sparse, nonpolar gases. The result is an imbalance of forces acting on the surface molecules and the observed high surface tension of water (72.8 mN/m). If the vapor phase is replaced by a condensed phase such as octane, which has a higher molecular density but still interacts only by nonpolar (i.e., dispersion) forces, the interfacial energy as given by the interfacial tension will be reduced significantly (52 mN/m). If the extent of molecular interaction between phases can be increased by the introduction of polar groups, as, for instance, in 1-octanol, the interfacial energy reduction will be even greater (8.5 mN/m). Clearly, any change in the number or the nature of the molecules occupying the surface would be expected to result in a change in the interfacial energy. Therein lies the explanation for the action of surfactants in lowering the surface and interfacial tension of aqueous solutions. It also explains why few surfactants will affect the surface tension of organic liquids—the molecular natures of the liquid and the surfactant are not sufficiently different to make adsorption particularly favorable or, if adsorption occurs, the energy gain is not sufficient to produce a measurable change in surface tension. The actions of fluorocarbon and silicone materials are exceptions, as will be seen later.

The molecular characteristics necessary for a material to perform as a surfactant in aqueous solution have already been extensively discussed. It is useful, however, to reiterate the basic functions of the molecular groups to promote an understanding of their effects on the liquid surface tension. The hydrophilic group, of course, serves the purpose of imparting sufficient water solubility to the molecule to provide a workable concentration of surfactant to produce the desired result. Monolayers of insoluble organic compounds will lower the surface tension of water, but such monolayers must be applied as a separate phase and are not practical for most applications. A significant exception might be the application of materials to prevent or retard evaporation of water from reservoirs or to combat mosquito infestations by "suffocating" the mosquito larva. In such cases, however, the desired result does not rely on the reduction of surface tension so much as on the formation of a barrier to retard the passage of water molecules from the liquid phase to the vapor phase or the respiration of the larva at the water surface. The hydrophobic group of a surfactant must play two essential roles in determining surfactant properties. It must provide the proper solubility properties, so that the molecules will be preferentially adsorbed at the water/vapor interface, and it must favorably alter the energy of interaction between the liquid interface and the contacting gas molecules. Each function will be directly related to the chemical nature of the hydrophobic group and, in some cases, to that of the hydrophile. The specifics of the relationships between chemical structure and activity at the water–vapor interface are discussed in more detail below.

When hydrocarbon-based surfactants are dissolved in polar organic media such as the lower alcohols, in which they have an appreciable solubility, the reduction in surface tension will be small or nonexistent, since the nature of the interface will not be changed significantly by the adsorption of surfactant molecules. Most nonpolar organic liquids are similarly unaffected by the presence of the more common surfactants, since the surfactants usually have very limited solubility in such solvents and the pure organic liquid will usually have an interfacial energy lower than that produced by the adsorption of surfactant molecules. The exceptions to this rule, as noted above, are the siloxane and fluorinated surfactants, which do, in some cases, produce significant surface tension lowering effects in organic media.

3.3.2. Efficiency, Effectiveness, and Surfactant Structure

In a discussion of the performance of a surfactant in lowering the surface tension of a solution, it is necessary to consider two aspects of the process: (1) the concentration of surfactant required to produce a given surface tension reduction and (2) the maximum reduction in surface tension that can be obtained, regardless of the concentration of surfactant present. The two effects may be differentiated by defining the surfactant "efficiency" as that bulk-phase concentration necessary to reduce the surface tension by a predetermined amount, commonly 20 mN/m, and its "effectiveness" as the maximum reduction that can be obtained by the addition of any quantity of surfactant.

The extent of reduction of the surface tension of a solution depends on the substitution of surfactant for solvent molecules at the interface. Therefore, the relative concentration of surfactant in the bulk and interfacial phases should serve as an indicator of the adsorption efficiency of a given surfactant and as a quantitative measure of the activity of the material at the solution–vapor interface. For a given homologous series of straight-chain surfactants in water

$$CH_3(CH_2)_n{-}S$$

where S is the hydrophilic group and n is the number of methylene groups in the chain, an analysis based on the thermodynamics of transfer of a surfactant molecule from the bulk phase to the interface leads to the conclusion that the so-called efficiency of adsorption will be directly related to the length of the hydrophobic chain. If the energy of such transfer is divided into components related to that for the terminal methyl group in a straight-chain molecule (ΔG_{trm} $-CH_3$), subsequent methylene groups (ΔG_{trn} $-CH_2-$), and the hydrophilic group (ΔG_{trs} –S), and a standard reduction level of 20 mN/m is chosen, the surfactant efficiency can be defined as the negative logarithm of the bulk-phase concentration required to produce a reduction of 20 mN/m, so that

$$-\log(C)_{20} = pC_{20} = \left[n\left(\frac{-\Delta G_{\mathrm{trn}}}{2.3RT}\right)\right] + \frac{-\Delta G_{\mathrm{trs}}}{2.3RT} + K \tag{3.18}$$

where $K = -\Delta G_{\mathrm{trm}}$. For a given head group S under constant conditions of temperature, pressure, solvent composition, and so on, the equation reduces to a direct dependence of efficiency on the length of the hydrocarbon chain n.

Since the surfactant efficiency is directly related to the thermodynamics of chain transfer from bulk to interface, it is reasonable to expect that chain modifications that alter that characteristic, such as changes in the hydrophobic character of the surfactant, will produce corresponding changes in the value of pC_{20}. The linear relationship between the number of $-CH_2-$ linkages in a chain and the adsorption efficiency for a variety of hydrophilic groups illustrated in Figure 3.8.

Branching in the hydrophobic group will result in a reduction in the hydrophobicity of a surfactant chain relative to that of a related straight-chain material with the same total carbon content. It is found, for example, that carbon atoms located on branch sites will contribute approximately two-thirds as much to the surface activity of a surfactant molecule as one located in the main chain. Similar results are observed for surfactants with two or more shorter-chain hydrophobes of equal total carbon content (e.g., internal substitution of the hydrophilic group) and for the presence of unsaturation in the chain. The phenyl group, $-C_6H_4-$, will usually contribute an effect equivalent to approximately 3.5 methylene groups.

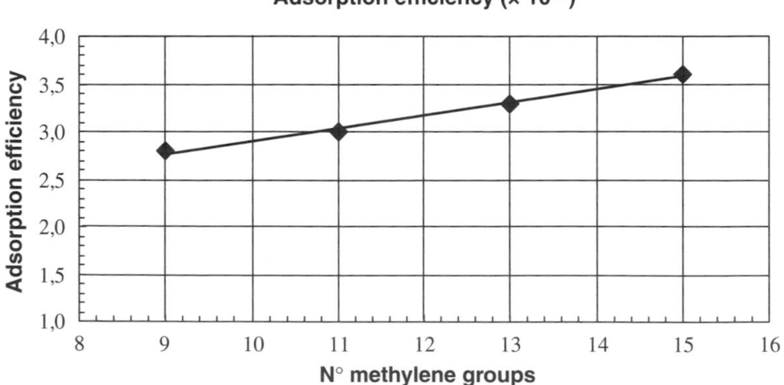

Figure 3.8. The effect of hydrocarbon chain length (as $-CH_2-$ groups) on the adsorption efficiency of sodium sulfonate surfactants.

If a surfactant possesses two polar groups, as do, for example, the taurines or sulfoesters

$$R\text{–}X\text{–}(CH_2\text{–})n\text{–}SO_3^-M^+$$

where R is a normal surfactant hydrophobe and X is a polar amide or ester linkage, the methylene groups lying between the two polar groups will contribute an effect equivalent to approximately half that found for such groups in R.

Although usually regarded as hydrophilic, the first oxyethylene group attached to the hydrophobic chain in surfactants of the type

$$R(OCH_2CH_2)_nOSO_3^-\ Na^+$$

where $n < 4$, actually appears to behave in a manner suggestive of the addition of approximately 2.5 methylene groups to the hydrophobic chain. Such anomalous behavior might be attributed to changes in the solvation of the ether linkage brought on by the close proximity of the highly solvated sulfate group. Succeeding OE groups appear to have little or no significant effect on the hydrophobicity of the molecule. Sulfated polyoxyethylene surfactants having $n > 4$, as do most of the commercially important members of this class, are usually composed of several POE chain lengths and do not lend themselves to easy analysis of the effects of the POE units.

In nitrogen-based cationic surfactants, it has generally been found that the presence of short-chain alkyl groups (fewer than four carbon atoms) attached to the nitrogen has little apparent effect on the efficiency of adsorption of the molecule. The dominant factor will always be the length of the primary hydrophobic chain. That effect is true whether the alkyl groups are attached to a quaternary ammonium group, an amine oxide, or a heterocyclic nucleus such as pyridine.

The nature of the charge on an ionic surfactant has a small effect on the efficiency of surfactant adsorption. It is primarily the hydrophobic group that dominates that characteristic. Some effect will, however, be seen if the counterion to the primary charge is one that is highly ion-paired: that is, one that is not highly solvated in the system and therefore produces a lower net electrical charge as the molecules are adsorbed at the interface. The addition of neutral electrolyte to a surfactant solution will produce a similar result in increasing the efficiency of adsorption of a given ionic surfactant.

When one considers the efficiency of adsorption of nonionic surfactants, it must be remembered that significant differences in the electrical nature of the hydrophilic group can be expected to result in considerable change in the manner in which they adsorb at a S/V interface. For POE surfactants with the same hydrophobic group and an average of 7–30 OE units, the efficiency of adsorption at the solution–vapor interface has been found to adhere to an approximately linear relationship of the form

$$pC_{20} = A_{\mathrm{tr}} + nB \tag{3.19}$$

where A_{tr} and B_{tr} are constants related to the free energy of transfer of —CH_2— and OE groups, respectively, from the bulk phase to the interface and n is the number of OE units in the POE chain. As is usually the case for POE nonionic surfactants, most data reported have been obtained using nonhomogeneous POE chains. The available data indicate that the efficiency of adsorption will decrease slightly as the number of OE units on the surfactant increases.

Up to this point, we have seen that the efficiency of surfactant adsorption at the solution–vapor interface is dominated by the nature of the hydrophobic group and is only slightly affected by the hydrophilic head group. It is often found that the second characteristic of the adsorption process, the so-called adsorption effectiveness, will be much more sensitive to other factors and will quite often not parallel the trends found for adsorption efficiency.

The choice of 20 mN/m as a standard value of surface tension lowering for the definition of adsorption efficiency is convenient, but arbitrary. When one discusses the effectiveness of adsorption, as defined as the maximum lowering of surface tension regardless of surfactant concentration, the value of σ_{min} is determined by the system itself and represents a more firmly fixed point of reference. The value of σ_{min} for a given surfactant will be determined by one of two factors: (1) the solubility limit or Krafft temperature (T_K) of the compound or (2) the critical micelle concentration (cmc). In either case, operationally the maximum amount of surfactant adsorbed will be reached at the maximum bulk concentration of free surfactant, assuming one can ignore the slight decreases in σ found for some surfactants above the cmc.

Because the activity of surfactants used below T_K cannot reach the theoretical maximum as determined by the thermodynamics of surfactant aggregation, the surfactants will also be unable to achieve their maximum degree of adsorption at the solution–vapor interface. It is therefore important to know the value of T_K for a given system before considering its application. Most surfactants, however, are employed well above their Krafft temperature, so that the controlling factor for the determination of their effectiveness will be the cmc.

When one examines the shape of the σ–ln c curve for a surfactant, it can be seen that the curve becomes approximately linear at some concentration below the cmc. It can be shown that the effectiveness of the adsorption of a surfactant, $\Delta\sigma_{cmc}$, the linear slope of the curve in that concentration range, can be quantitatively related to the concentration of surfactant at which the Gibbs equation becomes linear, C_1, the surface tension attained at C_1, namely, σ_1, and the cmc. The relationship has the general form

$$-\Delta\sigma_{cmc} = (\sigma_0 - \sigma_1) + 2.3\Omega\, RT\, \Gamma_m \log \frac{C_{cmc}}{C_1} \tag{3.20}$$

where σ_o is the surface tension of the pure solvent and Γ_m is the maximum in surface excess of adsorbed surfactant at the interface. The factor Ω in Eq. (3.20) is related to the number of molecular or atomic units that will be adsorbed at the interface with the adsorption of each surfactant molecule. For nonionic surfactants or ionic materials in the presence of a large excess of neutral electrolyte, $\Omega = 1$; for fully ionized ionic surfactants, $\Omega = 2$, since one counterion must be adsorbed for each surfactant molecule, giving a total of two species.

Equation (3.20) shows that the effectiveness of a surfactant at lowering the surface tension of a solution is related to three main factors: (1) the cmc of the surfactant; (2) the surfactant concentration required to attain the surface tension at which Gibbs equation linearity begins, C_1; and (3) the maximum surface excess concentration of the surfactant, Γ_m, at surface saturation. The effectiveness of a surfactant can be conveniently quantified by using a value of C_1 at which the surface tension has been reduced by 20 mN/m, assuming $\Gamma_{20} \approx \Gamma_m$, so that the two concepts of efficiency and effectiveness can be directly linked quantitatively.

Application of Eq. (3.20) allows for the calculation of a standard quantity, cmc/C_{20}, which serves as a useful measure of overall surfactant effectiveness. Some representative values that illustrate the effects of well-controlled changes in surfactant structure are given in Table 3.2.

It is often found that the efficiency and effectiveness of surfactants at lowering the surface tension of a solution do not run parallel; in fact, it is commonly observed that more efficient materials that produce significant lowering of the surface tension at low concentrations will be less effective or will have a smaller

TABLE 3.2. Experimental Values of cmc/C_{20}, Γ_{20}, and σ_{min} for Some Typical Surfactants in Aqueous Solution

Surfactant	Temperature (°C)	cmc/C_{20}	Γ_{20} ($\times 10^{10}$ mol/cm^2)	σ_{min} (mN/m)
$C_{12}H_{25}SO_4^-Na^+$	25	2.0	3.3	40.3
$C_{12}H_{25}SO_3^-Na^+$	25	2.3	2.9	40.8
$C_{12}H_{25}SO_4^-Na^+$	60	1.7	2.6	44.8
$C_{12}H_{25}SO_3^-Na^+$	60	1.9	2.5	43.9
$C_{16}H_{33}SO_4^-Na^+$	60	2.5	3.3	37.8
$C_{12}H_{25}C_6H_4SO_3^-Na^+$	70	1.3	3.7	47.0
p-$C_{12}H_{25}C_6H_4SO_3^-Na^+$	75	1.6	2.8	48.8
$C_{16}H_{33}C_6H_4SO_3^-Na^+$	70	1.9	1.9	45.0
$C_{12}H_{25}C_5H_5N^+Br^-$	30	2.1	2.8	42.8
$C_{14}H_{29}C_5H_5N^+Br^-$	30	2.2	2.8	41.8
$C_{12}H_{25}N(CH_3)_3^+Br^-$	30	2.1	2.7	41.8
$C_{10}H_{21}(POE)_6OH$	25	17.0	3.0	30.8
$C_{12}H_{25}(POE)_6OH$	25	9.6	3.7	31.8
$C_{16}H_{33}(POE)_6OH$	25	6.3	4.4	32.8
$C_{12}H_{25}(POE)_9OH$	23	17.0	2.3	36.8
$C_{16}H_{33}(POE)_9OH$	25	7.8	3.1	36.8
$C_{12}H_{25}(POE)_{12}OH$	23	11.8	1.9	40.8
$C_{16}H_{33}(POE)_{12}OH$	25	8.5	2.3	39.8
$C_{16}H_{33}(POE)_{15}OH$	25	8.9	2.1	40.8
$C_{16}H_{33}(POE)_{12}OH$	25	8.0	1.4	45.8
p,t-$C_8H_{17}C_6H_4(POE)_7OH$	25	22.9	2.9	30.8

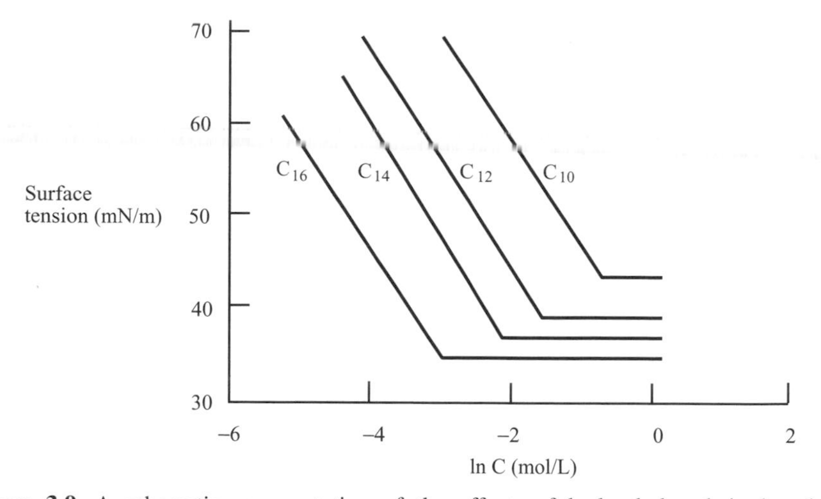

Figure 3.9. A schematic representation of the effects of hydrophobe chain length on surfactant cmc and σ_{min}.

Γ_m. This follows from the complex relationship between adsorption at the interface and micelle formation in the solution.

On a molecular basis, the conflicting factors can be seen conceptually as arising from the different roles of the molecular structure in the adsorption process. Surfactant efficiency is related to the extent of adsorption at the interface as a function of bulk surfactant concentration. At a concentration well below the cmc, efficiency can be structurally related to the hydrophobicity of the surfactant tail and the nature of the head group. For a given homologous series of surfactants, it will be a function of the thermodynamics of transfer of the hydrophobic tail from the bulk to the surface phase. A plot of σ–ln c for such a series will exhibit a relatively regular shift in the linear portion of the curve to lower concentration as methylene groups were added to the chain. An idealized example of such a series is shown in Figure 3.9.

While the role of molecular structure in determining surfactant efficiency is primarily thermodynamic, its effectiveness is more directly related to the size of the hydrophobic and hydrophilic portions of the adsorbing molecules. When one considers the adsorption of molecules at an interface, it can be seen that the maximum number of molecules that can fit into a given area will depend on the area occupied by each molecule. That area will, to a good approximation at least, be determined by either the cross-sectional area of the hydrophobic chain or the area required for the arrangement for closest packing of the head groups (Figure 3.10), whichever is greater. For straight-chain 1 : 1 ionic surfactants, it is usually found that the head group requirement will predominate, so that for a given homologous series, the surface tension minimum obtained will vary only slightly with the length of the hydrocarbon chain.

Since the decrease in surface tension obtained is directly related to the surface excess adsorption of the surfactant by the Gibbs equation, a reduction in the amount

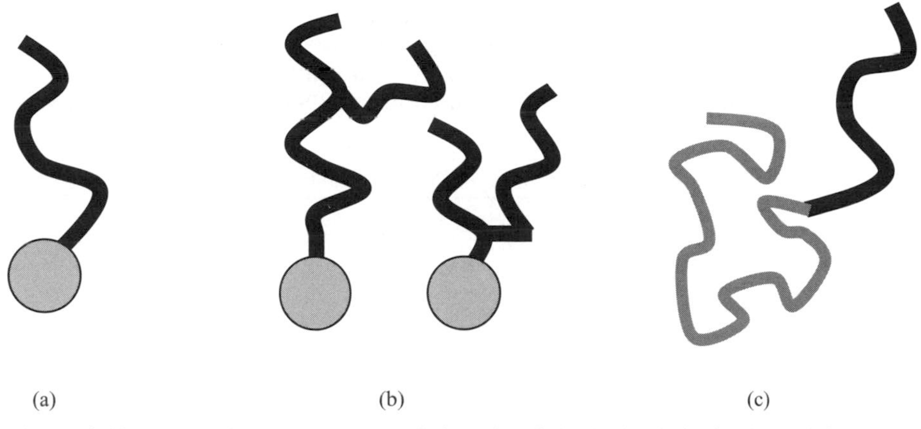

Figure 3.10. Schematic representation of the role of the hydrophobe in determining the effectiveness of surfactant adsorption: (a) *n*-alkyl—area determined by head group; (b) branched or double-tailed—area determined by bulk of tail relative to head; (c) polyoxyethylene nonionic—area determined by coiling of POE chain.

of material that can be adsorbed in a given surface area will reduce the ultimate surface tension lowering attained. The efficiency will change more or less regularly with the chain length. The sign of the charge on the ionic surfactant has only a minor effect on the ultimate surface tension attained, indicating that the geometric requirements (including electrostatic effects) are relatively constant from one head group to the next. In the presence of neutral electrolyte, of course, electrostatic repulsions between adjacent molecules will be reduced, making their effective areas smaller. The net result will be a slight increase in surfactant effectiveness.

The complex relationship between the molecular structure of a surfactant and its impact on surfactant efficiency and effectiveness can be illustrated by the case of a series of nonionic POE surfactants in which the number of OE units is held constant and the hydrocarbon chain length is increased (Table 3.3). In the series it is found that the surface excess at surface saturation Γ_m decreases regularly from 4.4×10^{-10} mol/cm^2 for C_{16} to 2.7×10^{-10} for C_6, while the σ_{min} remains relatively constant. This would indicate that while the efficiency of surfactant adsorption is increasing with the length of the hydrocarbon chain, the overall effectiveness of the material is relatively unchanged.

It can be seen from Table 3.3 that in the cases of the C_{16} and *p,t*-$C_8H_{17}C_6H_4$ hydrophobic groups, as the size of the hydrophilic group (n) increases, the effectiveness (as σ_{min}) decreases. This effect can be related to the fact that each additional OE group added to the head of the surfactant increases the total area required for adsorption of the molecule, reduces the packing density of hydrophobic groups at the interface, and therefore results in a smaller reduction in the surface tension of the system. If the area per molecule a_o required for the adsorption of the $C_{16}H_{33}(POE)_x$ series of surfactants is examined, it can be seen that the addition of each OE unit increases the requirement by an average of 5 Å^2 or 0.05 nm^2. From insoluble monolayer experiments, it has been shown that the surface tension

TABLE 3.3. Effect of Polyoxyethylene and Hydrocarbon Chain Length on the Efficiency and Effectiveness of Surface Tension Lowering for a Number of POE Nonionic Surfactants with the General Formula $R(POE)_nOH$

R	n	Γ_m (x 10^{10} mol/cm^2)	σ_{min} (mN/M)
C_6	6	2.7	32.8
C_{10}	6	3.0	30.8
C_{12}	6	3.7	31.8
C_{16}	6	4.4	32.8
	7	3.8	33.8
	9	3.1	36.8
	12	2.3	39.8
	15	2.1	40.8
	21	1.4	45.8
p,t-$C_8H_{17}C_6H_4$	7	2.9	30.8
	8	2.6	32.8
	9	2.5	34.3
	10	2.2	35.8

(or surface pressure $\pi = \sigma_0 - \sigma$) is related to the orientation of the adsorbed molecules at the interface, with maximum lowering resulting from an essentially perpendicular orientation between hydrophobe and interface (Figure 3.11). For soluble monolayers such as those in question here, that orientation will be directly affected by the proximity of the neighboring molecules; thus factors that cause an increase in molecular separation will also allow the adsorbed molecules to tilt more relative to the surface, producing a smaller effective surface tension reduction (Figure 3.11b).

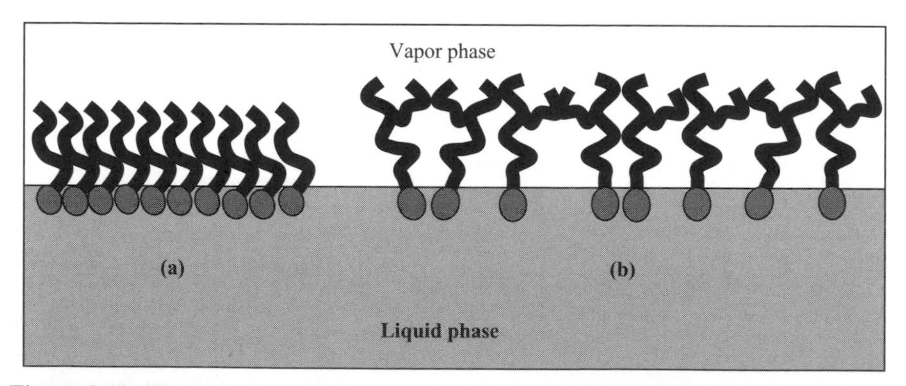

Figure 3.11. The role of surfactant structure and packing in the determination of packing efficiency and surface tension: (a) straight-chain hydrophobes, closest packing and maximum effectiveness—area per molecule determined by head group; (b) branched, unsymmetric, substituted, and other types of tails reduce packing efficiency—area determined by tail.

While an increase in the hydrocarbon chain length in a series of normal alkyl surfactants between C_8 and C_{20} will have a minor effect on the effectiveness of a surfactant, other structural changes can produce much more dramatic effects. We have seen that structural features such as branching and multiple-chain hydrophobes will generally result in increases in the cmc of surfactants with the same total carbon content. Those changes seem to have a much smaller effect on the efficiency of the surfactant (C_{20}) than on its effectiveness. For sodium *n*-dodecylbenzene sulfonate the cmc and σ_{min} are 1.2×10^{-3} M and 36 mN/m, while the same values for sodium(2-methylundecylbenzene)sulfonate are 2.3×10^{-3} M and 27 mN/m, respectively.

The introduction of slightly polar groups such as unsaturation, ether, ester, or amide linkages, or hydroxyl groups located well away from the head group will usually result in a significant lowering of both the efficiency and effectiveness of the surfactant as compared to a similar material with no polar units. Such a result has generally been attributed to changes in orientation of the adsorbed molecule with respect to the surface due to interactions between the polar group and the water (Figure 3.12). If the polar group is situated very near the primary hydrophilic group, its orientational effect will be less dramatic, although it may still have a significant effect on the cmc of the material.

Changes in the hydrophobic group in which fluorine atoms are substituted for hydrogen will usually result in significant increases in the efficiency and effectiveness of the surfactant. The substitution of fluorine for hydrogen in a straight-chain surfactant results in a relatively small increase in chain cross-sectional area, as compared to a methyl branch, for example, so that the changes must be related to the chemical nature of the substitution. As has already been pointed out, fluorinated organic materials have a relatively low cohesive energy density and therefore little interaction with adjacent phases, or themselves, for that matter. They therefore have very favorable thermodynamic driving forces for adsorption (leading to high efficiency), as well as low surface energies. Their effectiveness is reflected in the very low surface tension values produced (as low as 20 mN/m in some instances).

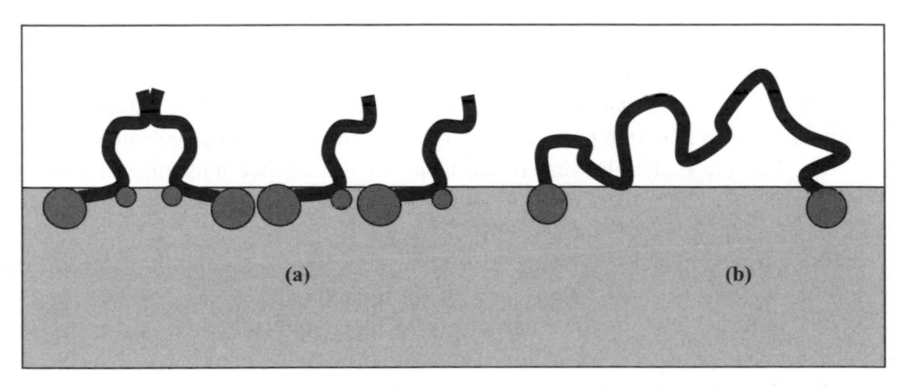

Figure 3.12. The effect of additional polar groups on the adsorption of surfactants: (a) additional polar groups along the hydrophobic chain; (b) multiple, separated head groups.

Thus for, the discussion has dealt primarily with the effects of changes in the hydrophobic group on the ability of a surfactant to reduce the surface tension of a solution. It was stated earlier that an ionic head group usually plays a relatively minor role in determining the efficiency and effectiveness of a surfactant. While that may be true for groups closely related in size and electrostatic character, alterations in those factors can produce significant changes in their activity at the solution–air interface. A class of surfactants well suited to the study of such effects is that of the quaternary ammonium salts in which three of the alkyl groups are short-chain units such as methyl, ethyl, and propyl. The substitution of propyl for methyl groups in *n*–tetradecyltrimethylammonium bromide resulted in a significant reduction in the efficiency of adsorption, while not affecting the minimum surface tension obtained. Presumably, the presence of the bulkier propyl substituents on the head group greatly increases its area per molecule and therefore reduces its adsorption efficiency.

In the case of anionic head groups, there appear to be only relatively minor variations in effect from one group to another. The difference in cross-sectional area between sulfate and sulfonate groups does not appear to influence greatly the activity of surfactants in lowering surface tensions, although some difference can be noted when differences in cmc are taken into consideration. The role of the counterion can be important when changes result in significant alterations in the ion binding properties of the molecule. Tight ion binding will reduce the extent of electrostatic repulsion between adsorbed molecules, allowing for tighter packing of surfactant at the interface and, in general, increases in both the efficiency and the effectiveness of the surfactant. A similar result is obtained by the addition of neutral electrolyte.

An interesting class of surfactants that has found practical application because of tight ion binding and its effects on surface activity are those in which both the anion and the cation of the pair are, individually, surface-active. Materials such as *n*-decyltrimethylammonium *n*-decylsulfate (written in reverse to illustrate the ion binding)

$$C_{10}H_{21}(CH_3)_3N^+ \ {}^-O_4SC_{10}H_{21}$$

have been found to be much more efficient and effective at surface tension reduction than either simple material alone. The very strong ion pairing nature of their association results in a high packing density at the interface and large values for σ_{20} and σ_{min}. Related materials were found to lower the surface tension of a solution to approximately 24 mN/m, one of the lowest surface tensions yet reported for totally hydrocarbon surfactant systems.

As has already been mentioned, the effect of changes in the hydrophobic chain length on the effectiveness of surface tension reduction in nonionic surfactants is relatively minimal. Increases in the length of the polyoxyethylene chain, on the other hand, lead to significant reductions in the effectiveness of a given surfactant hydrophobic group. It appears, then, that the primary factor involved in the efficiency and effectiveness of nonionics in this application is the length of the hydrophilic chain. A similar result could be expected for other types of nonionic material, although there are few good experimental data available to warrant excessive generalization.

The discussion above introduced some basic concepts related to the properties of fluid–fluid, and particularly liquid–vapor interfaces. The practical effects of surface tension lowering were not addressed because they generally appear in the context of phenomena such as emulsification, foaming, wetting, and detergency, to be discussed later. For further details on the subject of surface tension lowering and surfactant adsorption at fluid interfaces, the reader is referred to the works cited in the Bibliography.

PROBLEMS

3.1. Calculate the total reversible thermodynamic work required to produce a spray of water of droplet diameter 2000 nm from 40 L of water at 25°C. Take the surface tension of water as 72 mN/m. How many water drops would be produced, theoretically, if the droplets were all of equal size?

3.2. The surface tension of aqueous solutions of LiCl have the following values at 25°C:

(%)	σ	(%)	σ	(%)	σ	(%)	σ
5.46	74.2	7.37	75.1	10.2	76.3	13.9	78.1

Using the Gibbs adsorption equation, calculate the surface excess concentration, Γ_{LiCl}, in molecules/cm^2 for an 8% solution of LiCl. The surface tension of pure water at 25° is 72.0 mN/m.

3.3. Solutions of an unknown alcohol A in water have the following surface tensions at 20°C. Using a plot of surface tension–alcohol concentration, suggest whether the alcohol in question can be considered surface-active. If not, what do the results suggest about the solution characteristics of the mixture? The surface tension of water at 20°C is 72.8 mN/m.

% A	σ	% A	σ	% A	σ	% A	σ	% A	σ
7.5	60.9	10.0	59.0	25.0	46.4	50.0	33.0	50.0	27.3

3.4. A 5.2×10^{-5} -g sample of palmitic acid ($C_{15}H_{31}COOH$) is spread on a pure water surface as a solution in toluene and the solvent evaporated. Using a Langmuir trough apparatus, the monolayer is compressed to an area of 265 cm^2 at which point it is known to form a close-packed monolayer. Calculate the area (in nm^2) occupied by each molecule.

3.5. On the basis of the result obtained in Problem 3.4 and general principles of molecular geometry, which part of the molecule is the primary factor controlling the area occupied? According to your answer, what can you say about the areas that you would expect to be occupied by lauric acid, stearic acid, and butanoic acid?

3.6. Explain the theoretical significance of a negative surface excess Γ_i, in terms of the derivation of Eq. (3.1)

3.7. Assuming a constant value of the sticking coefficient α [Eq. (3.11)] of 0.5, estimate the surface residence time for an atom or molecule of the following materials at 50°C: ethanol, hexadecane, glycerol, gallium, lead, aluminum, yellow sulfur, and olive oil.

3.8. Assuming ideal solutions, calculate the theoretical surface tensions of the following mixtures of acetone in acetic acid at 25°C: 5%, 10%, 25%, 35%, 50%, and 75%. Plot the results and compare them with literature values. Do the two materials form ideal solutions?

3.9. Rank the following surfactants in terms of their expected effectiveness of adsorption: sodium tetradecyl sulfate, sodium di(2-ethylhexylsulfosuccinate), sodium triisopropylnaphthalene sulfonate, $(C_{16}H_{33})_2(CH_3)_2N^+Cl^-$, and $(CH_3)_3$-$(C_{12}H_{25})N^+Br^-$.

3.10. The following interfacial tensions (in mN/m) with water have been determined at 20°C: *n*-octane (50.8), *n*-octanol (8.5), and octanoic acid (7.0). Explain the observed results in terms of general concepts of molecular interactions at interfaces.

4 Surfactants in Solution: Monolayers and Micelles

The amphiphilic nature of surfactants causes them to exhibit many properties that appear on first sight, to be contradictory. Because of their special molecular structures, they possess something of a "love–hate" relationship in most solvents, resulting in a tug of war among competing forces striving for a comfortable (energetically speaking) accommodation within a given environment. Surfactants, one might say, appear to feel to some extent that the grass is greener on the other side of the fence, and as a result, they spend much of their time sitting on that "fence" between phases. Some of the basic characteristics of those "fences" were introduced in Chapter 3. This chapter will begin the process of expanding on the specifics of how surfactant molecular structures affect their surface activity. Specific topics on the adsorption of surfactants at specific interfaces will be discussed in later chapters. At this point, it is important to understand some of the more important aspects of the solution behavior of surfactants and some of the circumstances that can affect that behavior.

In their energetic "need" to minimize unfavorable interactions or to maximize favorable interactions with their environment, surfactants spend much of their time at interfaces or associating with others of their own kind. The purpose of the discussions here is to introduce some of the more important and useful fundamental concepts of surfactants in solution, as we currently understand them. A basic understanding of these concepts can help guide a prospective surfactant user in understanding a given phenomenon and choosing a material that may suit a particular need. This chapter is concerned primarily with the more simple—if that term can be applied in the present context—aspects of surfactant activity in terms of self-assembled or spontaneous, thermodynamically driven aggregate structures in solution. As is usually the case with surfactant-related discussions, the primary emphasis will be placed on aqueous systems. Chapter 5 gives a broad introduction to the more complex and highly ordered self-assembled structures such as vesicle, bi- and multiplayer membranes, and the new darlings of the field, continuous bilayer systems. All of those areas are becoming more important in current and potential technological and research applications of surfactants and other amphiphiles.

Surfactant Science and Technology, Third Edition by Drew Myers

4.1. SURFACTANT SOLUBILITY

The specific structures of surfactant molecules, having well-defined lyophilic and lyophobic components, is responsible for their tendency to concentrate at interfaces and thereby reduce the interfacial free energy of the system in which they are found. A molecule with the same elemental composition but a different structural distribution of its constituent atoms may show little or no surface activity. The primary mechanism for energy reduction in most cases will be adsorption at the available interfaces. However, when all interfaces are or begin to be saturated, the overall energy reduction may continue through other mechanisms as illustrated in Figure 4.1. The physical manifestation of one such mechanism is the crystallization or precipitation of the surfactant from solution—that is, bulk-phase separation such as that seen for a solution of any solute that has exceeded its solubility limit. In the case of surfactants, alternative options include the formation of molecular aggregates such as micelles and liquid crystal mesophases that remain in solution as thermodynamically stable, dispersed species with properties distinct from those of the monomeric solution. Before turning our attention to the subject of micelles, it is necessary to understand something of the relationship between the solubility of a surfactant or amphiphile in the solvent in question and its tendency to form micelles or other aggregate structures.

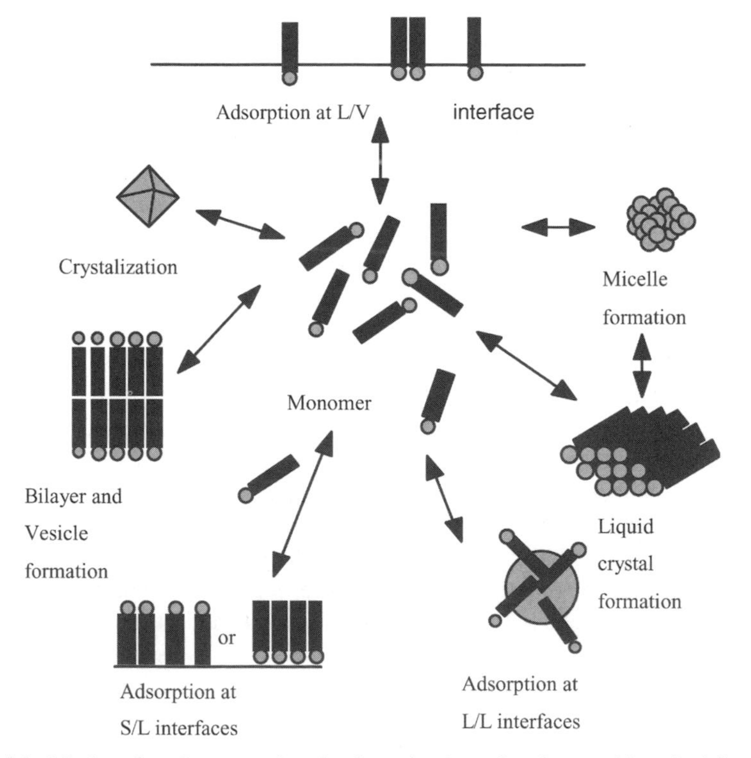

Figure 4.1. Modes of surfactant action for the reduction of surface and interfacial energies.

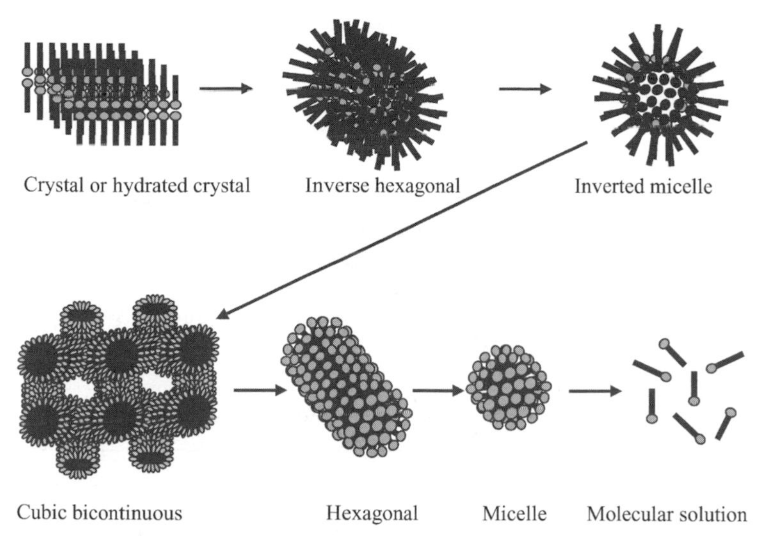

Figure 4.2. A hypothetical spectrum of surfactant mesophases in aqueous solution as formed with increased water content.

For most pure solutes, solubility is a more-or-less "yes or no" question. Under a given set of conditions of solvent and temperature, and sometimes pressure, the solute has a specific solubility limit which, when passed, results in the formation of crystals or at least a distinct separate phase that can hypothetically be separated from the solvent or supernatant liquid by physical means. While crystalline hydrates may be separated from water solutions, they will normally have specific compositions that make them unique and subject to characterization by chemical analysis, for example. Surfactants and other amphiphiles, on the other hand, can exhibit a number of intermediate or mesophases in going from a dilute solution of individual or "independent" molecules to crystalline hydrates or anhydrous structures. A hypothetical "spectrum" of surfactant mesophases in aqueous solution is given in Figure 4.2.

As pointed out in Chapter 1, a primary driving force for the industrial development of synthetic surfactants was the problem of the insolubility of the fatty acid soaps in the presence of multivalent cations such as calcium and magnesium or at low pH. While most common surfactants have a substantial solubility in water, that characteristic changes significantly with changes in the length of the hydrophobic tail, the nature of the head group, the electrical charge of the counterion, the system temperature, and the solution environment. For many ionic materials, for instance, it is found that the overall solubility of the material in water increases as the temperature increases. That effect is the result of the physical characteristics of the solid phase—that is, the crystal lattice energy and heat of hydration of the material being dissolved.

For ionic surfactants, the solubility of a material will often be observed to undergo a sharp, discontinuous increase at some characteristic temperature, commonly

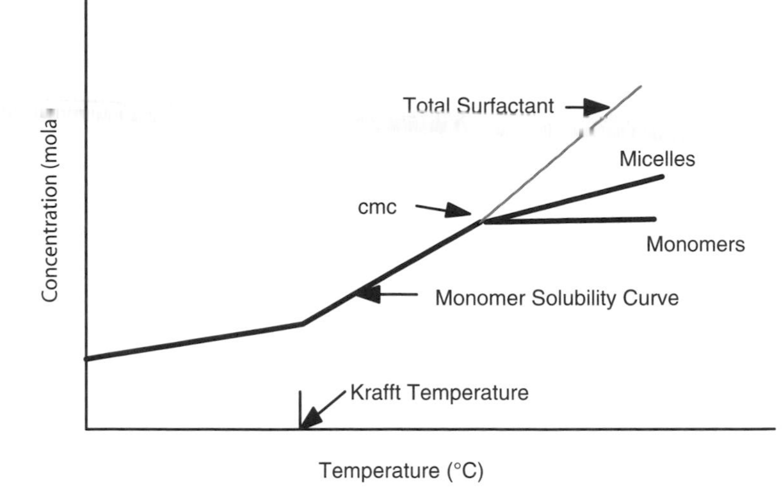

Figure 4.3. Temperature–solubility relationship for typical ionic surfactants.

referred to as the *Krafft temperature*, T_K. Below that temperature, the solubility of the surfactant is determined by the crystal lattice energy and the heat of hydration of the system. The concentration of the monomeric species in solution will be limited to some equilibrium value determined by those properties. Above T_K, the solubility of the surfactant monomer increases to the point at which aggregate formation may begin, and the aggregated species (e.g., a micelle) becomes the thermodynamically favored or predominant form in solution.

The micelle may be viewed, to a first approximation, as structurally resembling the solid crystal or a crystalline hydrate, so that the energy change in going from the crystal to the micelle will be less than the change in going to the monomeric species in solution. Thermodynamically, then, the formation of micelles favors an overall increase in solubility. The concentration of surfactant monomer may increase or decrease slightly at higher concentrations (at a fixed temperature), but micelles will be the predominant form of surfactant present above a critical surfactant concentration, the critical micelle concentration (cmc). The apparent solubility of the surfactant, then, will depend on not only the solubility of the monomeric material but also the solubility of the micelles or other aggregate structures. A schematic representation of the temperature–solubility relationship for ionic surfactants is shown in Figure 4.3.

The Krafft temperatures of a number of common ionic surfactants are given in Table 4.1. It can be seen from the data that T_K can vary as a function of both the nature of the hydrophobic group and the character of the ionic interactions between the surfactant and its counterion. It should be noticed that no data are listed for nonionic surfactants. Nonionic surfactants, because of their different mechanism of solubilization, do not exhibit a Krafft temperature. They do, however, have a characteristic temperature–solubility relationship in water that causes them to become

TABLE 4.1. The Krafft temperatures T_K of Typical Ionic Surfactants

Surfactant	T_K (°C)
$C_{12}H_{25}SO_3^-Na^+$	38
$C_{14}H_{29}SO_3^-Na^+$	48
$C_{16}H_{33}SO_3^-Na^+$	57
$C_{12}H_{25}OSO_3^-Na^+$	16
$C_{14}H_{29}OSO_3^-Na^+$	30
$C_{16}H_{33}OSO_3^-Na^+$	45
$C_{10}H_{21}CH(CH_3)C_6H_4SO_3^-Na^+$	32
$C_{12}H_{25}CH(CH_3)C_6H_4SO_3^-Na^+$	46
$C_{14}H_{29}CH(CH_3)C_6H_4SO_3^-Na^+$	54
$C_{16}H_{33}CH(CH_3)C_6H_4SO_3^-Na^+$	61
$C_{16}H_{33}OCH_2CH_2OSO_3^-Na^+$	36
$C_{16}H_{33}(OCH_2CH_2)_2OSO_3^-Na^+$	24
$C_{16}H_{33}(OCH_2CH_2)_3OSO_3^-Na^+$	19
$C_{10}H_{21}COOC(CH_2)_2SO_3^-Na^+$	8
$C_{12}H_{25}COOC(CH_2)_2SO_3^-Na^+$	24
$C_{14}H_{29}COOC(CH_2)_2SO_3^-Na^+$	36
$C_{10}H_{21}OOC(CH_2)_2SO_3^-Na^+$	12
$C_{12}H_{25}OOC(CH_2)_2SO_3^-Na^+$	26
$C_{14}H_{29}OOC(CH_2)_2SO_3^-Na^+$	39
n-$C_7F_{15}SO_3^-Na^+$	56
n-$C_8F_{17}SO_3^-Li^+$	<0
n-$C_8F_{17}SO_3^-Na^+$	75
n-$C_8F_{17}SO_3^-K^+$	80
n-$C_8F_{17}SO_3^-NH_4^+$	41
n-$C_7F_{15}COO^-Li^+$	<0
n-$C_7F_{15}COO^-Na^+$	8
n-$C_7F_{15}COO^-K^+$	26
n-$C_7F_{15}COO^-NH_4^+$	2

less soluble as the temperature increases. In some cases, phase separation occurs, producing a cloudy suspension of surfactant. The temperature (usually a range of temperature) at which the phases separate is referred to as the "cloud point" for that surfactant. This will be disscussed in more detail later.

The intimate relationship between the Krafft temperature and the solid state of the surfactant is confirmed by the good correlation between T_K for a surfactant of a given chain length and the melting point of the corresponding hydrocarbon material. Such correlations can also be found for the appearance of other structural changes in surfactant solutions. As we shall see in later chapters, good practical use can be made of such temperature-related phenomena. Note that in Table 4.1 fluorinated surfactants have Krafft temperatures in roughly the same temperature range as hydrocarbon materials containing twice as many carbon atoms. That tendency is seen, not surprisingly, in comparing most surfactant properties of hydrocarbon versus fluorocarbon materials.

As indicated above, an important characteristic of a surfactant in solution is its solubility relative to the critical concentration at which thermodynamic considerations result in the onset of molecular aggregation or micelle formation. Since micelle formation is of critical importance to many surfactant applications, the understanding of the phenomenon relative to surfactant structures constitutes an important element in the overall understanding of surfactant structure–property relationships.

4.2. THE PHASE SPECTRUM OF SURFACTANTS IN SOLUTION

Most academic discussions of surfactants in solution concern relatively low concentrations, so the system contains what may be called "simple" surfactant species such as monomers and their basic aggregates or micelles. Before entering into a discussion of micelles, however, it is important to know that although they have been the subject of exhaustive studies and theoretical considerations, they are only one of the several states in which surfactants can exist in solution. A complete understanding of surfactant solution systems, including correlations between chemical structures and surface properties, requires a knowledge of the complete spectrum of possible states of the surfactant. While no attempt is made here to provide a detailed discussion of surfactant phase behavior, it is important that the subject be at least introduced into any description of the solution behavior of surface-active agents.

When one considers the wide range of possible environments for surfactant molecules in the presence of solvents, it is not surprising that the subject can appear overwhelming to the casual observer. As illustrated in Figure 4.1, the possibilities range from the highly ordered crystalline phase to the dilute monomeric solution, which, although not completely without structure, has order only at the level of molecular dimensions. Between the extremes lie a variety of phases whose natures depend intimately on the chemical structure of the surfactant, the total bulk-phase composition, and the environment of the system (temperature, pH, cosolutes, etc.). Knowledge of those structures, and of the reasons for and consequences of their formation, influences both our academic understanding of surfactants and their technological application.

Pure, dry surfactants, like most materials, can be made to crystallize relatively easily. Because of their amphiphilic nature, however, the resulting structures always appear to be lamellar with alternating head-to-head and tail-to-tail arrangements (Figure 4.4). The energy of the surfactant crystal, as reflected by its melting point, for example, will be determined primarily by the chemical structure of the molecules. Terminally substituted, *n*-alkyl sulfates, for example, will have higher melting points than will the corresponding branched or internally substituted materials basically due to the more compact and ordered packing structures available to

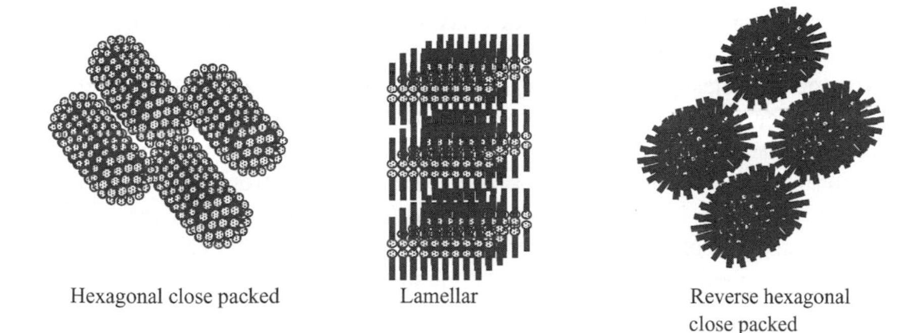

Figure 4.4. Typical arrangement of surfactant molecules in crystalline lattices.

the straight-chain materials. Additionally, highly polar, small hydrophilic groups will provide enhanced crystal stability over bulky, more polarizable functionalities.

The packing of long hydrocarbon chains into a crystalline alignment is difficult because of the many possible variations in configuration for the units of the chain due to rotation about the four bonds to each carbon atom (rotational isomers). That difficulty is reflected in the relatively low melting points and poorly defined crystal structure of most hydrocarbons under normal conditions. When members of a homologous series or structural isomers are present as a mixture, a situation common to many important surfactant systems, the difficulty of crystal formation, is magnified. The crystallization of pure surfactants from a mixture, therefore, can be difficult, especially if the mixture is that of a series of homologs. For that reason, crystals of natural fatty acid soaps, commercial polyoxyethylene (POE) nonionic surfactants, and other surfactants containing homologous species or branched isomers are rare, and even relatively pure samples may exhibit a variety of crystal structures depending on the conditions of crystallization.

When surfactants are crystallized from water and other solvents that are strongly associated with the polar head group, it is common for the crystalline form to retain a small amount of solvent in the crystal phase. In the case of water, the material would be a hydrate. The presence of solvent molecules associated with the head group allows for the existence of several unique compositions and morphological structures that, although truly crystalline, are different from the structure of the anhydrous crystal.

As water or other solvent is added to a crystalline surfactant, the structure of the system will undergo a transition from the most highly ordered crystalline state to one of greater disorder usually referred to as a liquid crystalline or "mesophase." In such phases some structure is retained in one molecular region of the system, while a more liquid or amorphous structure is developed in the other. Such crystalline/amorphous phases, 18–20 of which have been reported for some molecular structures, are characterized by possessing some physical properties of both crystalline and fluid phases. These phases will have at least one highly ordered dimension and, as a result, will exhibit relatively sharp X-ray diffraction patterns and optical birefringence. In other dimensions, the phases will behave in a manner more similar to that for nonstructured or minimally structured fluids.

Two general classes of liquid crystalline structures or mesophases are encountered whether one is considering surfactants or other types of material. These classes are the thermotropic liquid crystals, in which the structure and properties are determined by the temperature of the system, and lyotropic liquid crystals in which the structure is determined by specific interactions between the surfactant molecules and the solvent. With the exception of the natural fatty acid soaps, experimental data support the view that almost all surfactant liquid crystals are lyotropic in nature.

Theories on liquid crystal formation predict the existence of at least 18 distinct liquid crystalline structures for a given molecular composition and structure. Nature, however, appears to have been kind in that only three of those possibilities have been identified in simple, two-component surfactant–water systems. The same is often true for three-component surfactant-oil-water systems. The three liquid crystalline phases usually associated with surfactants are the lamellar, hexagonal, and cubic. Of the three, the cubic phase is the most difficult to define and detect. It has been invoked to describe the new kid on the block in surfactant aggregates, the so–called "cubosomes." It may have a wide variety of structural variations that involve components of the other mesophases. The remaining two types are more easily characterized and, as a result, are better understood.

The lamellar liquid crystal can be viewed as a mobile or "plasticized" derivative of the basic surfactant crystalline phase. The hydrophobic chains in these structures possess a significant degree of freedom or randomness, unlike the case in the crystalline phase, in which the chains are usually locked into the all–trans configuration (for terminally substituted *n*-alkyl hydrophobes). The level of disorder of the lamellar phase may vary smoothly or change abruptly as solvent is added, depending on the specific system. It is therefore possible for a surfactant to pass through several distinct lamellar phases. Because the basic unit is a bilayer structure, lamellar phases are usually uniaxial. The lamellar phase resembles the bilayer and multilayer membranes to be discussed later, although they are formed as a result of changes in solvent concentration rather than the specific molecular structural features of the surfactant.

The hexagonal liquid crystal is a high-viscosity fluid phase composed of a close-packed array of cylindrical assemblies of theoretically unlimited size in the axial direction. The structures may be "normal" (in water) in that the hydrophilic head groups are located on the outer surface of the cylinder, or "inverted," with the hydrophile located internally.

As mentioned before, surfactant liquid crystalline phases are normally lyotropic. The characteristics of the system, then, are highly dependent on the nature and amount of solvent present. In a phase diagram of a specific surfactant, the liquid crystalline phases may span a broad range of compositions, and may constitute by far the major fraction of all possible compositions. With the continued addition of water or other solvent, the system will eventually pass through the regions of the mesophases into the more familiar isotropic solution phase. The liquid state is the most highly random condition of condensed matter and, as a result, tends to have fewer easily detected structural features. Surfactant solutions, however, are far from devoid of structure, it is only the scale of the structure that changes as dilution occurs.

The addition of a third (or fourth, etc.) component such as a water-immiscible oil, electrolyte, or a polar nonsurfactant solute to an aqueous surfactant solution can lead to the formation of new phases with distinct properties. These new components will alter the thermodynamic balance of the system and, as a result, may alter the nature of the aggregated species present. The extent of their effect will vary with concentration, structural features, and specific interactions; however, under the proper conditions they will give rise to important new structures and phases not encountered in a simple two-component system. The factors controlling the nature of the system will no longer be simply those of the solvent and solute, but will involve complex three-way (or more) interactions.

The natures and properties of the multicomponent systems are discussed in somewhat more detail in the following chapters. They are introduced here only to complete the discussion of the "spectrum" of surfactant structures commonly encountered. One thermodynamically interesting phase consisting of surfactant, water, oil, and sometimes a fourth "cosurfactant" component, is referred to as the "microemulsion" region of the multicomponent phase diagram. There is still some controversy as to the exact nature of microemulsions, since structures in this region seem to span the size range from conventional micelles (with diameters of a few nanometers) to the more easily defined emulsions (several hundred nanometers). Should microemulsions be considered a new, separate phase or simply an extension of the micellar phase in which the basic structure is enlarged by the presence of an additive incorporated into the micelle? There are reasonably compelling arguments that support the idea of microemulsions as a phase separate from the normal swollen micelle, particularly the question of size. Other factors make it less easy to separate the two systems into distinct classes. In the final analysis, trying to define a boundary between micelles and microemulsions is much like trying to define the wavelength at which light changes from red to orange. It will probably be a question of convenience or individual personal preference (or the pronouncement of some august international body).

Conventional emulsions, unlike the microemulsions, are easily identified as dispersions of one liquid phase in another. In such systems, the energetics of surfactant aggregation is not a major factor in their formation. Conventional emulsions, therefore, are only indirectly related to the subject of this chapter. They are, however, related in the sense that a direct line of evolution can be drawn from the crystalline surfactant phase, through the mesophases, micelles, and microemulsions, to emulsions, all resulting from changes in the composition of the system.

In addition to facilitating our understanding of the fundamental principles of surfactant solution behavior, knowledge of the details of the solution properties of a surfactant can be of immense practical importance. From a practical standpoint, a great deal can be achieved by understanding solution phase diagrams of surfactants. Of particular importance is understanding of structure–solubility relationships, selection of optimum components for a given product application, understanding of the details of surface and interfacial activity, and design of new surfactant molecular structures for optimal performance in both old and new applications. The major drawbacks of such studies are that they require a significant amount of time and experimental effort, and they are really useful, in general, only for pure, single-component, well-characterized surfactant systems—something not very common in most technological applications.

Although the study of surfactant solution properties throughout the complete concentration range is of obvious theoretical and occasional practical importance, no attempt is made to cover in detail those phases more structured than the simplest aggregates of surfactants in dilute solution. For more information on surfactant phase diagrams, the reader is referred to the excellent works of Laughlin cited in the Bibliography for this chapter.

4.3. THE HISTORY AND DEVELOPMENT OF MICELLAR THEORY

The aggregation of surfactants into clusters or micelles in dilute solutions, as we will see, is a direct consequence of the thermodynamic requirements of the particular surfactant–solvent system under consideration. It has been suggested that phases occurring between the simplest micelles and true crystals are natural consequences of the removal of water from the micellar system, but do not constitute thermodynamically distinct states. In other words, the factors determining the structures of the mesophases are identical to those that control the formation of micelles in the first place. The same would be true of aggregates other than micelles, which do not fall under the classification of mesophases.

The number of publications related to micelles, micelle structures, and the thermodynamics of micelle formation is enormous. Extensive interest in the phenomenon of the self-association of surface-active species is evident in such wide-ranging chemical and technological areas as organic and physical chemistry, biochemistry, polymer chemistry, pharmaceuticals, petroleum recovery, minerals processing, cosmetics, and food science. In addition to the general scientific literature, hundreds of patents have been issued covering new materials and uses related to micelle formation and the effects of those structures on different phenomena of potential commercial interest. Even with the vast amount of experimental and theoretical work devoted to understanding of the aggregation of surface-active molecules, no theory or model has emerged that can unambiguously satisfy all the evidence and all the interpretations of that evidence.

Early in the twentieth century, it was recognized that aqueous solutions of surface-active agents do not follow the patterns of solution behavior common to most solutes as their concentration is increased. It was suggested that the unusual behavior of surfactants could be attributed to the aggregation of individual molecules into clusters in solution above a fairly well-defined concentration. Although that somewhat radical idea received a rather cool initial reception, the concept of molecular aggregation in solution eventually began to develop a significant following and today everyone accepts it as fact. Micelles have now been studied with almost every technique devised by modern science, including X-ray diffraction (XRD), nuclear magnetic resonance (NMR), electron spin resonance (ESR), small-angle neutron scattering (SANS), light scattering, fluorescence, calorimetry, and many other solution and spectroscopic techniques. Despite being probed, prodded, and picked apart, however, micelles have still refused to yield the ultimate data, the interpretation of which is universally accepted and that unequivocally defines the true nature and characteristics of the aggregated species. It is possible, of course, that the diversity of surfactant structures and micellar and related aggregate species (vesicles, bilayers, microemulsions, etc.) are such that only very general laws will be found to be applicable to all; perhaps each system will have its specific twists, which preclude the existence of a "universal theory of everything" for surfactant aggregation, although in the generally ordered scheme of natural phenomena, such a prospect is unlikely. However, in science, as in many other human endeavors, it is as much the thrill of the hunt as the final capture that supplies the driving force for our activities.

It is generally accepted that most surface-active molecules in aqueous solution can aggregate into structures or clusters averaging 30–200 monomeric units in such a way that the hydrophobic portions of the molecules are closely associated and mutually protected from extensive contact with the bulk of the water phase. Not so universally accepted are some of the ideas concerning micellar shapes, the nature of the interior of a micelle, the "roughness" of the aggregate surface, the sites of adsorption of additional solutes into (or onto) micelles, and the size distribution of micelles in a given system. Although sophisticated experimental techniques continue to provide new insights into the nature of micelles, we still have things to learn. Given the inherent tendency of scientists to question and refine experimental procedures and to offer alternative interpretations for the results, it seems likely that questions concerning the theory of micelle formation and a complete model of the molecular nature of micelles will remain "fair game" for some time to come.

4.3.1. Manifestations of Micelle Formations

Early in the study of the solution properties of surface-active materials, it became obvious that the bulk solution properties of such materials were unusual and could change dramatically over very small concentration ranges. The measurement of properties such as surface tension, electrical conductivity, or light scattering as a function of surfactant concentration produces property curves that normally exhibit relatively sharp discontinuities at comparatively low concentration (Figure 4.5). The sudden change in a measured property is interpreted as indicating a significant change in the nature of the solute species affecting the measured quantity. In the case of the measurement of equivalent conductivity (top curve), the break may be associated with an increase in the mass per unit charge of the conducting species. For light scattering (bottom curve), the change in solution turbidity indicates the appearance of a scattering species of significantly greater size than the monomeric

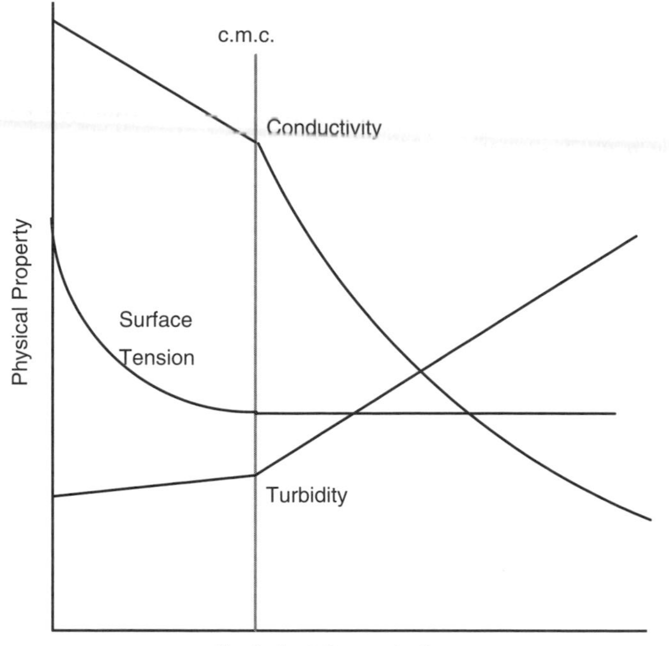

Figure 4.5. Some important manifestations of micelle formation: abrupt changes in solution conductivity, a discontinuity in the surface tension–concentration curve; a sudden increase in solution turbidity.

solute. These and many other types of measurement serve as evidence for the formation of aggregates or micelles in solutions of surfactants at relatively well-defined concentrations.

In 1920 it was reported that the osmotic activity of solutions of potassium stearate indicated the presence of a considerable degree of aggregation and suggested that the aggregated species should be termed micelles. To explain both the osmotic data and corresponding changes in conductance, it was suggested that two distinct types of micelle were being formed, a spherical species composed of ionized salt molecules and a nonionic lamellar aggregate structure involving unionized acid molecules (Figure 4.6a,b). Subsequent interpretations of the results of such studies were made in terms of a single type of structure. The single-structure or Hartley model called for essentially spherical micelles with a diameter equal to approximately twice the length of the hydrocarbon chain (Figure 4.6c). It was suggested that the structure was composed of 50–100 molecules and that the association should occurred over a relatively narrow concentration range. The interior of the micelle was described as being essentially hydrocarbon in nature, while the surface consisted of the charged head groups. The close proximity of the head groups required that some fraction be tightly bound to their counterions, thereby reducing repulsions between neighbors and reducing the overall mobility of the aggregated

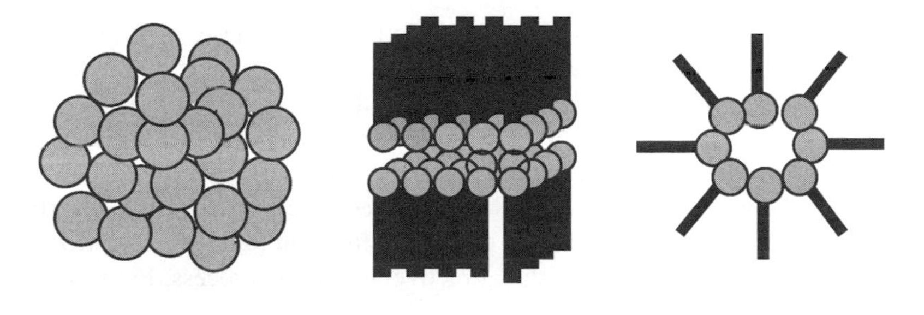

Figure 4.6. Five of the proposed micelle shapes, as interpreted from experimental data: (a) spherical; (b) lamellar; (c) inverted (or reversed); (d) disk; (e) cylindrical or rodlike.

species in an electric field. The classical Hartley micelle successfully described the characteristics of many, if not all, surfactant systems and achieved a deservedly honored place in the history of surface and colloid science. Other proposed structures (are illustrated in Fig. 4.6 b-e). More recent modifications to the basic micellar model have served to fill in the fine points but have not significantly changed the basic picture of micellar structures.

Early discussions of the micellization phenomenon emphasized that the "dislike" of the hydrophobic portion of a surfactant molecule for water was not a repulsive interaction, but rather an attractive preference of water for water and hydrocarbon for hydrocarbon. It was not suggested that there existed a particularly strong attraction among the hydrophobic chains in the molecules, since their interactions are nonpolar and, therefore, relatively small. That idea was reflected in the low melting and boiling points of hydrocarbons relative to polar materials of similar or lower molecular weight. Because of its chemical nature, however, water possesses a very strong cohesive force, which results in many of its unusual properties.

When a molecule containing both a hydrophobic group and a hydrophilic group is introduced into water, a distortion of the water structure to accommodate the solute molecules occurs, disrupting the happy accommodation of the water molecules and requiring them to orient around the hydrophobic tail in a more icelike structure. That more structured arrangement increases the free energy (basically the entropy) of the system. The physical result of such an energy increase is a tendency for the surfactant molecules to adsorb at available interfaces where preferred molecular orientations may serve to reduce the total free energy of the solution, or for the formation of molecular aggregates with their hydrophobic portions directed toward the interior of the micelle. Micellization, therefore, is an alternative mechanism to adsorption for the reduction of solution free energy by the minimization of the distortion of the structure of the bulk water. Although the removal of the hydrophobe from the water environment results in a decrease in energy, the adsorbed or aggregated hydrophobe may experience a loss of freedom (decrease in entropy) that would thermodynamically reduce the attractiveness of the process. In this case, the water wins out, overall.

It must also be remembered that the surfactant molecule also possesses a hydrophilic group, the interaction of which with water may decrease the free energy of the system. The partial removal of that group from the solution through adsorption or micelle formation can result in an increase in free energy. Additionally, the hydrophilic group may possess an electrostatic charge so that the process of adsorption or micellization can introduce electrostatic repulsions, which act to inhibit the removal of the molecule from solution. The situation, then, becomes a tug of war between the opposing free-energy considerations. The occurrence of micellization in a given surfactant system, and the concentration at which micelle formation occurs, will therefore be determined by the relative balance of the forces favoring and retarding the molecular aggregation process. Since the magnitudes of the opposing forces are determined by the chemical compositions of the solute molecules, where all other aspects (temperature, pressure, solvent, etc.) are held constant, it is the chemical constitution of the surface-active species that ultimately controls events. It should be possible, then, to make reasonable generalizations about the micellization characteristics of surfactants and their chemical structures.

Using the Hartley concept as a starting model, modern studies using techniques unimagined by the earlier workers have produced more detailed pictures of the submicroscopic nature of micelles. Micelles, of course, are not static species. They are very dynamic in that there is a constant, rapid interchange of molecules between the aggregates and the solution phase. It is also reasonable to assume that surfactant molecules do not pack into a micelle in such an orderly manner as to produce a smooth, uniform surface structure. If one could photograph a micelle with ultrahigh-speed film, freezing the motion of the molecules, the picture would almost certainly show an irregular molecular cluster more closely resembling a cocklebur than a golf ball.

The simplicity of the Hartley micelle has left it open to criticism, since it fails to adequately explain many experimentally observed phenomena. Models of micelles have been suggested that appear to differ substantially from those of Hartley. Particularly significant differences are a much greater degree of penetration of water into the micelle interior and a relatively smaller interior or core radius. Some of those models appear to better explain some of the solubilization data for hydrophobic additives and the measured or inferred microviscosities of micellar interiors. They have also been used to better explain some results in micelle-catalyzed reactions. The various models of micelles are, of course, just that—models that assist the investigator in visualizing what may be occurring in a surfactant solution to produce the observed experimental results. The relationship of models to reality may be questionable, but their utility as "tools of the trade" is quite real.

Although the classical picture of a micelle is that of a sphere, most evidence indicates that spherical micelles are not the rule and may in fact be the exception. As a result of geometric packing requirements and analyses (to be discussed below), ellipsoidal, disk-shaped, and rodlike structures may be the more commonly encountered shapes. However, from the standpoint of providing a concept of micelles and micelle formation for the nonspecialist, the Hartley spherical model remains a useful and meaningful tool.

4.3.2. Thermodynamics of Dilute Surfactant Solutions

As stated above, it is unlikely that a single "theory of everything" for the aggregation of surfactant molecules into micelles or other structures will be developed soon. It is all something like models for predicting the weather—we have a pretty good grasp of what is going on in general terms, but when we throw in a seemingly small variation (the famous fluttering butterfly in China, for example), things can rapidly begin to fall apart. Such models may "predict" the global weather 50 or 100 years hence, but we still don't know for sure whether it will rain 3 days from now! While classical approaches based on phase separation and mass action models have proved extremely useful, they do not possess adequate flexibility to extend their utility to explain such phenomena as the existence of cylindrical micelles, vesicles, and bilayer structures. In particular, they have not been able to theoretically quantify the role played by molecular geometry in predicting the shapes and structures that may result from a given molecular architecture.

In attempting to devise a comprehensive theory for micelle formation, there are two possible approaches. One may, if so inclined, begin with basic statistical mechanics, taking into account complex interactions between surfactant molecules and water, as well as solute–solute and solvent–solvent interactions. However, since the fundamental principles of the hydrophobic interactions between small molecules in water are still not clearly defined, there seems to be little hope that such an approach will produce a satisfactory result. However, even if a theoretically satisfying model did result, the mathematical complexities would possibly obscure any clear insight based on chemical realities.

Occupying the other end of the theoretical spectrum are approaches that ignore the statistical mechanical details in favor of an overall thermodynamic understanding. Pure thermodynamics, however, tends to be somewhat obscure (to those not skilled in the art) and must eventually fall back on some aspects of molecular interaction to validate its conclusions. Whatever favorite theory an author may champion, in order to be useful it must be able to not only explain experimental observations but also successfully predict as yet unobserved phenomena—as does any valid scientific theory. If too many unknown or variable parameters are included, theory tends to become a numerical game, an exercise in curve fitting that may lose its predictive capacity and credibility.

On the other hand, a theoretical assumption may start out as an acceptable or comfortable explanation of experimental data, but if the "truth" of that assumption cannot be tested or if its shortcomings are not obvious, it may be promoted to the rank of being an article of faith causing contradictory data to be discarded or relegated to a category of "bad" experimental technique. In the absence of a firm basis for choosing one approach to the theory of micelle formation, or any other theory for that matter, over others, and in line with the stated goal of keeping things as simple as possible, the following discussion is limited to a brief summary of the classical concepts of micelle formation.

4.3.3. Classical Theories of Micelle Formation

In the literature on micelle formation, two primary models have gained general acceptance as useful (although not necessarily accurate) for understanding the energetic basis of the process. The two approaches are the *mass action model*, in which the micelles and monomeric species are considered to be in a kind of chemical equilibrium, and the *phase separation model*, in which the micelles are considered to constitute a new phase formed in the system at and above the critical micelle concentration. In each case, classical thermodynamic approaches are used to describe the overall process of micellization.

In the mass action model, it is assumed that equilibrium exists between the monomeric surfactant and the micelles. For the case of nonionic (or un-ionized) surfactants, the monomer–micelle equilibrium can be written

$$nS \leftrightarrow S_n \tag{4.1}$$

with a corresponding equilibrium constant, K_m, given by

$$K_m = \frac{[S_n]}{[S]_n} \tag{4.2}$$

where brackets indicate molar concentrations and n is the number of monomers in the micelle or the aggregation number. Theoretically, one must use activities rather than concentrations in Eq. (4.2); however, the substitution of concentrations for activities is generally justified by the fact that the cmc occurs at such low concentrations that activity coefficients can be assumed to be unity.

It is usually observed that the cmc for a surfactant is relatively sharp and characteristic of a given surfactant. Although the detailed theory of micelle formation can become quite complex, the sharpness of the cmc can be explained conceptually in terms of the law of mass action. If C_t denotes the total concentration of surfactant in solution, C_m the fraction of monomer units in the aggregated or micellar state, and C_s that of free molecules, Eq. (4.2) may be written

$$K_m = \frac{C_m}{C_s} \tag{4.3}$$

In the process of micelle formation, there will be some value of C, C_{eq}, at which the number of surfactant molecules in the micellar form will be equal to that in the form of free surfactant molecules. At that concentration, $C_m = C_s = C_{eq}/2$. Using Eq. (4.3), one can then write

$$K_m = \left(\frac{C_{eq}}{2}\right)^{-(n-1)} \tag{4.4}$$

At any value of C_t, the relationship between C_s and C_m can be found by substitution of Eq. (4.4) into Eq. (4.2)

$$\frac{C_m}{(C_s)_n} = \left(\frac{C_{eq}}{2}\right)^{-(n-1)} \tag{4.5}$$

where $C_t = C_s + C_m$. Rearrangement of Eq. (3.5) gives

$$\frac{C_m}{C_{eq}} = \frac{1}{2}\left(\frac{2C_s}{C_{eq}}\right)_n \tag{4.6}$$

Using Eq. (4.6) as a starting point, one can estimate how the individual concentrations vary in the area of $C_t = C_{eq}$ for a given aggregation number, n. Aggregation numbers for many surfactants lie in the range of 50–100; Table 4.2 gives the percentages of molecules in the associated state for $n = 50$, 75, and 100, calculated according to Eq. (4.6). The results indicate that, while the cmc for a given system may not represent a truly sharp change in conditions, once the formation of micelles has begun, any increase in surfactant concentration will be directed almost

TABLE 4.2. Percentage of Total Surfactant Molecules in Micellar Form Near $C_s = C_{eq}$ Calculated According to Eq. (4.6)

	Percent C_t in Micellar Form (%)		
C_s/C_{eq}	$n = 50$	$n = 75$	$n = 100$
0.45	0.57	0.04	0.003
0.47	4.6	1.01	0.22
0.49	27	18	12
0.495	38	32	27
0.50	50	50	50
0.505	62	68	73
0.51	73	81	88
0.52	87	95	98
0.53	95	99	99.7
0.54	98	99.7	99.95
0.55	99.1	99.9	99.99

completely to the formation of more micelles. It is also obvious that the larger the aggregation number for a given system, the sharper will be the transition from monomolecular solution to predominantly micelles.

The alternative approach to modeling micelle formation is to think in terms of a phase separation model in which, at the cmc, the concentration of the free surfactant molecules becomes constant (like a solubility limit or K_{sp}) and all additional molecules go into the formation of micelles. Analysis of the two approaches produces the same general result in terms of the energetic balance of micelle formation (with some slight differences in detail), so that the choice of model is really a matter of preference and circumstances. There is evidence that the activity of free surfactant molecules does increase above the cmc, which tends to support the mass action model; however, for most purposes, that detail is of little consequence.

4.3.4. Free Energy of Micellization

Using Eq. (4.2) as a basis, the standard free energy for micelle formation per mole of micelles is given by

$$\Delta G^\circ_{\mathrm{m}} = -RT \ln K_{\mathrm{m}} = -RT \ln S_{\mathrm{n}} + nRT \ln S \tag{4.7}$$

while the standard free-energy change per mole of free surfactant is

$$\frac{\Delta G^\circ_{\mathrm{m}}}{n} = -\left(\frac{RT}{n}\right) \ln \cdot S_n + RT \ln S \tag{4.8}$$

As shown above, at (or near) the cmc $S \approx S_n$, so that the first term on the right side of Eq. (4.8) can be neglected, and an approximate expression for the free energy of micellization per mole of surfactant will be

$$\Delta G^\circ_{\mathrm{m}} = RT \ln(\mathrm{cmc}) \tag{4.9}$$

The situation is complicated somewhat in the case of ionized surfactants because the presence of the counterion and its degree of association with the monomer and micelle must be taken into consideration. For an ionic surfactant, the mass action equation would be

$$nS_x + (n - m)C_y \leftrightarrow S_n m(x) \tag{4.10}$$

The degree of dissociation of the surfactant molecules in the micelle α, the micellar charge, is given by $\alpha = m/n$. The ionic equivalent to Eq. (4.2) would then be

$$K_{\mathrm{m}} = \frac{[S_n]}{[S_x]_n [C_y]^{n-m}} \tag{4.11}$$

where m is the concentration of free counterions, C (e.g., those not bound to the micelle). The standard free energy of micelle formation will be

$$\Delta G^{\circ}_{\mathrm{m}} = \frac{\mathrm{RT}}{n}\{n \ln [S_x] + (n - m) \ln [C_y] - \ln [\mathrm{S}_n]\} \tag{4.12}$$

At the cmc $[\mathrm{S}^{-(+)}] = [\mathrm{C}^{+(-)}] =$ cmc for a fully ionized surfactant, and Eq. (4.12) can be approximated as

$$\Delta G^{\circ}_{m} = RT\left[1 + \left(\frac{m}{n}\right)\right] \ln \mathrm{cmc} \tag{4.13}$$

When the ionic micelle is in a solution of high electrolyte content, the situation described by Eq. (4.12) reverts to the simple nonionic case given by Eq. (4.9).

In general, but not always, micelle formation is found to be an exothermic process, favored by a decrease in temperature. The enthalpy of micellization, ΔH_{m}, given by

$$-\Delta H_{\mathrm{m}} = RT^2\left[\frac{d \ln \mathrm{cmc}}{dT}\right] \tag{4.14}$$

may therefore be either positive or negative, depending on the system and conditions. The process, however, always has a substantial positive entropic contribution to overcome any positive enthalpy term, so that micelle formation is primarily an entropy-driven process.

More elaborate models employ more complicated mathematical treatments with more rigorous statements of the physical phenomena involved. However, they yield little information of value as far as understanding a given practical system is concerned. A different, and perhaps more conceptually useful, approach emphasizes the importance of molecular geometry in defining the characteristics of an aggregating system. Such a geometric approach would seem to be especially useful for applications in which the chemical structure of the surfactant is of central importance.

4.4. MOLECULAR GEOMETRY AND THE FORMATION OF ASSOCIATION COLLOIDS

The theoretical developments based on the effects of geometry on molecular aggregation have shown that physical characteristics of surfactants such as cmc, aggregate size and shape, and micellar size distribution (polydispersity) can be quantitatively described without relying on a detailed knowledge of the specific energetic components of the various molecular interactions. It is also useful in that it applies equally well to micelles, vesicles, and bilayer membranes; the latter lie outside the normal models of association processes. For that reason, the geometric approach warrants a somewhat closer look.

The classical picture of micelles formed by simple surfactant systems in aqueous solution is that of a sphere with a core of essentially liquidlike hydrocarbon surrounded by a shell containing the hydrophilic head groups along with associated counterions, water of hydration, and other matter. Regardless of any controversy surrounding the model, it is usually assumed that there are no water molecules included in the micellar core, since the driving force for micelle formation is a reduction of water–hydrocarbon contacts. Water will, however, be closely associated with the micellar surface; as a result, some water–core contact must occur at or near the supposed boundary between the two regions. The extent of that water–hydrocarbon contact will be determined by the surface area occupied by each head group and the radius of the core. It seems clear from a conceptual viewpoint that the relative ratio between the micellar core volume and surface area must play an important role in controlling the thermodynamics and architecture of the association process. Equally important is the need to understand the constraints that such molecular geometry places on the ability of surfactants to pack during the aggregation process to produce micelles, microemulsions, vesicles, and bilayers.

Israelachvili (1992) and others have shown that the geometric factors that control the packing of surfactants and lipids into aggregated structures can be conveniently given by what is termed a "critical packing parameter" or shape factor given by $v/a_o l_c$, where v is the volume of the hydrophobic portion of the molecule, a_o is the optimum head group area, and l_c is the critical length of the hydrophobic tail, effectively the maximum extent to which the chain can be stretched under the specific conditions imposed by molecular structure, environment, and other factors. The value of the packing parameter will determine the type of association structure formed in each case. A summary of some of the structures to be expected from molecules falling into various "critical packing" categories are listed in Table 4.3.

TABLE 4.3. Expected Aggregate Characteristics in Relation to Surfactant Critical Packing Parameter, $v/a_o l_c$

Critical Packing Parameter	General Surfactant Type	Expected Aggregate Structure
<0.33	Simple surfactants with single chains and relatively large head groups	Spherical or ellipsoidal micelles
0.33–0.5	Simple surfactants with relatively small head groups, or ionic surfactants in the presence of large amounts of electrolyte	Relatively large cylindrical or rod-shaped micelles
0.5–1.0	Double-chain surfactants with large head groups and flexible chains	Vesicles and flexible bilayer structures
1.0	Double-chain surfactants with small head groups or rigid, immobile chains	Planar extended bilayer structures
>1.0	Double-chain surfactants with small head groups, very large, bulky hydrophobic groups	Reversed or inverted micelles

While examples almost surely will be found of materials that do not fit neatly into such a scheme, the general concepts are usually valid. For surfactants and other amphiphilic materials that form bilayer structures, Israelachvili has offered several generalizations that make it easier to understand the geometric consequences of the surfactant structure, including

1. Molecules with relatively small head groups, and therefore large values for v/a_0l_c, will normally form extended bilayers, large (low curvature) vesicles, or inverted micellar structures. Such structures can be created in anionic systems by changes in pH, high salt concentrations, or the addition of multivalent cations, especially Ca^{2+}.
2. Molecules containing unsaturated hydrocarbon chains, especially multiple cis double bonds, will have smaller values for l_c, and thus will tend toward the formation of larger vesicles or inverted structures.
3. Multichained molecules held above the melting temperature of the hydrocarbon chain may undergo increased chain motion, allowing trans–gauche chain isomerization, reducing the effective value of l_c, and resulting in changes in aggregate structures. This effect may be of particular importance in understanding the effects of temperature on biological membranes.

The foregoing generalizations on bilayer assemblies of surfactants and other amphiphilic molecules offer a broad view of the types of structure that may be formed as a result of the self-assembly process. They consider only the fundamental relationships between structure and the geometric characteristics of the molecules involved. Not considered are any effects on the systems that may exist because of curvature or other distortions of the molecular packing. The interested reader can obtain more in-depth discussions in the works of Israelachvili and Fendler cited in the Bibliography for this chapter. The unique characteristics of the bilayer and vesicle assemblies have attracted the attention of scientists in many disciplines for both theoretical and practical reasons. The following brief discussion skims the surface of what is sure to become an even more interesting and important area of surfactant-related surface science.

Although it is convenient to visualize the micellar core as a bulk hydrocarbon phase, the density may not be equal to that of the analogous true bulk material. X-Ray evidence indicates that the molecular volumes of surfactants in micelles are essentially unchanged by the aggregation process. If a molecular volume for a hydrocarbon chain in the micellar core equal to that of a normal hydrocarbon is assumed, the core volume v can be calculated from

$$v = n'(27.4 + 26.9n_c') \times 10^{-3}(\text{nm}^3) \tag{4.15}$$

where n' is an effective micellar aggregation number and n_c' is the number of carbon atoms per chain in the core. In general, the value of n_c' will be one less than the total number of carbons in the hydrocarbon chain n_c since the first carbon after the head group is highly solvated and may be considered to be a part of it. For normal surfactants with a single hydrocarbon tail, n' will be equal to the aggregation number n, while for those that possess a double tail, $n' = 2n$.

If one assumes that the micellar core has no "hole" at its center, one dimension of the aggregate species will be limited by the length of the hydrocarbon chain when extended to its fullest. That maximum length can be calculated by assuming a distance of 0.253 nm between alternate carbon atoms of the extended chain and adding the value of the van der Waals radius of the terminal methyl group (≈ 0.21 nm) and half the bond distance between the first carbon in the core and that bonded to the head group (≈ 0.06 nm). The maximum extended length l_{max} for a normal hydrocarbon chain with n_c' core carbon atoms, therefore, is given by

$$l_{max} = 0.15 + 0.1265 n_c' \text{ nm} \tag{4.16}$$

Since hydrocarbon chains in the liquid state are never fully extended, a dimension, l_{eff}, can be defined that gives the statistically most likely extension as calculated by the same procedure used for the calculation of polymer chain dimensions. For a chain with $n_c' = 11$, the ratio of l_{max} to l_{eff} will be approximately 0.75. In the micellar core, due to restrictions imposed by the attachment of the hydrocarbon tail to the head group bound at the surface, the mobility of the chains may be significantly limited relative to that of bulk hydrocarbon chains. The presence of "kinks" or gauche chain conformations, which may be imposed by packing considerations, will result in a calculated l_{max} amounting to only about 80% of the theoretical maximum.

Since hydrocarbon chains possess restricted bond angles as well as bond lengths, additional restrictions on the maximum extension of the chain arise beyond those mentioned previously. Chain segments located at the transition region from core to shell, for example, cannot assume arbitrary conformations in order to produce a perfectly smooth surface. The micellar surface, therefore, must be assumed to be rough or irregular, although the dynamic nature of the aggregate may obscure any practical effect of such roughness.

Several groups have considered in detail some of the geometric restrictions that govern micelle sizes, shapes, and size dispersity. Their analysis of the geometric and thermodynamic factors appears to allow for the prediction of most aspects of the aggregation of surface-active molecular species, including the cmc, average aggregation number, polydispersity of micelle sizes, and the most likely shape of the aggregated species.

Although the geometric approach shows great promise, it has not worked its way into the general thinking on micelles. It is, however, finding wide acceptance in areas related to biological membranes and aggregates. It can be expected that as more experimental data can be correlated with the predictions of geometric considerations, such an approach will gain ground as a basis for the design of surfactant molecules with specific desirable aggregation characteristics. With the preceding concepts in mind, we now turn to some of the experimental results that have helped bring us to our current state of understanding of surfactant aggregation or micelle formation.

4.5. EXPERIMENTAL OBSERVATIONS OF MICELLAR SYSTEMS

The preceding sections presented a brief review of some of the basic theoretical concepts pertaining to the formation of surfactant micelles. The following sections are devoted to the presentation of some experimental results from the literature that illustrate many of the various effects of surfactant chemical structure and solution environment on the aggregation process. While the previously described models of micelle formation serve as a useful basis for the interpretation of the results given below, it must be remembered that in all cases they represent a very simplified picture and involve a number of assumptions, many of which are not fully justified by experiment.

4.5.1. Micellar Aggregation Numbers

Several references have been made to the number of surfactant monomers aggregating to form a micelle—the aggregation number, n. The classical method for determining n is to use elastic light scattering. It is possible to determine a weight-average molecular weight (Mw) for a micellar solution, and therefore the average number of surfactant molecules in the structure, from the intensity of light scattered at a given angle at surfactant concentrations above the cmc relative to that of the pure solvent (or solvent plus surfactant below the cmc). Newer techniques such as laser light scattering and fluorescence quenching produce data that allow for the determination of the aggregation number and the distribution of micellar sizes, as well as giving some idea of their approximate shapes. Typical aggregation numbers for various surfactant types are given in Table 4.4.

Because the size and size distribution of micelles are sensitive to many internal (hydrophobic structure, head group type, etc.) and external (temperature, pressure,

TABLE 4.4. Aggregation Numbers for Some Surfactants in Water

Surfactant	Temperature (°C)	Aggregation Number
$C_{10}H_{21}SO_3^- Na^+$	30	40
$C_{12}H_{25}SO_3^- Na^+$	40	54
$(C_{12}H_{25}SO_3^-)_2Mg^{2+}$	60	107
$C_{12}H_{25}SO_4^- Na^+$	23	71
$C_{14}H_{29}SO_3^- Na^+$	60	80
$C_{12}H_{25}N(CH_3)_3{}^+Br^-$	23	50
$C_8H_{17}O(CH_2CH_2O)_6H$	30	41
$C_{10}H_{21}O(CH_2CH_2O)_6H$	35	260
$C_{12}H_{25}O(CH_2CH_2O)_6H$	15	140
$C_{12}H_{25}O(CH_2CH_2O)_6H$	25	400
$C_{12}H_{25}O(CH_2CH_2O)_6H$	35	1400
$C_{14}H_{29}O(CH_2CH_2O)_6H$	35	7500

pH, electrolyte content, etc.) factors, it is sometimes difficult to place too much significance on reported values of n. However, some generalizations can be made that are usually found to be true. They include

1. In aqueous solutions, it is generally observed that the greater the length of the hydrophobic chain of a homologous series of surfactants, the larger will be the aggregation number, n.
2. A similar increase in n is seen when there is a decrease in the "hydrophilicity" of the head group—for example, a higher degree of ion binding for an ionic surfactant or a shorter polyoxyethylene chain in a typical nonionic material.
3. External factors that result in a reduction in the "hydrophilicity" of the head group such as high electrolyte concentrations will also cause an increase in the aggregation number.
4. Changes in temperature will affect nonionic and ionic surfactants differently. In general, higher temperatures will result in small decreases in aggregation numbers for ionic surfactants but significantly large increases for nonionic materials. The effect on nonionic surfactants is related to the cloud point phenomenon discussed previously.
5. The addition of small amounts of nonsurfactant organic materials of low water solubility will often produce an apparent increase in micelle size, although that may be more an effect of solubilization (see Chapter 6) than an actual increase in the number of surfactant molecules present in the micelle.

While the question of the size of micelles is of great theoretical interest, it is rarely very significant (as far as we know) in most surfactant applications. Of more general importance is the concentration at which micelle formation occurs, the critical micelle concentration, since it is then when many of the most useful surfactant properties come into play.

4.5.2. The Critical Micelle Concentration

Because many factors have been shown to strongly affect the observed critical micelle concentration of surfactant systems, the following discussion is divided to isolate (as much as possible) the various important components, including the nature of the hydrophobic tail, the head group type, including counterion, and the role of external factors not directly related to the chemical structure of the surfactant.

There are a number of relatively easy (with practice) experimental methods for determining the cmc of a surfactant. One compendium, for instance, contains 71 possibilities, along with a critical discussion of each. The method of choice will depend on the availability of the various techniques, the relationship between the technique and the ultimate application, and the personal preferences of the investigator.

Because of uncertainties in micellar thermodynamics and the nature of micellar species, each procedure for the determination of critical micelle concentrations will carry with it the need to make a somewhat arbitrary decision as to the exact value to be reported. The chosen value may depend on the way in which the data are plotted and lines are extrapolated, the characteristics of the measurements, the specific techniques (and care) used in the procedure, the accuracy of experimental controls, and the judgment of the investigator. Light scattering measurements, for example, can be used to determine the cmc and yield a value related to the weight-average molecular weight, the size of the micelle, and even some shape factor. An older technique such as dye solubilization, on the other hand, will produce a number-average value that will always be smaller than the corresponding weight average for a polydisperse system. In addition, procedures such as the use of dyes introduce foreign materials that may actually alter the thermodynamics of the system through molecular interactions with the surfactant, so that even the choice of dye may influence the results. Even time-honored techniques that measure some physical characteristic of the solution such as surface tension and conductivity will often produce results differing in numerical value, if not order of magnitude.

It is obvious, therefore, that any discussion of cmc data must be tempered with the knowledge that the reported values must not be taken to be absolute; rather, they reflect certain variable factors inherent in the procedures employed for their determination. The variations found for nominally the same material under supposedly identical conditions in the literature should be accepted as minor "noise" that does not significantly affect the overall picture of the system. With those caveats in mind, our attention now turns to some of the trends that have been identified over the years that relate surfactant critical micelle concentrations to molecular structures.

4.5.3. The Hydrophobic Group

The length of the chain of a hydrocarbon surfactant has been shown to be a major factor determining its cmc. It is known that the cmc decreases logarithmically as the number of carbons in the chain of a homologous series n_c increases. For straight-chain hydrocarbon surfactants of 16 carbon atoms or less bound to a single terminal head group, the cmc is usually reduced by approximately one-half with the addition of each $-CH_2-$ group. For nonionic surfactants, the effect can be much larger, with a decrease by a factor of 10 following the addition of two carbons to the chain. The insertion of a phenyl and other linking group, the branching of the alkyl group, and the presence of a polar substituent groups on the hydrophobic chain can produce different effects on the cmc, as discussed below. For now we confine the subject to simple alkyl hydrophobic groups.

Mathematically, the relationship between the hydrocarbon chain length and cmc can be expressed by the so-called Klevens constant as

$$\log_{10}(\text{cmc}) = A - Bn_c \qquad (4.17)$$

where A and B are constants specific to the homologous series under constant conditions of temperature, pressure, and other parameters. Values of A and B for a wide variety of surfactant types have been determined, and some are listed in Table 4.5.

TABLE 4.5. Klevens Constants [Eq. (4.17)] for Common Surfactant Classes

Surfactant Class	Temperature (°C)	*A*	*B*
Carboxylate soaps (Na^+)	20	1.85	0.30
Carboxylate soaps (K^+)	25	1.92	0.29
n-Alkyl-1-sulfates (Na^+)	45	1.42	0.30
n-Alkyl-2-sulfates(Na^+)	55	1.28	0.27
n-Alkyl-1-sulfonates	40	1.59	0.29
p, *n*-Alkylbenzene sulfonates	55	1.68	0.29
n-Alkylammonium chlorides	25	1.25	0.27
n-Alkyltrimethylammonium bromides	25	1.72	0.30
n-Alkylpyridinium bromides	30	1.72	0.31

It has generally been found that the value of *A* is approximately constant for a particular ionic head group, while *B* is constant and approximately equal to $\log_{10} 2$ for all paraffin chain salts having a single ionic head group. The value of *B* will change, however, in systems having two head groups, or for nonionic systems.

A number of empirical relationships between cmc and hydrocarbon chain length consistent with the relationship in Eq. (4.17) have been developed by various researchers. The fundamental principle on which the derivations are based stems from the fact that the cmc is related to the free-energy change on micelle formation through Eqs. (4.13) and (4.14). Rearrangement of Eq. (4.13) yields

$$\ln(\text{cmc}) = \frac{\Delta G}{RT} + \ln 55.4 \tag{4.18}$$

Converting to $\log_{10}$ gives

$$\log_{10}(\text{cmc}) = \frac{\Delta G}{2.3RT} + \log_{10} 55.4 \tag{4.19}$$

Because the aggregation process represents a balance of the forces tending to favor micelle formation and those in opposition, the free-energy term can be divided into its components as

$$\Delta G_{\text{m}} = \Delta G_{\text{mh}} + \Delta G_{\text{mw}} \tag{4.20}$$

where the superscripts h and w indicate the hydrophobic forces driving the system to aggregation and the work required to bring the hydrophilic head groups into close proximity at the micellar surface, respectively. Data on the solubility of hydrocarbons in water indicate that the contribution to ΔG_{mh} of each $-CH_2-$ group added to the chain is a constant. In addition, the contribution of the terminal $-CH_3$ differs only by the addition of a constant *K*, so that

$$\log_{10}(\text{cmc}) = \frac{\Delta G_{\text{mw}}}{2.3RT} + n_{\text{c}} \frac{\Delta G_{\text{mh}}}{2.3RT} + \text{K}' \tag{4.21}$$

where

$$K' = \log_{10} + \frac{K}{2.3RT} \tag{4.22}$$

and n_c is the number of carbon atoms in the hydrocarbon tail. If it is assumed that the contribution of the head group ΔG_{mw} is independent of n_c, then for a homologous series Eq. (4.21) can be abbreviated in the form of Eq. (4.17) where

$$A = \frac{\Delta G_{mw}}{2.3RT} + K' \tag{4.23a}$$

and

$$B = \frac{-\Delta G_{mh}}{2.3RT} \tag{4.23b}$$

It is apparent that such an analysis qualitatively describes the empirical observations related to the fact that the constant A is relatively invariable for a given head group and B shows only small changes in different homologous series of ionic surfactants.

For nonionic surfactants, in the absence of electrical contributions to the aggregation process, the relative importance of the tail and head groups to the system changes. An empirical relationship between the cmc and the number of oxyethylene $(OE)_y$ groups in several nonionic surfactants series has the form

$$\ln(\mathrm{cmc}) = A' + B'y \tag{4.24}$$

where A' and B' are constants related to a given hydrophobic group. Examples of A' and B' for several commonly encountered hydrophobic groups are given in Table 4.6. In each case, the results are for one temperature and can be expected to vary significantly, given the known sensitivity of such systems to changes in T. In all the cases, not surprisingly, the cmc is found to decrease as the hydrophobicity of the molecule increases.

TABLE 4.6. Empirical Constants Relating cmc and OE Content for Various Hydrophobic Groups in Nonionic Surfactants [(Eq. (4.24)]

Hydrophobic Group	A	B
$C_{12}H_{25}O$–	3.60	0.048
$C_{13}H_{27}O$–	3.59	0.091
$C_{18}H_{35}O$– (oleyl)	3.67	0.015
$C_{18}H_{37}O$– (stearyl)	2.97	0.070
$C_9H_{19}C_6H_4O$–	3.49	0.065

TABLE 4.7. Critical Micelle Concentrations of Some Surfactants Having Two or More Ionic Groups Compared, Where Possible, to That of the Corresponding Fatty Acid

Surfactant ($R = n$-Alkyl)	cmc (mM)	Ratio of cmc's (for Corresponding Fatty Acid)
$R_8CH(COOK)_2$	350	3.5
$R_{10}CH(COOK)_2$	130	5.2
$R_{12}CH(COOK)_2$	48	7.7
$R_{14}CH(COOK)_2$	17	10
$R_{16}CH(COOK)_2$	6.3	≈ 15
$R_{12}CH(SO_3H)(COOH)$	2.4	—
$R_{14}CH(SO_3H)(COOH)$	0.6	—
$R_{14}CH(SO_3Na)COOCH_2CH_2SO_3Na$	8.0	—
$R_{16}CH(SO_3Na)COOCH_2CH_2SO_3Na$	2.5	—

Most of the ionic surfactants investigated have been of the simple 1 : 1 electrolyte type. With the appearance of surfactants with two or three ionic groups at one end of the hydrocarbon tail, it has become of interest to determine the cmc-related consequences of such structures. A number of studies of materials such as α-sulfonated fatty acids and their esters, alkyl malonates, and alkyl tricarboxylates exhibit a linear relationship similar to Eq. (4.17). Such surfactants have generally been found to have a lower Krafft point than the corresponding surfactant with a single head group. Because of the large size of the head group, members of these classes of materials exhibit other properties significantly different from those of more common surfactants of similar characteristics. A comparison of the cmc's of alkyl malonates with those of the corresponding fatty acid indicates that the extra head group produces a large increase in the cmc (Table 4.7). The difference results mainly from the larger electrical contribution to the thermodynamics of micelle formation. If, instead of a simple straight hydrophobic tail, a surfactant has a branched structure, with the head group attached at some point other than the terminal carbon, as for instance, in sodium tetradecane-2-sulfonate

$$CH_3(CH_2)_{11}CH(SO_3^-Na^+)CH_3$$

or if there are two independently attached hydrophobes, as in the diesters of sulfosuccinates

$$C_8H_{17}O_2CCH_2CH(SO_3^-Na^+)CO_2C_8H_{17}$$

the aggregation process can be expected to differ from that of normal hydrocarbon surfactants.

When the hydrophobe is branched, the additional carbon atoms off the mainchain contribute a factor equivalent to about half that for a mainchain carbon. The critical micelle concentrations of a series of sodium alkyl sulfates in which the total number

TABLE 4.8. Effects of Sulfate Substitution and Total Carbon Content on Cmc's of Sodium Alkyl Sulfates at 40°C

Total Carbon Number	Sulfate Position	cmc (mM)
8	1	136
8	2	180
10	2	49.5
11	3	28.9
11	6	83
14	1	2.4
14	2	3.3
14	3	4.3
14	4	5.15
14	5	6.75
14	7	9.70
15	2	1.71
15	3	2.20
15	5	3.4
15	8	6.65
16	1	5.8
16	4	1.72
16	6	2.35
16	8	4.25
18	1	1.65
18	2	2.6
18	4	4.5
18	6	7.2

of carbon atoms ranged from 8 to 18 and the position of the sulfate substitution changed from the terminal to the middle carbons are listed in Table 4.8. The data show that, except for the lower carbon number samples, the homologous series follows a linear relationship similar to the Klevens equation.

The values of the constants A and B for the different series are given in Table 4.9. From the data it can be seen that the value of A decreases slightly as the sulfate group is moved toward the interior of the hydrocarbon chain. Such an effect may be interpreted as indicating a smaller thermodynamic contribution to micelle formation from added methylene groups with internal sulfation. The values of B remain relatively constant within a series with different sulfate locations. For a series of alkyl sulfates with the same number of carbon atoms, the cmc increases as the sulfate group is moved internally. In the example of sodium tetradecylsulfate, the measured cmc varies from 9.7 mM for the 7-sulfate to 2.4 mM for the terminal sulfate. By analogy to straight-chain sulfates, such a change would correspond to a decrease of about two carbon atoms.

Many surfactants of commercial interest have nonterminal hydrophilic substitution. However, because such materials are seldom of sufficient purity to warrant

TABLE 4.9. Effects of Sulfate Substitution on Klevens Relationship [Eq. (3.17)] for C_{20} Alkyl Sulfate Surfactants at 40°C

Sulfate Position	A	B
1	0.294	4.49
2	0.286	4.53
3	0.280	4.55
4	0.266	4.47
5	0.258	4.44
6	0.270	4.72
7	0.256	4.59
8	0.251	4.58
9	0.245	4.55
10	0.240	4.52

detailed thermodynamic analysis, extensive micellization data have not been reported. The effects of other types of branching and hydrophobic substitution on the micellization of 2-*n*-alkylbenzene sulfonates have been reported. Other reports have dealt with the cmc's of alkylbenzene sulfonates with various points of attachment to the benzene ring. A comparison of the results for sodium alkyl sulfonates with those for alkylbenzene sulfonates (Tables 4.10 and 4.11), reflects the more hydrophilic nature of the aromatic ring as well as some branching-related effects. Analysis of the data indicates that, in terms of micellization effects, the benzene ring is approximately equivalent to 3.5 carbon atoms.

Surfactants that contain two hydrophobic chains, such as the sodium dialkylsulfosuccinates, exhibit a number of interesting and useful properties that make it advantageous to understand their structure–property relationships. The work done on such materials has been limited by the general lack of isomeric purity of the commercial materials. The cmc's of a few such materials that were carefully puri-

TABLE 4.10. The cmc's of Some Simple Sodium Alkyl Sulfonate Surfactants

Surfactant	Temperature (°C)	cmc (mM)
$C_8H_{17}SO_3^-$	40	160
$C_{10}H_{21}SO_3^-$	40	41
$C_{12}H_{25}SO_3^-$	40	9.7
$C_{14}H_{29}SO_3^-$	40	2.5
$C_{16}H_{33}SO_3^-$	50	0.7
$C_8H_{17}OOC(CH_2)_2SO_3^-$	30	46
$C_{10}H_{21}OOC(CH_2)_2SO_3^-$	30	11
$C_{12}H_{25}OOC(CH_2)_2SO_3^-$	30	2.2
$C_{14}H_{29}OOC(CH_2)_2SO_3^-$	40	0.9

TABLE 4.11. The cmc's of Typical Sodium Alkylbenzene-sulfonate Surfactants

Alkylbenzene Group	Temperature (°C)	cmc (mM)
p-Hexyl	75	37
p-Heptyl	75	21
p-Octyl	60	15
o-Octyl	55	19
p-Nonyl	75	6.5
p-Decyl	50	3.8
p-Dodecyl	30	1.2
p-Tetradecyl	75	0.66
p-1-Methyldecyl	35	2.53
p-1-Methyldodecyl	35	0.72
p-1-Methyltetradecyl	40	0.31
p-1-Methylhexadecyl	50	0.13

fied, have been reported, however (Table 4.12). It can be seen that the cmc values for the straight-chain esters follow the Klevens relationship, although the value of B is slightly smaller than that found for single-chain surfactants. As might be expected, the cmc's for the branched esters of equal carbon number occur at higher concentrations.

As mentioned above, the addition of a benzene ring to the hydrophobic chain has an effect on the cmc similar to the addition of 3.5 $-CH_2-$ groups. Since many natural fatty acids serve as starting materials for synthetic surfactant manufacture, it is of interest to know what effect might be expected from the presence of ethylenic unsaturation in the chain. Some data for the cmc's of saturated and unsaturated fatty acid analogs are given in Table 4.13. As can be seen, the presence of a single double bond in the chain increases the cmc by as much as a factor as 3–4 compared to the saturated compound. In addition to the electronic presence of the double bond, the isomeric configuration (cis or trans) will have an effect, with the cis isomer usually

TABLE 4.12. The cmc's of Typical Surfactants Containing Two Hydrophobic Groups (Not Branched)

Surfactant	cmc (mM)
Sodium di-*n*-butylsulfosuccinate	200
Sodium di-*i*-butylsulfosuccinate	200
Sodium dipentylsulfosuccinate	53
Sodium dihexylsulfosuccinate	12.4
Sodium dioctylsulfosuccinate	6.8
$(C_8H_{17})_2(CH_3)_2N^+Cl^-$	26.6
$(C_{10}H_{21})_2(CH_3)_2N^+Cl^-$	2.0
$(C_{12}H_{25})_2(CH_3)_2N^+Cl^-$	0.18

TABLE 4.13. Effects of Ethylenenic Unsaturation and Polar Substitution on the cmc of Potassium Soaps

Carboxylate Salt	Temperature (°C)	cmc (mM)
Octadecanoate (stearate)	60	0.5
Octadecenoate (oleate)	25	1.5
trans-9-Octadecenoate (elaidate)	50	1.5
$C_{19}H_{29}COO$ (abietate)	25	12
9,10-Dihydroxystearate	60	7.5
Ricinoleate	55	3.6

having a higher cmc, presumably because of the more difficult packing requirements imposed by the isomer.

If polar atoms such as oxygen or nitrogen are added to the hydrophobic chain (but not associated with a head group), the usual result is an increase in the cmc. The substitution of an –OH for hydrogen, for example, reduces the effect of the carbon atoms between the substitution and the head group to half that expected in the absence of substitution. If the polar group and the head group are attached at the same carbon, that carbon atom appears to make no contribution to the hydrophobic character of the chain.

A number of commercial surfactants are available in which all or most of the hydrophobic character is derived from the presence of polyoxypropylene (POP) groups. The observed effect of such substitution on aggregation has been that each propylene oxide group is equivalent to approximately 0.4 $-CH_2-$ groups.

As demands placed on surfactants have become more stringent, new classes of materials have been developed that do not conveniently fit into the classical groupings of conventional hydrocarbon-based materials. These newer classes include those in which fluorine replaces hydrogen atoms on the carbon chain, silicone-based surfactants, and more recently so called biosurfactants that can be a witch's brew of protein, carbohydrate, and/or hydrocarbon units.

The hydrophobic unit of the silicone-based surfactants consists of low-molecular-weight polyorganosiloxane derivatives, usually polydimethylsiloxanes. Possibly because of their somewhat ill-defined structure and composition, they have received relatively little attention in the general scientific literature, although their unique surface characteristics have proved them useful in many technological applications, especially in nonaqueous systems.

Probably the most rapidly developing of the nonhydrocarbon surfactants families are the fluorocarbons. The substitution of fluorine for hydrogen on the hydrophobic chain has produced surfactants of several types with extremely interesting and useful properties. The presence of the fluorine atoms results in large decreases in critical micelle concentration relative to the corresponding hydrocarbon. The addition of one $-CF_2-$ group to a chain produces a much greater effect on the cmc than that of the $-CH_2-$ group. From the point of view of the general character of fluorinated materials, that observation is totally consistent with the lower cohesive energy density of fluorocarbon materials. One can say that fluorine atoms, once bound to carbon, are extremely happy and are reluctant to undergo extensive interactions with their neighbors.

Because of the electronic character of the carbon–fluorine bond, fluorinated materials have been found to exhibit much lower surface energies and surface tensions than do conventional materials. The ability of polytetrafluoroethylene (PTFE or Teflon) to act as a nonstick coating, for example, results from the fact that the atoms in such a material do not interact strongly with those in adjoining phases, or themselves, for that matter. At the molecular level, the dispersion interactions among molecules are too weak to provide sufficient attraction for significant adhesion or cohesion to occur. That weakness of lateral molecular interactions is also reflected in the fact that fluorinated materials have boiling points much lower than would be expected based on molecular weight considerations, and that uncrosslinked PTFE has very low mechanical strength. It has been the introduction of chemical crosslinks into the technology that has brought us to the current state of the art in fluorocarbon coatings.

Fluorocarbons, hydrocarbons, and silicones exhibit widely differing cohesive energies as a result of the different natures of the C–C–F, Si–C–H, and C–C–H bonds. Those differences are a reflection of differences in the electronegativities of the various bond types. Surface energies and surface tensions can be related to those cohesive energy parameters. Liquids having the same number of carbon atoms show decreasing surface tensions in the order hydrocarbons > silicones > fluorocarbons.

A perfluorinated surfactant will have all carbon–hydrogen bonds replaced by carbon–fluorine, so that the simplest formula for the saturated materials will be $CF_3(C_nF_{2n})S$, where S signifies any of the possible surfactant head groups. Since various degrees of branching along the fluorocarbon chain are common, especially in commercial samples, care must be taken in the evaluation of such materials and the interpretation of experimental results. If hydrogen is substituted for a terminal fluorine, there will be a increase in the cmc and the minimum surface tension the surfactant can produce in aqueous solution (Table 4.14).

TABLE 4.14. Effects of Hydrogen Substitution on the cmc's of Fluorinated Surfactants

Surfactant	cmc (mM)
$H(CF_2)_6COO^-NH_4^+$	250
$H(CF_2)_8COO^-NH_4^+$	38
$H(CF_2)_{10}COO^-NH_4^+$	9
$H(CF_2)_6COOH$	150
$H(CF_2)_8COOH$	30
$C_5F_{11}COOH$	51
$C_7F_{15}COOH$	9
$C_9F_{19}COOH$	0.8
$C_5F_{11}COO^-K^+$	500
$C_7F_{15}COO^-K^+$	27
$C_7F_{15}COO^-K^+$	0.9

When substitution occurs internally, leaving the terminal CF_3– intact, the effects are much less significant. This is in agreement with general observations that indicate that the surface energies of polymers and adsorbed monolayers are determined primarily by the top atomic layers, as they are less dependent on the chemical nature of succeeding layers. The fact that all CF_3-terminated commercial surfactants do not produce the same low surface tension is a reflection of molecular packing defects introduced by fluorocarbon chain branching, the presence of homologs and other impurities, and the steric demands of the various linking groups and head groups employed. In essentially all cases, however, the fluorinated materials will be significantly more surface-active than the analogous hydrocarbon.

4.5.4. The Hydrophilic Group

As seen from the preceding section, the nature of the hydrophobic group has a major effect on the critical micelle concentration of a surfactant. The effect of the hydrophilic head group on the cmc's of a series of surfactants with the same hydrocarbon chain may also vary considerably, depending on the nature of the change. In aqueous solution, for example, the difference in cmc for a C_{12} hydrocarbon with an ionic head group will lie in the range of 0.001 M while a nonionic material with the same hydrocarbon chain will have a cmc in the range of 0.0001 M.

The cmc's of several "model" surfactants are given in Table 4.15. It is evident from the data that the nature of the ionic head group has a rather small effect

TABLE 4.15. Effects of Hydrophilic Group on the cmc's of Surfactants with Common Hydrophobes

Hydrophobe	Hydrophile	Temperature (°C)	cmc (mM)
$C_{12}H_{25}$	$COO^- K^+$	25	12.5
	$SO_3^- K^+$	25	9.0
	$SO_3^- Na^+$	25	8.1
	$H_3N^+ Cl^-$	30	14
	$(CH_3)_3N^+ Cl^-$	30	20
	$(CH_3)_3N^+ Br^-$	25	16
$C_{16}H_{23}$	$H_3N^+ Cl^-$	55	0.85
	$(CH_3)_3N^+ Cl^-$	30	1.3
	$(CH_3)_3N^+ Br^-$	60	1.0
	$(CH_3)_2C_2H_4OH\ N^+Cl^-$	30	1.2
	$(CH_3)(C_2H_4OH)_2N^+ Cl^-$	30	1.0
	$(C_2H_4OH)_3N^+ Cl^-$	30	1.0
C_8H_{17}	OCH_2CH_2OH	25	4.9
	$(OCH_2CH_2)_2OH$	25	5.8
C_9H_{19}	$COO(CH_2CH_2O)_9CH_3$	27	1.0
	$COO(CH_2CH_2O)_{16}CH_3$	27	1.8
$C_{10}H_{21}$	$O(CH_2CH_2O)_8CH_3$	30	0.6
	$O(CH_2CH_2O)_{11}CH_3$	30	0.95
	$O(CH_2CH_2O)_{12}CH_3$	30	1.1

compared to that seen for changes in the hydrocarbon chain. Such a result is not surprising in view of the fact that the primary driving force in favor of micelle formation is the energy gain due to reduction of water–hydrophobe interactions, while the effect of the ionic group, beyond its impact on water solubility, is to work against the aggregation process. For all 1 : 1 electrolyte/paraffin chain salts that are completely dissociated, the electrical contribution to dissolution and to micelle formation will be relatively constant. Of course, differences in ionic radius, degree of hydration, and other nonelectrical contributions will result in small differences among the various groups.

As noted earlier, the location of the head group along the hydrophobic chain can greatly affect micellization. It has been shown, for example, that as the charge on the hydrophilic group is moved away from the α carbon of the hydrophobe, the cmc will decrease. Such a result has been attributed to the increased work required to move the charge toward a medium or low dielectric constant (the micellar core). Of the more common ionic head groups, the order of decreasing cmc values for a given hydrocarbon chain is found to be carboxylates (containing one more carbon atom) > sulfonates > sulfates.

The sulfated alkyl- and alkylbenzene polyoxyethylenes is a class of ionic surfactants that has become increasingly important from an applications point of view. The basic structure of the family is

$$R{-}(OCH_2CH_2)_xOSO_3^-M^+$$

where x is the number of OE groups in the chain. (The use of x to denote the degree of polymerization of the POE chain, instead of n as in Chapter 2, is necessary to avoid confusion with the micellar aggregation number n defined earlier in this chapter.) As discussed in Chapter 2, these materials have found extensive use in many applications and are second only to the alkylbenzene sulfonates in total consumption worldwide. From the viewpoint of the physical chemistry of surfactants, it might be expected that the addition of oxyethylene groups adjacent to the ionic sulfate would increase the hydrophilic character of the molecule and modify the solution properties of the materials accordingly. In fact, when the terminal hydroxyl group of such nonionics is sulfated, a number of changes in solution characteristics do occur—but not necessarily those that might be expected from analogy with the related nonionic structures. The critical micelle concentrations of a series of dodecyloxyethylene sulfates with x varied from 1 to 4 are given in Table 4.16.

It can be seen that, contrary to what might be expected, the cmc's of the materials decrease as each OE group is added, indicating an increase in the overall hydrophobicity of the molecules. Early speculation concerning such results focused on some effect that reduced the ionic character of the sulfate group in the presence of the oxyethylene linkages. Investigations into the degree of dissociation of alkyl ether sulfate micelles have shown, however, that in fact dissociation increases with x, the opposite to what would be expected for a decrease in cmc.

Additional anomalies can be found in the effect of added electrolyte on the cmc's of these materials. It is usually found that the addition of electrolyte to

TABLE 4.16. Effects of Oxyethylene (OE) Introduction between Hydrocarbon Chain and Sulfate Group in $C_{12}H_{25}(OCH_2CH_2)_xOSO_3Na$ Surfactants at 25°C

Number of EO (x)	cmc (mM)
0	8.0
1	0.5
2	0.3
4	0.2

solutions of ionic surfactants results in a decrease in the cmc, while nonionic materials are only slightly affected. In the presence of 0.1 N NaCl, the materials in Table 4.16 show a shift in cmc even greater than that found for the analogous dodecylsulfate salt. It seems clear, then, that the combined effects of oxyethylene groups and ionic sulfates cannot be analyzed by simple analogy to either simple class of surfactants. In the absence of a clear-cut explanation for the unusual micellization properties of these materials, it has been suggested that the results might stem from either a reduction of the degree of hydration of the OE groups due to the presence of the sulfate or to a decrease in the work required to bring the ionic groups into close proximity during micelle formation as a result of the greater space requirements of the combination of hydrophiles. Data on the adsorption of alkyl ether sulfates onto solid surfaces and at the liquid–vapor interface, and data on micellar aggregation numbers, tend to support the importance of head group size and sulfate spacing to the micellization of these materials, especially for $x > 2$.

4.5.5. Counterion Effects on Micellization

As indicated by Eq. (4.12), the free energy of micelle formation for ionic surfactants contains a term related to the interactions of solvent, in most cases water, with the ionic head group. The degree of ionization of the ionic group, in terms of tight ion binding, solvent-separated ion pairing, or complete ionization, might be expected to greatly influence the magnitude of ΔG_{mw} and consequently the cmc and aggregation number of the system. Since electrostatic repulsions among the ionic groups would be greatest for complete ionization, it is not surprising to find that the cmc of surfactants in aqueous solution decreases as the degree of ion binding increases.

From regular solution theory it is found that the extent of ion pairing in a system will increase as the polarizability and valence of the counterion increase. Conversely, a larger radius of hydration will result in greater ion separation. It has been found that, for a given hydrophobic tail and anionic head group, the cmc generally decreases in the order $Li^+ > Na^+ > K^+ > Cs^+ > N(CH_3)_4^+ > N(CH_2CH_3)_4^+ > Ca^{2+} \approx Mg^{2+}$. In the case of cationic surfactants such as dodecyltrimethylammonium halides, the cmc's are found to decrease in the order $F^- > Cl^- > Br^- > I^-$.

TABLE 4.17. The cmc's of Various Metal Salts of Dodecylsulfate

Counterion	Temperature (°C)	cmc (mM)
Li^+	25	8.8
	40	10.5
Na^+	25	8.1
	40	8.9
K^+	40	7.8
Cs^+	40	6.9
$(CH_3)_4N^+$	25	5.6
$(\frac{1}{2}\ Ca^{2+})$	54	2.6
$(\frac{1}{2}\ Mg^{2+})$	25	1.6
$(\frac{1}{2}\ Zn^{2+})$	60	2.1

Although within a given valency the size of the hydrated counterion will have some effect on the micellization of an ionic surfactant, a more significant effect is produced by changes in valency. As the counterion is changed from monovalent to di- and trivalent, the cmc is found to decrease rapidly. The cmc's of various salts of dodecylsulfate are listed in Table 4.17. As discussed earlier, the divalent and higher salts of carboxylic acid soaps generally have very low water solubility and are not useful as surfactants in aqueous solution. They have found use in nonaqueous solvents because of their increased solubility in those systems, especially in the preparation of water-in-oil emulsions. As we will see, the presence of ions in aqueous surfactant solutions beyond the stoichiometric concentration can produce a more significant effect than changes within a valency group.

4.5.6. The Effects of Additives on the Micellization Process

Most industrial applications of surfactants involve the presence in the solution of cosolutes and other additives that can potentially affect the micellization process through specific interactions with the surfactant molecules (thereby altering the effective activity of the surfactant in solution) or by altering the thermodynamics of the micellization process by changing the nature of the solvent or the various interactions leading to or opposing micelle formation. Examples of specific interactions between surfactant molecules and cosolutes are common when the system contains polymeric materials. Because of the growing importance of such systems, they are treated as a special topic below.

In the absence of specific interactions, which can be quite complex, it is useful to be able to rely on laboratory cmc and aggregation number data to predict the characteristics of micellization of a surfactant in use. The use of such data, however, must be tempered by the knowledge that it is really valid only if the conditions of use parallel those under which the measurements are made. The reality of surfactant life is that in many applications, such parallelism may not, in fact, apply.

The solution changes commonly encountered in use that might be expected to impact the process include (1) the presence of electrolyte, (2) changes in pH, (3) the presence of organic materials that may be essentially water-insoluble (e.g., hydrocarbons), (4) the presence of water-miscible cosolvents, (5) the presence of materials that have low water solubility but contain polar groups that impart some surface activity although they are not classified formally as surfactants, and (6) the presence of polymeric materials. In the brief discussion of each category that follows, it must be remembered that each surfactant system can exhibit characteristics different from the general observations noted here.

4.5.6.1. Electrolyte Effects on Micelle Formation

In aqueous solution, the presence of electrolytes causes a decrease in the cmc of most surfactants, with by far the greatest effect found for ionic materials. For such materials, the effect of addition of electrolyte on the cmc can be empirically quantified with the relationship

$$\log_{10}(\text{cmc}) = -a \log_{10} c_i + b \tag{4.25}$$

where a and b are constants for a given ionic head group at a particular temperature and c_i is the total concentration of monovalent counterions in moles per liter.

The lowering of the cmc is due primarily to a reduction in the electrostatic repulsion between head groups and, consequently, a smaller contribution of those groups to the free-energy term opposing micellization [ΔG_{mw}, Eq. (4.20)]. For nonionic and zwitterionic materials, the impact of added electrolyte is significantly less, and the relationship in Eq. (4.20) does not apply. For such surfactants, a relationship of the form

$$\log_{10}(\text{cmc}) = -KC_s + \text{constant} \quad (\text{for } C_s < 1) \tag{4.26}$$

has been suggested, where K is a constant for a particular surfactant, electrolyte, and temperature, and C_s is the concentration of added electrolyte in moles per liter. For alkyl betaines it has been found that the value of K in Eq. (4.26) increases with an increase in the length of the hydrophobic chain and with the charge on the anion of the electrolyte.

The observed changes in the cmc of nonionic and zwitterionic materials with the addition of electrolytes cannot be attributed to the same electrostatic effects as for fully ionic surfactants. The most generally accepted explanation of such effects has been developed in the context of changes in the solvent properties of the aqueous solution for the hydrophobic group and the degree of solvation of the hydrophilic. It is well known that the solubility of many materials in water can be significantly altered by the addition of neutral ions. The result of such addition can be a reduced solubility, commonly referred to as "salting out," or increased solubility or "salting in." The specific effect will depend on the nature of the added electrolyte.

It has also been suggested that the changes in cmc found for nonionic materials with electrolyte addition are related to the amount of work required to disrupt the structure of the aqueous solvent by the insertion of the surfactant molecule. If added electrolyte acts to enhance structure, that is, if it increases the organization of the water molecules through the action of the added ions, the introduction of the monomeric surfactant molecule will require an additional amount of work to overcome that added structural energy. The net result will be a "salting out" of the surfactant and a decrease in the cmc. If, on the other hand, the added electrolyte acts as a structure breaker, the required work will be less, the surfactant will be "salted in," and the cmc will increase.

Nonionic surfactants that have highly hydrogen bonding interactions in water such as POE ethers and sugar esters will seldom show significant salt effects until the electrolyte concentration reaches the level at which the activity of the water becomes affected. At that point, the competition between the dissolved salts and the hydrophilic group for the available water becomes intense and the cmc will be found to decrease. The same might be expected for the addition of non-ionized additives that have a similar interaction with water such as sugars and other polyhydroxy materials, organic acids and amines, and water-soluble polymers.

In the case of the POE nonionic surfactants, there exists an additional possible phenomenon that may help to explain the effect of certain cations on their properties. It is well known, for example, that the cyclic polyoxyethylene or "crown" ethers can form very strong complexes with many appropriately sized ions such as Na^+ and K^+. When the ionic radius of the ion is properly matched to the size of the "basket" formed by the cyclic ether, interactions between the ether oxygen atoms and the ion produce complexes with exceedingly large stability constants. It seems reasonable to expect that linear POE chains of intermediate length, relatively free to assume various configurations in solution, could do so in such a way as to form a "pseudocrown" ether capable of forming complexes with cationic ions. Should such a phenomenon occur with either the monomer or the micelle, we would expect the overall thermodynamics of the system to be affected, including that of the micellization process. Such a scenario is presently somewhat speculative, but it represents an interesting potential field for further research.

The effectiveness of a given ion at altering the micellization process can be qualitatively related to the radius of hydration of the added ions, and the contribution of the cations and anions will be approximately additive. In general, the smaller the radius of hydration of the ion, the greater is its effect on the cmc. The approximate order of effectiveness of anions at decreasing the cmc is the following:

$$\frac{1}{2}SO_4^{2-} > F^- > BrO_3^- > Cl^- > Br^- > NO_3^- > I^- > CNS^-$$

For cations, the order is

$$NH_4^+ > K^+ > Na^+ > Li^+ > \frac{1}{2}Ca^{2+}$$

It has been found that the tetraalkylammonium salts of surfactants exhibit an increase in the cmc in the order

$$(C_3H_7)_4N^+ > (C_2H_5)_4N^+ > (CH_3)_4N^+$$

4.5.6.2. The Effect of pH

Since most modern, industrially important surfactants consist of long-chain alkyl salts of strong acids, it might be expected that solution pH would have a relatively small effect, if any, on the cmc of the materials, an expectation generally borne out by experience. In solutions of sulfonate and sulfate salts, where the concentration of acid or base significantly exceeds that of the surfactant, the excess will act as if it were simply neutral electrolyte with roughly the same results as discussed above.

Unlike the salts of strong acids, the carboxylate soap surfactants exhibit a significant sensitivity to pH. Since the carboxylate group is not fully ionized near or below the pK_a, the electrostatic interactions between head groups retarding micelle formation will vary with the solution pH, resulting in significant changes in the cmc. A similar result will be observed for the cationic alkylammonium salts near and above the pK_b, resulting in a decrease in the cmc. When the surfactant is in the fully ionized form, excess acid or base will act as neutral electrolyte as mentioned above.

It is to be expected that pH will have no effect on the cmc of nonionic surfactants, and that is generally found to be the case. However, at very low pH it is possible that protonation of the ether oxygen of OE surfactants could occur. Such an event would, no doubt, alter the characteristics of the system. Little can be found in the literature pertaining to such effects, however.

A number of amphoteric surfactant systems show pH sensitivity related to the pK values of their substituent groups. At low pH, materials containing carboxyl and amine groups would act as cationic surfactants, while at high pH the activity would be anionic, by analogy to the action of amino acids. If the cation is a quaternary ammonium salt, no pH sensitivity would be expected, as would be the case for a strong-acid anionic group. The pH sensitivity of amphoteric surfactants, therefore, will vary according to the specific structure of the materials. The possibilities can be grouped in the following way:

1. Quaternary ammonium/strong-acid salt with no significant pH sensitivity

$$RR'_3N^+XSO_3^-$$

where R is a long-chain alkyl group, R′ is a short-chain alkyl, and X is a linking carbon chain usually of one or two carbons

2. Quaternary ammonium/weak acid, which will be zwitterionic at high pH and cationic below the pK_a of the acid

$$RR'_3N^+XCOOH$$

where R, R′, and X are as defined in item 1

3. Amine/weak acid, which will be anionic at high pH, cationic at low pH, and zwitterionic at some pH between the respective pK values of the groups

$$RR'R''_2N^+XCOOH$$

where at least one of R′ or R″ is a hydrogen

4. Amine/strong acid, which will be anionic at high pH and zwitterionic below the pK_b of the amine

$$RR'R''_2N^+XSO_3^-$$

4.5.6.3. The Effects of Added Organic Materials

Organic materials that have low water solubility can be solubilized in micelles to produce systems with substantial organic content where no solubility would occur in the absence of surfactant. The details of the phenomenon of solubilization in surfactant micelles are presented in Chapter 6. Here we are concerned only with the possible effects of the phenomenon on the micellization process itself.

In the process of solubilization of water-immiscible organics in micelles, the size of the aggregate, and therefore the curvature of its surface, can change significantly. In the presence of such changes, it can be expected that there will be changes in the energetic requirements of interactions among the component parts of the surfactants in the micelle, especially the head groups at the micelle surface. Changes in the hydrophobic interactions among the hydrocarbon tails due to the insertion of additive molecules into the core may also occur. The combined effect of the presence of the solubilized material is usually to produce a slight decrease in the measured cmc of the system. The effect, however, is usually substantially smaller than that observed for the addition of electrolyte or changes due to structural changes in the surfactant molecule.

The effects of added organic materials on the cmc of some ionic surfactants are shown in Table 4.18. From the results it is evident that the effects, while relatively small, can be experimentally significant, especially as the length of the hydrocarbon

TABLE 4.18. Effects of Organic Additives on the cmc's of Ionic Surfactants

Surfactant	Additive (mM)		cmc (mM)
$C_{12}H_{25}NH_3^+\ Cl^-$	Cyclohexane	(0)	1.45
		(1.54)	1.34
		(2.5)	1.30
		(4.2)	1.20
	Heptane	(0.87)	1.34
		(2.12)	1.31
		(2.72)	1.28
	Toluene	(0.78)	1.40
		(2.19)	1.35
		(2.96)	1.31
$C_{10}H_{21}SO_3^-\ Na^+$	Benzene	(0)	41
		(34)	38
$C_{12}H_{25}SO_3^-\ Na^+$	Benzene	(0)	9.7
		(7.5)	9.2
$C_{14}H_{29}SO_3^-\ Na^+$	Benzene	(0)	2.5

tail of the surfactant increases. Because of these effects, it becomes necessary to assess the results of cmc determinations using the dye solubilization technique with the notion that the solubilization process could conceivably alter the cmc of the system.

Organic additives with substantial water miscibility such as the lower alcohols, dioxane, acetone, glycol, and tetrahydrofuran would not be expected to partition into the interior of the micelle when present in small amounts. The effect of such materials on the cmc, therefore, would be expected to be relatively minor. As the carbon number of the additive goes beyond C_2, the inherent surface activity of the alcohol can start to become significant. Otherwise, it will be only at high concentrations, where the additive may be considered to be a cosolvent, that major effects on cmc will be evident. Those effects will be the result of changes in the bulk solvent properties of the system. The energy requirements for bringing the hydrophobic tail into solution may decrease, leading to an increase in the cmc. Conversely, the added organic material will result in a reduction in the dielectric constant of the solvent mixture. Such an effect would tend to decrease the cmc of ionic surfactants as a result of their lower solubility and reduced repulsion between adjacent head groups at the micellar surface. The net effect on the cmc will therefore depend on the relative magnitudes of the two opposing trends. In any case, the effects are usually found to be relatively minor until substantial additive concentrations are reached.

The properties of a surfactant solution are found to change much more rapidly with the introduction of small amounts of long-chain alcohols, especially C_4 and greater. Because so many classes of surfactants of importance academically and industrially are derived from alcohols or raw materials containing alcoholic impurities, the recognition of the effects of such materials can be very important. Most of the observed effects can be attributed to the inherent surface activity of the longer alcohols—preferential adsorption at interfaces and a high proclivity for mixed micelle formation.

A number of studies have been carried out to determine the relationships among such factors as the number of carbon atoms in the alcohol chain, that of the surfactant tail, the alcohol concentration, and the observed cmc of the system. A useful logarithmic relationship has the form

$$\ln\left\{-\left[\frac{\delta\,(\mathrm{cmc})}{\delta C_a}\right]\right\} = -0.69\,m + 1.1\,m_a + \text{constant} \qquad (4.27)$$

where m and m_a are the number of carbon atoms in the surfactant chain and the alcohol, respectively, and C_a is the molar concentration of alcohol in the mixture. The term $-[\delta\,(\mathrm{cmc})/\delta C_a]$ is derived from a plot of cmc versus C_a. It is generally found that alcohols of carbon numbers 2–7 follow Eq. (4.27) rather well, while deviations begin to occur for the higher alcohols (Table 4.19), presumably because of the difficulty of packing the longer alcohol chains into the micellar structure.

The interactions between surfactants and alcohols have become of greater importance as a result of the intense interest in microemulsions and their potential

TABLE 4.19. Effects of Added Alcohol Chain Length [as Change in cmc with Alcohol Concentration, δ(cmc)/δ(Ca)] on Potassium Carboxylate Surfactant cmc's

	$\delta(\text{cmc})/\delta C_a$			
Alcohol	C_7COOK (10°C)	C_9COOK (10°C)	$C_{11}COOK$ (10°C)	$C_{13}COOK$ (18°C)
C_3OH	0.14	0.065	0.012	0.0032
C_4OH	0.38	0.19	0.038	0.0098
C_6OH	3.6	1.3	0.37	0.098
C_8OH	23	8.3	3.5	1.0
$C_{10}OH$	112	55	18	8.1

application in various areas of technological importance. Because of the rapid rate and volume of publications in the area of microemulsion technology, often emphasizing surfactant–additive interactions, Chapter 6 includes a summary of some current thinking on the subject.

4.5.7. The Effect of Temperature on Micellization

The effects of temperature changes on the cmc of surfactants in aqueous solution have been found to be quite complex. It is found, for example, that the cmc of most ionic surfactants passes through a minimum as the temperature is varied from about 0°C through 60–70°C. Nonionic and zwitterionic materials are not quite so predictable, although it is has been found that some nonionic materials reach a cmc minimum around 50°C.

When one considers the possible reasons for the observed temperature effects on cmc, it seems clear that one area of impact is in the degree of hydration of the head group, since the structuring of water molecules is known to be very temperature-sensitive. As the temperature of the system is increased, the degree of water–solute hydration will decrease, as will the cohesive interactions among water molecules. On one hand, the result will be a decrease in the energy factors favoring solution and an increase in the tendency toward micelle formation. On the other hand, the factors that reduce head group hydration at higher temperatures will also reduce energy increase caused by the structured water molecules around the hydrophobic portion of the surfactant molecule. The result from that scenario will be a reduction of the magnitude of the free-energy component attributable to the hydrophobic or entropic effect [ΔG_{mh}, Eq. (4.18)]. Such an effect will increase the "solubility"—or more accurately, decrease the "insolubility"—of the tail in water, a result that is in opposition to micelle formation. Since the two temperature effects act in opposite directions, the net effect of increasing or decreasing the cmc will depend on the relative magnitudes of the two.

The temperature dependence of the cmc's of polyoxyethylene nonionic surfactants is especially important, since the head group interaction is essentially

TABLE 4.20. Cloud Points (1% Solution) of Representative (Average) POE Nonionic Surfactants

Surfactant	Cloud Point (°C)
$C_9H_{19}COO(CH_2CH_2O)_7CH_3$	44
$C_9H_{19}COO(CH_2CH_2O)_{10}CH_3$	65
$C_9H_{19}COO(CH_2CH_2O)_{12}CH_3$	74
$C_9H_{19}COO(CH_2CH_2O)_{16}CH_3$	>100
$C_{11}H_{23}COO(CH_2CH_2O)_6CH_3$	31
$C_{11}H_{23}COO(CH_2CH_2O)_8CH_3$	53
$C_{11}H_{23}COO(CH_2CH_2O)_{10}CH_3$	74
$C_{11}H_{23}COO(CH_2CH_2O)_{12}CH_3$	79

totally hydrogen bonding in nature. Materials relying solely on hydrogen bonding for solubilization in aqueous solution are commonly found to exhibit an inverse temperature–solubility relationship. A major manifestation of such a relationship is the presence of the so-called cloud point for many nonionic surfactants.

The "cloud point," as defined in Chapter 1, is the temperature at which the solubility of the nonionic surfactant is not sufficient to provide the solubility necessary for effective surfactant action. In essence, it is a lower critical solution temperature for the low-molecular-weight POE chain. At the cloud point, a normally transparent solution of nonionic surfactant becomes cloudy and bulk-phase separation occurs. That is not to say that the material precipitates from solution; rather, a second swollen phase containing a high fraction of the POE surfactant appears, and its domains are significantly larger than those of a normal micelle.

Because of the role of the POE units in providing solubility to nonionic surfactants, it is not surprising to find that such surfactants with relatively short POE chains possess cloud points in easily accessible temperature ranges (Table 4.20). They seldom pass through a minimum in the cmc–T curve. If one considers a surfactant with 10 OE units, the loss of one hydrogen bond represents a loss of 10% of the energy contributing to solution of the monomeric species. For a surfactant with 20 OEs, such a change will represent a loss of only 5%. Clearly, then, the situation of the nonionic surfactants is complicated by the possibility of phase separation occurring before the increased "solubility" of the hydrophobe can bring about an balancing increase in the cmc.

4.6. MICELLE FORMATION IN MIXED SURFACTANT SYSTEMS

When one discusses the solution behavior of many, if not most, industrial surfactants, it is important to remember that experimental results must be interpreted in the context of a surfactant mixture rather than a pure homogeneous material. Studies of such systems are important both academically, assuming that the mixture can be properly analyzed as to its composition, and practically, since most detergents and soaps contain homologs of higher or lower chain length than that of the primary or "average" component.

Determinations of the cmc of well-defined, binary mixtures of surfactants have shown that the greater the difference in the cmc between the components of the mixture, the greater is the change in the cmc of the more hydrophilic member of the pair as the chain length of the more hydrophobic member is increased. Early analyses of the solution behavior of binary surfactant mixtures seemed to indicate that even homologous surfactants gave nonideal mixing in solution. It has since been recognized, however, that analysis of the results must account for the fact that at the cmc, the mole fractions of the monomeric surfactants in solution are not equal to the stoichiometric mole fractions; each value must be decreased by the amounts of each mole fraction incorporated into the micellar phase.

Theories describing the phenomena relating to the cmc of surfactant mixtures and their respective compositions have been developed by considering the effective mole fraction of each species in a binary mixture at the cmc. Although there is generally good agreement between the theory and experiment, in cases in which there is a sufficient difference between the chain lengths of the two surfactants, significant deviations have been found. The observed differences have been explained by such effects as (1) relatively small changes in the mole fraction of the smaller chain component due to preferential aggregation of the more hydrophobic material (i.e., homogeneous micelle formation) and (2) the difficulty of inclusion of the longer chain into micelles of the shorter material. In some cases, where the difference is very large, the shorter component may well act as an added electrolyte, with the consequent effects to be expected from such an addition, rather than becoming directly involved in the micellization process. When ternary surfactant mixtures are considered, it is usually found that the cmc of the mixture will fall somewhere between that of the highest and lowest values determined for the individual components.

The critical micelle concentrations of mixtures of POE nonionic surfactants are of particular interest, since the synthesis of such materials on a commercial basis will always produce a rather broad range of POE chain lengths. Because they contain no electrostatic contribution to the free energy of micelle formation, they can be treated theoretically with a simpler relationship between composition and cmc. In a mixture of nonionic surfactants in which the average POE chain lengths are approximately the same and the hydrocarbon chains different, there was a smooth decrease in the cmc of the mixture as the mole fraction of the more hydrophobic material (lower cmc) was increased, reminiscent of the surface tension–mole fraction curves found for miscible organic materials mixed with water.

The critical micelle concentrations of mixtures of ionic and nonionic surfactants has not been as fully explored as that of mixtures of structurally related materials, although it appears as if such systems are reasonably well behaved. Using the assumption that the mole fraction of ionic and nonionic surfactants in the micelle is the same as the bulk ratio, a good correlation can be made between micellar aggregation number, cmc, and the composition of the mixture.

The presence of an ionic surfactant in mixture with a nonionic usually results in an increase in the cloud point of the nonionic component. In fact, the mixture may not show a cloud point, or the transition may occur over a broader temperature range, indicating that the ionic component is forming mixed micelles with the nonionic surfactant, thereby increasing its "solubility" at higher temperatures. As a result, it is possible to formulate mixtures of ionic and nonionic surfactants for use at temperatures and under solvent conditions (electrolyte, etc.) in which neither component alone would be effective.

As we shall see in subsequent chapters, many mixtures of surfactants, especially ionic with nonionic, exhibit surface properties significantly better than do those obtained with either component alone. Such synergistic effects greatly improve many technological applications in areas such as emulsion formulations, emulsion polymerization, surface tension reduction, coating operations, personal care and cosmetics products, pharmaceuticals, and petroleum recovery, to name only a few. The use of mixed surfactant systems should always be considered as a method for obtaining the optimal performance for any practical surfactant application.

A more unique and less extensively researched class of mixed surfactant systems is that in which the two components are of opposite charge, that is, a mixture of a cationic and an anionic surfactant. In aqueous solvents such mixtures will often result in precipitation of stoichiometric amounts of the two materials due to ion pairing of the two surface-active components. A very careful combination of the two ionic classes can produce interesting results in terms of surface tension lowering (effectiveness) due to the formation of close ion pairs in the surface monolayer. In nonaqueous solvents, on the other hand, interesting and useful results may be obtained since the ion-paired combination may be significantly soluble in the organic solvent while still retaining useful properties in terms of aggregation and adsorption. Some such combinations have shown promise as phase transfer catalysts in which one or both components act as "mules" to shuttle reactants and products between aqueous and organic phases.

As fluorinated surfactants become more widely used throughout industry, regulatory constraints permitting, there often arise needs for a mixture of hydrocarbon and fluorocarbon materials to meet system performance requirements. For example, fluorocarbon surfactants are excellent at lowering the surface tension of aqueous systems at very low concentrations, but are normally of little use for forming or stabilizing emulsions. In a system requiring a low surface tension and emulsion stabilization it may be convenient to use both classes of materials. In such instances it must be determined whether the two types of surfactant will form mixed micelles, or whether two different types of homogeneous micelles will result. The presence of two distinct micellar types in a single solution, with all the accompanying characteristic differences, poses many interesting theoretical and practical questions. One might expect that given the known immiscibility of heavily fluorinated materials with hydrocarbons, comicellization would not be the rule. Although the experimental data are limited, there seems to be good evidence that the formation of two micellar species does occur in many instances.

4.7. MICELLE FORMATION IN NONAQUEOUS MEDIA

The formation of surfactant aggregates in nonaqueous solvents has received far less attention than the related phenomenon in water. In the present context, the term "nonaqueous" refers primarily to organic solvents such as hydrocarbons, aromatics, halogenated materials, and other liquids of low polarity and low dielectric constant. It does not generally include, for example, such solvents as dimethylformamide (DMF), dimethylsulfoxide (DMSO), and the lower alcohols ($C < 4$), although they will be briefly mentioned below.

In the past there was some controversy as to whether such a phenomenon as surfactant aggregation in fact occurs in the same sense as in aqueous solutions. There is no doubt that some chemical species, many surfactants included, do undergo an aggregation process in hydrocarbon and other nonpolar solvents. It is well known, for example, that carboxylic acids will dimerize in benzene, as evidenced by the fact that the molecular weights determined by boiling point elevation are routinely twice that of the monomeric species. Overwhelming experimental evidence, however, points to the fact that many other chemical types not only dimerize but also form relatively large aggregates in nonpolar solvents that must be considered to be related, if not identical, to the micelles formed in aqueous systems. Whether one prefers to call such species *micelles*, *reversed micelles*, or *inverted micelles*, or to use some other terminology, the characteristics and applications of those species warrant their inclusion in the current discussion.

4.7.1. Aggregation in Polar Organic Solvents

In polar solvents such as glycerol, ethylene glycol, dimethylsulfoxide, and formamide, many nonionic surfactants such as the POE alkylphenols aggregate to form micelles resembling those in aqueous solution. No micelles appear to be formed, however, in solutions of the lower alcohols or related solvents. Some attempts have been made to correlate the logarithms of the cmc's of such nonionic surfactants with the solubility parameters of the solvents employed. In several polar solvents, the free energy of micellization of $C_{12}H_{25}$–$(OE)_n$ surfactants with $n = 4,6,8$ has been resolved into the hydrocarbon and polar group contributions. In all the systems studied, the nature of the hydrocarbon chain has been identified as the most important factor driving the system toward micellization.

Many surfactant applications, especially those related to pharmaceuticals, require the presence of mixed solvent systems. The nonionic $C_{12}H_{25}$–$(OE)_6$ was found to form micelles in solutions of water, formamide, and mixtures of the two. Increasing the amount of formamide in the mixture led to an increase in the cmc. In aqueous solutions of sodium dodecylsulfate, the cmc is initially decreased by the addition of ethyl alcohol, reaches a minimum at about 5% alcohol, and increases from that point. The initial decrease in cmc is attributed to the preferential association of the ethanol with the surfactant micelle, followed by an increase in the solubility of the surfactant monomer as the solvent became less aqueous in character. Similar behavior has been observed for aqueous solutions containing nonionic surfactants and other water-miscible organic solvents.

The forces leading to micelle formation in polar organic solvents are not well understood, but they probably lie somewhere between the classical aqueous driving forces and the more-or-less opposite phenomena operating in nonpolar systems. There is undoubtedly a spectrum of mechanisms to be explored on theoretical and practical grounds for the ambitious graduate student or industry intern.

4.7.2. Micelles in Nonpolar Solvents

One reason for the scarcity of information on nonaqueous micelles is the relative difficulty of obtaining good, reproducible data. In water, micellization can be relatively easily followed using laboratory techniques such as surface tension measurements, conductivity, light scattering, and dye solubilization. In organic media, the two classic workhorses of surfactant studies, surface tension and conductivity measurements, are pretty useless. Dye solubilization has been used, but tends to be difficult to repeat quantitatively. A number of spectroscopic techniques have also been used with varying degrees of success. However, because the aggregation process in organic solvents is apparently not a sudden-onset phenomenon as in water, identification of the exact concentration of surfactant present when the process occurs is often subject to a wide range of interpretations. Precise data, therefore, are hard to come by.

The forces and changes involved in surfactant aggregation in nonpolar nonaqueous solvents differ considerably from those already discussed for water-based systems. The orientation of the surfactant relative to the bulk solvent will be the opposite that in water (hence the term "reversed" micelle; see Figure 4.6c). In addition, the micelle, regardless of the nature of the surfactant, will be un-ionized in solvents of low dielectric constant, and thus will have no significant electrical properties relative to the bulk solvent. Electrostatic interactions may, as we shall see, play an important role in the aggregation process, but in a sense opposite that in aqueous solution where strong head group repulsion tended to work against micelle formation.

As pointed out previously, the primary driving force for the formation of micelles in aqueous solution is the unfavorable entropic effect—also referred to as the "hydrophobic effect" in some literature—of ordering water molecules around the surfactant tail. In nonaqueous solvents, that effect would not be expected to be important since there would be little energy difference between solvent–solvent, solvent–tail, and tail–tail interactions. That is the case even if the solvent is aromatic or halogenated, rather than a simple hydrocarbon. Systems containing fluorinated materials or silicones are possible exceptions, as indicated by the fact that the surface tensions of some organic liquids is lowered by such surfactants. A more significant energetic consequence of nonaqueous micelle formation is the reduction of unfavorable interactions between the polar or ionic head groups of the surfactant molecules and the nonpolar solvent molecules. By analogy, such an effect might be called a "hydrophilic effect."

Unlike the situation for aqueous micelles, in which interactions among the hydrophobic tails contribute little to the overall free energy of micelle formation, ionic, dipolar, or hydrogen bonding interactions between head groups in reversed micelles may be the primary driving forces driving micelle formation. In the face of factors favoring aggregation, there seem to be few obvious factors opposing the formation of nonaqueous micelles, such as head group repulsion. The possible exception is unfavorable entropy losses as a result of fewer degrees of freedom for monomers in the micelle relative to those free in solution.

Of the many possible reasons for the relative scarcity of experimental data on nonaqueous micelles versus the aqueous variety, one of the most important findings is the failure of the easy and straightforward techniques applicable in water to work in most nonaqueous situations. Particularly important are the measurements of conductivity and surface tension. The ionization of charge-carrying species in solvents of low dielectric constant is, of course, difficult at best, and very high potentials are required to perform electrochemical measurements in such systems. In addition, since most surfactants possess hydrocarbon tails, their adsorption at the solution–air interface, if it occurs, will be such that the polar head group will be directed outward, a situation that could actually result in an increase in measured surface tension. Materials that can produce a lowering of the surface tension of organic solvents, namely, fluorocarbons and silicones, usually do so in a smooth decrease over a few mN/m, so that a phenomenon such as a cmc cannot be readily defined.

Unlike aqueous surfactant solutions in which micellar size and shape may vary considerably, small spherical micelles appear to be the most favored, especially when the reduction of solvent–polar group interactions is important. In much the same way as in water-based systems, geometric considerations often play an important role in determining micelle size and shape. Many materials that commonly form nonaqueous micellar solutions possess large, bulky hydrocarbon tails with a cross-sectional area significantly greater than that of the polar head group. Typical examples of such materials are sodium di-2-ethylhexylsulfosuccinate and sodium dinonylnaphthalene sulfonate:

$$C_8H_{17}OOCCH_2CH(SO_3^-Na^+)COOC_8H_{17}$$

$$(C_9H_{19})_2\text{—}C_{10}H_5SO_3^-Na^+$$

Since unambiguous experimental data are much less available on micelle formation in nonaqueous solvents, it is far more difficult to identify trends and draw conclusions concerning the relationships between chemical structures and critical micelle concentrations and aggregation numbers. Some compilations of such data are given in Tables 4.21 and 4.22. Because of the difficulties of obtaining precise data and the limited number of systems available, the numbers cited should be taken as approximate values that can change significantly if the conditions vary. For example, in hydrocarbon solvents, the nature of the polar head group is extremely important in the aggregation process. It has generally been found that ionic surfactants form larger nonaqueous micelles than do nonionic ones, with anionic sulfates surpassing the cationic ammonium salts. The aggregation number for

TABLE 4.21. Critical Micelle Concentrations for Some Surfactants in Organic Media

Surfactant	T (°C)	Solvent	cmc (mM)
Na^+ di-2-ethylhexylsulfosuccinate	30	Cyclohexane	1.6
Na^+ di-2-ethylhexylsulfosuccinate	30	Benzene	3
Na^+ di-2-ethylhexylsulfosuccinate	30	Dodecane	3
Li^+ dinonylnaphthalenesulfonate	25	Cyclohexane	1.1
Na^+ dinonylnaphthalenesulfonate	25	Benzene	0.1–1
Ba^{2+} dinonylnaphthalenesulfonate	25	Benzene	0.1–1
n-$C_{12}H_{25}NH_3^+$ $C_2H_5COO^-$	25	Benzene	2.2
n-$C_{12}H_{25}NH_3^+$ $C_3H_7COO^-$	25	Benzene	1.8
n-$C_{12}H_{25}NH_3^+$ $C_4H_9COO^-$	25	Benzene	2.0
n-$C_{12}H_{25}NH_3^+$ $C_3H_7COO^-$	25	Cyclohexane	2
n-$C_{12}H_{25}NH_3^+$ $C_3H_7COO^-$	25	CCl_4	1.6
n-$C_{18}H_{37}NH_3^+$ $C_2H_5COO^-$	25	Benzene	8
n-$C_{18}H_{37}NH_3^+$ $C_3H_7COO^-$	25	Benzene	2.7
n-$C_{12}H_{25}(OCH_2CH_2)_4OH$	—	Benzene	1.6
n-$C_{13}H_{27}O(CH_2CH_2)_6OH$	—	Benzene	2.6
n-$C_{12}H_{25}(OCH_2CH_2)_4OH$	—	Benzene	1.6
n-$C_{13}H_{27}O(CH_2CH_2)_6OH$	—	Benzene	2.6
$(n\text{-}C_7COO^-)_2\ Zn^{2+}$	—	Toluene	6
$(n\text{-}C_9COO^-)_2\ Zn^{2+}$	—	Toluene	5
$(n\text{-}C_{11}COO^-)_2\ Zn^{2+}$	—	Toluene	4

dinonylnaphthalene sulfonate in benzene was found to be essentially constant for a series of 10 different counterions, indicating a lack of sensitivity to the nature of the cation.

The effect of the hydrocarbon tail length in a homologous series of surfactants was found to be relatively small when compared to that in water. It was shown, however, that the aggregation numbers of the micelles decreased as the carbon number increased for a series of quaternary ammonium halides and metal carboxylates.

The presence of small amounts of water in a nonaqueous surfactant environment can have a significant effect on some systems. Particularly large effects have been found in solutions of sodium di-2-ethylhexylsulfosuccinate in toluene and phenylstearate soaps in benzene. It can be presumed that the effects of water and other impurities on nonaqueous micelle formation stem from alterations in the dipolar interactions between head groups induced by the additive or impurity. Although cmc values for a number of systems are listed in Table 4.21, the uncertainties associated with the nature of the aggregation process and of the aggregated species make a lengthy discussion of the type given for the aqueous systems complicated and somewhat more speculative than aqueous systems. That is not to say, however, that such phenomena are not theoretically and practically important. While the micellization process occurs in water, with some exceptions, over a fairly limited concentration range, the same is not always true for the nonaqueous process. Solu-

TABLE 4.22. Aggregation Numbers for Several Surfactants in Nonaqueous Solvents

Surfactant	T (°C)	Solvent	Aggregation Number
$n\text{-}C_9COOCH_2CHOHCH_2OH$	30	Benzene	41
$n\text{-}C_9COOCH_2CHOHCH_2OH$	30	Cl benzene	16
$n\text{-}C_{11}COOCH_2CHOHCH_2OH$	30	Benzene	73
$n\text{-}C_{11}COOCH_2CHOHCH_2OH$	30	Cl benzene	47
$n\text{-}C_{13}COOCH_2CHOHCH_2OH$	30	Benzene	15
$n\text{-}C_{15}COOCH_2CHOHCH_2OH$	30	Benzene	15
$n\text{-}C_{15}COOCH_2CHOHCH_2OH$	30	Cl benzene	9
$n\text{-}C_{17}COOCH_2CHOHCH_2OH$	30	Benzene	11
$n\text{-}C_{17}COOCH_2CHOHCH_2OH$	30	Cl benzene	3–4
$n\text{-}C_{12}H_{25}(OCH_2CH_2)_4OH$	25	Benzene	40–50
$n\text{-}C_{13}H_{27}O(CH_2CH_2)_6OH$	25	Benzene	70–80
Lecithin	25	Benzene	73
Lecithin	40	Benzene	55
Li^+ Phenyl stearate	25	Benzene	20
Na^+ Phenyl stearate	25	Benzene	25
K^+ Phenyl stearate	25	Benzene	22
$n\text{-}C_{12}H_{25}NH_3^+\ C_2H_5COO^-$	25	Benzene	2
$n\text{-}C_{12}H_{25}NH_3^+\ C_3H_7COO^-$	25	Benzene	6
$n\text{-}C_{12}H_{25}NH_3^+\ C_3H_7COO^-$	25	Cyclohexane	3
$n\text{-}C_{12}H_{25}NH_3^+\ C_3H_7COO^-$	25	CCl_4	10
Na^+ di-2-ethylhexylsulfosuccinate	30	Dodecane	32
Na^+ di-2-ethylhexylsulfosuccinate	30	Cyclohexane	32
Na^+ di-2-ethylhexylsulfosuccinate	30	Benzene	32
$(n\text{-}C_7COO^-)_2\ Zn^{2+}$	—	Toluene	6
$(n\text{-}C_9COO^-)_2\ Zn^{2+}$	—	Toluene	5
$(n\text{-}C_{11}COO^-)_2\ Zn^{2+}$	—	Toluene	4
Na^+ dinonylnaphthalenesulfonate	25	Benzene	10
Li^+ dinonylnaphthalenesulfonate	25	Cyclohexane	8

tion physical properties that are measured to determine micelle formation often undergo a smooth, continuous transition over several orders of magnitude in concentration. As a result, the designation of a given concentration as the cmc may become more a matter of the judgment of the investigator than the sensitivity of the technique. The same potential problems are present in aqueous systems, of course, but experience has shown them to be of only minor significance when due care is taken in the experiments.

PROBLEMS

4.1. Estimate the size of spherical micelles that would be formed by a series of single-chain hydrocarbon surfactants with chains of 10, 12, 14, 16, and 18 carbons.

4.2. Assuming that the head groups of the examples in Exercise 4.1 are sodium carboxylate, what would be the cmc of each material? Repeat the calculation for sodium sulfonate materials.

4.3. The following data for the cmc and aggregation number N were obtained for a typical straight-chain anionic hydrocarbon surfactant in solutions of various salt concentrations. Assuming a spherical geometry for the micelles, calculate for each system the volume of the hydrocarbon core, the effective radius of the core, and the cross-sectional area per chain at the aggregate surface

C (NaCl, M)	cmc (mM)	N
0.00	8.1	58
0.01	5.7	64
0.03	3.1	71
0.10	1.5	93
0.30	0.71	123

4.4. Given that the cmc of sodium dodecylsulfate is 8.9 mM at 45°C, calculate the expected cmc's for the C_{14}, C_{16}, and C_{18} members of the homologous series.

4.5. A nonionic surfactant with the formula $C_{12}H_{25}(CH_3)_2NO$ was found to have the following micellar characteristics as a function of temperature:

Temperature (°C)	cmc (mM)	N
1	0.124	77
27	0.092	76
40	0.080	78
50	0.076	73

(a) Assuming that the micellar radius is equal to the fully extended length of the hydrocarbon chain, calculate the area occupied by the head group.

(b) Using the equilibrium model for micellization, calculate the values of ΔG_{mic}, ΔH_{mic}, and ΔS_{mic} at 25°C.

4.6. It is usually found that the cmc of a homologous series of surfactants decreases by a factor of ~ 2 for every CH_2 added to the hydrophobic chain. Traube's rule states that adding a CH_2 to the chain changes the surface activity by a factor of 3, in the sense that a concentration one-third as large is required to obtain the same decrease in surface tension. What is the relation, if any, between these two phenomena?

4.7. Give (qualitatively) experimental evidence that the process of micelle formation in water is primarily an entropic rather than enthalpic phenomenon.

4.8. The different responses of ionic and nonionic surfactants to changes in temperature can have significant effects on their applications under unusual environmental conditions. It is often found by experience that mixtures of ionic and nonionic materials produce results superior to those from either pure material at the same total surfactant concentration. Propose an explanation for the common synergistic effect of surfactant combinations.

4.9. Calculate the critical packing parameters for the following surfactants and suggest what type of micelle or aggregate they will most likely form just above their critical micelle concentrations: sodium dodecyl sulfate, sodium laurate, sodium dodecylbenzenesulfonate, glycerol monostearate, glycerol monooleate, di-2-ethylhexyl sulfosuccinate, $CH_3COO^- \ ^+N(C_4H_9)_3(C_{12}H_{25})$, and sodium triisopropylnaphthalene sulfonate.

4.10. Industrial nonionic surfactants of the POE family are composed of a range of POE chain lengths, making them subject to possible batch-to-batch variations in some surfactant properties. Suggest a possible physical method for reducing the polydispersity of the POE chain distribution based on the known surfactant properties of the family.

5 Higher-Level Surfactant Aggregate Structures: Liquid Crystals, Continuous Biphases, and Microemulsions

As already described, a surfactant molecule contains a polar head group attached to a nonpolar, usually hydrocarbon, tail, which, when dissolved or dispersed in water or another liquid or mixture, is "driven" toward and adsorbed at almost any available interface. It is well known that surfactants and other amphiphiles adsorb very strongly to most solid surfaces and often require extreme measures such as concentrated nitric acid or potassium permanganate/sulfuric acid cleaning to ensure complete removal. When two essentially immiscible liquids are present, a preferred location for adsorption is at the interface between the two. When no more interfaces are available, self-assembled structures or aggregates begin to appear. The effects of such adsorption and aggregation are vitally important in many areas of biology and technology, but for the moment, the discussion will be limited basically to solvent (usually water)–amphiphile systems. Ternary systems will be addressed in the appropriate contexts.

Chapter 4 discussed the formation of relatively small, uniform, or isotropic association structures or micelles in dilute surfactant solutions. We know, however, that surfactants and related amphiphilic molecules, including the naturally occurring lipids, some proteins, and a variety of combined natural chemical species, tend to associate into structures more extensive than "simple" micelles in both aqueous and nonaqueous environments. In many cases, such assemblies can transform from one type to another as a result of sometimes subtle changes in solution conditions such as (1) changes in the concentration of the amphiphilic components, (2) the addition of new active components, (3) changes in solvent composition, (4) the addition of electrolytes, (5) temperature changes, (6) changes in solution pH, and (7) unspecified influences from internal and external sources—such as the phase of the moon, or so it seems at times.

The basic concepts that govern surfactant self-association or aggregation into micelles discussed previously also apply to the formation of larger, more extended

Surfactant Science and Technology, Third Edition by Drew Myers

aggregate systems. Such higher systems may be generally divided into "intermediate" and "bicontinuous" structures. The intermediate structures may be defined roughly as normal two-phase systems including liquid crystals, vesicles, bilayers, and membranes, and more disordered, but still structured microemulsions. In this chapter, attention will center on essentially two-component surfactant–water systems. In the real world, however, three or more component systems are those of most technological and academic interest. The inclusion of microemulsions in the category of self-assembled aggregate structures is a rather new concept to many.

Historically, microemulsions were discussed as a separate "family" of colloids that formed essentially spontaneously and were thermodynamically stable. However, microemulsions must, by definition, contain at least three components—solvent, amphiphile, and dispersed phase—and quite often contain a fourth, the so-called cosolvent. More recent experimental and theoretical work has tended to move them into the larger family of surfactant aggregates, their complex composition notwithstanding. That convention will be followed here, although there still remain a number of points of contention that need to be resolved on their classification as surfactant mesophases on a par with classical liquid crystals.

5.1. THE IMPORTANCE OF SURFACTANT PHASE INFORMATION

Phase diagrams, the basic science of obtaining information on the phase behavior of mixtures, including surfactants and related amphiphiles, is notoriously tedious and fraught with pitfalls. For that reason, along with a probable lack of attention at the instructional level, the potential utility of such information is quite often overlooked in the general course of a chemical education. That can be a potentially unfortunate situation, however, since phase information can represent some of the most useful physical data for determining the function and utility of surfactants and other materials in important applications.

For example, the physical behavior of the components in a physical or chemical process will almost certainly change with changes in the temperature, concentration, and chemical composition of the process mixture. Changes in viscosity, among other characteristics, will usually accompany changes in process temperature, as will changes in solvent concentration. Prior knowledge of such information can be important for process engineers in finalizing process designs.

At the chemical level, where process reactions depend critically on the chemical potentials of reactants, a phase change for one or more components may be accompanied by a change in reaction rates, impacting the apparent kinetics of a process for reasons other than strictly chemical reactivity; that is, the phase change may result in a misinterpretation of the chemistry of the process.

Many industrial separation technologies also rely on accurate knowledge of phases. One of the oldest chemical technologies, soapmaking, depends directly on recognizing and inducing phase changes in order obtain the desired product characteristics. The refining of fats also relies on the controlled precipitation of amphiphilic materials.

Detailed phase data provide a complete description of the solubility of surfactants and other amphiphiles. Such data can be important in the selection and design of complex systems involving surfactant activity. A surfactant candidate, seemingly interesting for economic reasons, for example, may be found inappropriate under some conditions of use if a complete understanding of its phase data is available. If such data were not available, the surfactant failure could possibly become apparent later in the development process, resulting in a very expensive reengineering process.

The optimum functioning of some surfactants, such as in foams and emulsions to be discussed in later chapters, have been found to depend on the phase behavior of the surfactant in the interfacial region. Foam stability, for example, has been correlated with the presence of more than one surfactant phase, a small amount of a liquid crystalline phase significantly improving the stability of the system.

Finally, although most ionic industrial surfactants are really mixtures of a homologous series of hydrophobic groups, a quantitative chemical analysis of the actual composition of nominally identical materials from batch to batch represents a significant burden. The same can be said for both the hydrophilic and hydrophobic groups of nonionic materials. The phase behavior of surfactants has been found to be so sensitive to the characters of both the head and tail groups that significant deviations or differences can be relatively easily detected from the phase behavior of the material relative to a control material. In fact, phase behavior can be used as an analytical tool to differentiate surfactants from nonsurfactant amphihilic materials.

5.2. AMPHIPHILIC FLUIDS

In some areas of research and technology, the term "amphiphilic fluids" is used with reference to multicomponent intermediate and bicontinuous aggregate systems. While the terminology is not "classical" in surface science, it is useful in terms of our new knowledge and understanding of complex systems containing amphiphilic materials. Systems falling into the category of amphiphilic fluids are not only important in physical chemistry but also form the basis of our understanding of structural biology. They have even inserted themselves into the "foreign" fields of soft-matter physics and materials science from both fundamental and practical perspectives. Their applications are also widespread, encompassing, for example, the actions of detergents in mammalian respiration. Cell membranes, we know, are complex macromolecular assemblies of, in large part, self-assembled phospholipids and associated protein and glycoprotein molecules.

Application of the concepts of higher-order amphiphilic self-assembled structures is not always as obvious in the more complex systems described here as one would like. At times it requires what seems to be a leap of faith, if such things are allowed in science, to get from a simple micelle to the complex structures to be discussed. This chapter presents some of the general aspects of the larger aggregate structures formed spontaneously by surfactants, and other amphiphilic species that may not technically be surfactants, but exhibit many of the same solution characteristics. The discussion is not meant to be comprehensive, but rather is intended to introduce some basic concepts along with some images that will help the reader grasp some of the why's and how's of the processes involved. With a basic conceptual understanding, potential users will hopefully be in a position to delve deeper into the menagerie of molecular aggregation processes in the search for a solution to their needs.

5.2.1. Liquid Crystalline, Bicontinuous, and Microemulsion Structures

The selective chemical affinity between the parts of a surfactant molecule or other amphiphile and the solvent or other components of a mixture is the driving mechanism responsible for aggregate structure formation. After decades of study, higher-order self-assembled surfactant and lipid structures continue to present a number of theoretical and experimental questions. Although this chapter presents some of the more fundamental principles and unresolved issues surrounding lyotropic self-assembly in equilibrium, the literature is far too extensive to be treated adequately here. Attention is focused on observations related to the more classical "intermediate" phases of complex aggregate structures. More involved membrane fusion intermediates and developments in the theory of phase stability of mesh, sponge, and cubic bicontinuous structures are addressed only briefly. Also generally excluded are geometrically deformed hexagonal phases, such as the so-called rectangular ribbon intermediates. The interested reader is referred to the works cited in the Bibliography for more details.

Micellar, lamellar, rodlike (usually hexagonal), mesh, and bicontinuous mesophases can form structured, equilibrium phases, in addition to the disordered mesophases now generally recognized to be dominated by microemulsions and, more recently, the so-called sponge mesophases. Theoretically, more recent advances in the understanding of lyotropic self-assembly has come about through the fusion of thermodynamics and molecular geometry (see Chapter 4), which in turn evolved from earlier concepts such as the classical hydrophile–lipophile balance (HLB), and the work of Tanford (1980; see General Readings list in Bibliography) related to the principles of hydrophobic interactions. Newer, more complete theories explain observations on the more complex and rigid systems based on the concept of membrane packing and bending energy as the dominating factors in the self-assembly process.

A great deal of the interest and activity in self-assembly processes today goes beyond the classical studies of equilibrium aggregation and phase studies, with a great deal of effort being devoted to such dynamic features as solution rheology and phase transformation kinetics. Application studies also look toward the synthesis of novel mesostructured inorganic materials using lyotropic systems as templates. Nevertheless, it is still important to increase our understanding of near-equilibrium self-assembly processes, since they are fundamental to understanding nonequilibrium, dynamic processes in amphiphilic systems and impact so many other fields, particularly the development of new materials and systems, and membrane function and fusion in biological systems.

The experimental–theoretical study of mesophase formation in amphiphilic systems emphasizes the basic chemical, physical, and materials science aspects of the systems. The most commonly discussed mesophases, beyond the simple micelles discussed in Chapter 4, are lamellar; aggregated micellar (packed in various cubic and hexagonal close-packed arrays), columnar or ribbon phases (rod-shaped micelles stacked in a two-dimensional hexagonal or rectangular array); microemulsions, and the cubic bicontinuous mesophases. The experimental techniques normally used to identify these mesophases are NMR lineshape analysis, diffusion measurements, small-angle neutron and X-ray scattering, and optical texture analyses. In addition, reconstruction of electron density profiles and very low temperature transmission electron microscopy (TEM) have been used to elucidate the details of these mesostructures.

The current theoretical understanding of these structures is based on the combined concepts of a preferred molecular shape and a preferred membrane curvature resulting from basic molecular geometry considerations. In essence, it is assumed that deviations from a preferred shape or curvature results in an increase in bending energy, and the resulting membrane or aggregate shapes are the result of the minimization of that energy.

Taking a spring as a conceptual model, there will be an equilibrium or rest configuration that represents an energy minimum that is determined by the chemical composition of the spring and the history of its formation. If that rest configuration is distorted by compression, extension, or bending, a restoring force or energy is introduced that will, if possible, restore the spring to its rest state. If the distortion becomes excessive, the spring may become permanently distorted (e.g., develop a new rest configuration) or break completely (Figure 5.1).

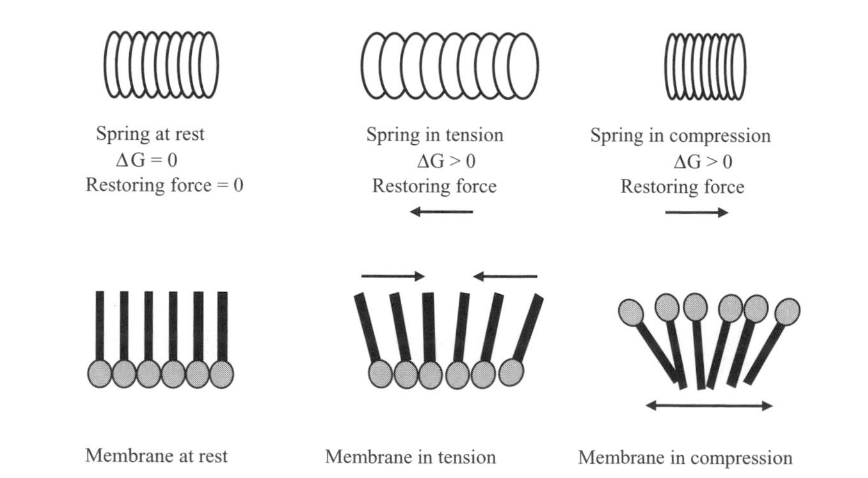

Figure 5.1. A spring model of bilayer membranes and related packing and bending energy characteristics.

A similar effect is invoked for amphiphilic membranes in that they will have a rest state, or curvature in this case, that is determined by thc geometric packing requirements of the constituent molecules that will try to maintain the minimum-energy state. If the membrane is distorted, the packing energy of constituents will try to force the membranes back into its minimum-energy state. Like the spring, if the restoring capacity of the membrane structure is exceeded, or if a component modification changes the net energy of the system, irreversible membrane rupture may occur. In biological systems, such rupture may be involved in such vital processes as cell division and the entry and expulsion of specific components such as ions, hormones, and enzymes. It may also be involved in unwanted activities such as the penetration of viruses or toxins into target cells.

Interfaces formed by amphilic molecules can take on three configurations: (1) they can be curved toward the hydrophobic region of the molecule, a conformation usually referred to as "type 1 curvature"; (2) they can curve toward the polar or hydrophilic region, the "type 2 curvature"; (3) or they can be essentially planar. The three basic categories are illustrated in Figure 5.2.

The effect of molecular geometry on the probable aggregation structures of amphiphiles was introduced in Chapter 4. As indicated, type 1 curvature is favored by amphiphiles for which the head group, S, is relatively more bulky than the average cross-sectional area of the tail, R, such as single-chained charged detergents. Type 2 is favored by molecules with bulky hydrophobic chain regions, such as double-chain surfactants or single-chain materials in high salt solution. Planar, or approximately so, assemblies are found for particular situations in which the critical packing parameter for the system is essentially unity. Most biological lipids, for example, must be delicately balanced in their hydrophobic–hydrophilic geometries, so that their phase diagrams at biologically relevant physical conditions favor almost flat or planar lamellar mesophases, since cell membranes require such structures for cell integrity and function. In a real system, however, the concept of a well-defined defect-free, planar lamellar state is an idealization, and such systems are usually delicately balanced and ready for transformation from one curved type to the other as required by the proper functioning of the system.

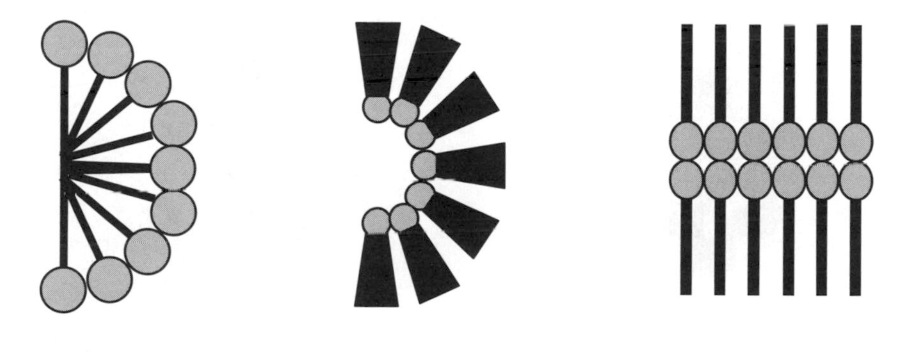
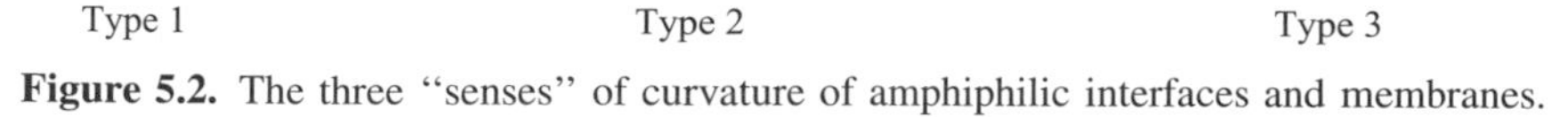

Figure 5.2. The three "senses" of curvature of amphiphilic interfaces and membranes.

5.2.2. "Classical" Liquid Crystals

The spontaneous formation of surfactant aggregate structures or micelles at relatively low concentrations opens the door to a veritable "zoo" of larger, more structurally complex, and certainly more theoretically complex self-assembled structures that inhabit our natural world, and make life as we know it possible. While the concepts presented in Chapter 4 to explain micelle formation in aqueous and nonaqueous solutions are relatively straightforward, chemically speaking, it should be obvious that there is a great deal of room for complications to set in as a system becomes more complex.

Simple micellar systems may be characterized generally as being dilute isotropic phases that show little structure beyond that of the localized micellar aggregate. That definition is a bit tenuous, however, since the broad spectrum of aggregate structures (Figure 5.3) can be continuous from the "simple" spherical, ellipsoidal, and disk micelles through the larger, intermediate nonisotropic structures such as liquid crystals, bicontinuous mesophases, vesicles, microemulsions, and extended membranes already mentioned. Transitions between such structures can result from subtle changes in amphiphile concentration, electrolyte concentration and pH changes, nonionic solute addition, temperature changes and other variables. This chapter introduces some of the structures and properties of the so-called lyotropic mesophases that may be present in more concentrated surfactant solutions. The subject of surfactant phase behavior is quite complex and has been

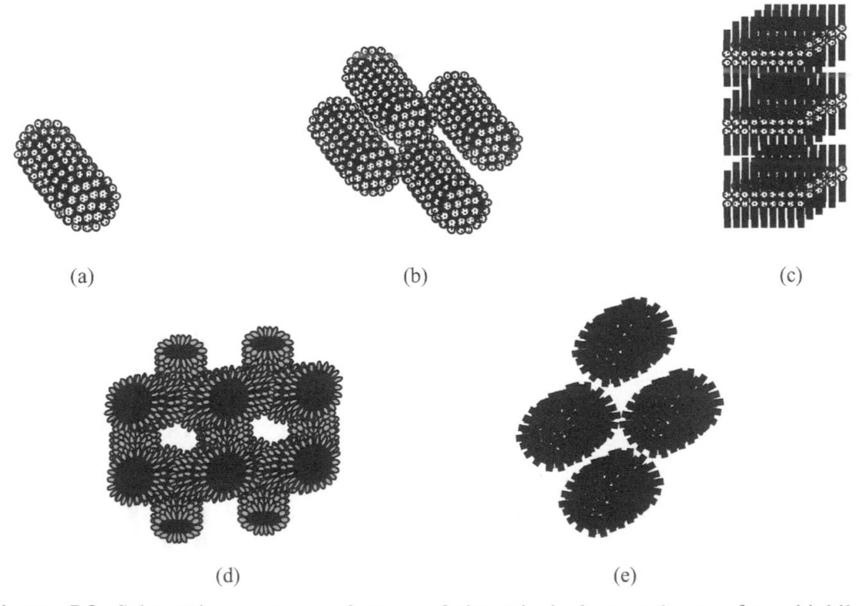

Figure 5.3. Schematic structures of some of the principal mesophases of amphiphiles: (a) rod-shaped micelles; (b) hexagonal close-packed rods; (c) bilayer or multilayer sheets; (d) a cubic bicontinuous phase; (e) reversed hexagonal rods.

reviewed in various publications cited in the Bibliography. The discussion here is limited to a general descriptive introduction to water-based systems. The possible role of mesophases in various surfactant applications will be mentioned in the appropriate chapters.

5.2.3. Liquid Crystalline Phases in Simple Binary Systems

It has been recognized for many years that surfactant solutions with concentrations well above the critical micelle concentration can exhibit physical properties that indicate the presence of various degrees of structure above the "simple" micellar level. Such structure formation may be manifested in bulk by abrupt changes in viscosity, conductivity, and other transport phenomena, birefringence, the existence of characteristic X-ray diffraction and radiation scattering patterns, or spectroscopic analyses. As the surfactant concentration is increased, changes in the physical characteristics of the solution signal corresponding changes in the nature of the aggregated solute as illustrated in Figure 4.2.

We know that the existence of mesophases in aqueous solutions of pure or reasonably homogeneous surfactants is a direct result of the nature of water–surfactant interactions. As the amount of solvent available between the simple dilute solution micelles decreases, interactions between adjacent structures increase to the point that a form of "second phase" coalescence can be invoked, leading to formation of larger disk-shaped or cylindrical micelles (Figure 5.3a). As the concentration process continues, hexagonal close-packed arrays of cylinders may appear (Figure 5.3b), producing the hexagonal or normal middle phase. The next step in the process is the coalescence of the adjacent, mutually parallel cylinders to produce the neat phase, characterized by lamellar bilayer structures separated by solvent phases (Figure 5.3c). Under some circumstances, especially in the presence of nonaqueous solvents, a reversed hexagonal phase will appear that contains close-packed cylindrical arrays, but with the internal region of the cylinders containing the aqueous phase (Figure 5.3d).

In some surfactant systems, more complex phase behavior involving one or more viscous isotropic structures will appear. Such phases usually exhibit an X-ray pattern characteristics of a cubic lattice. Such phases are now recognized as the primarily cubic bicontinuous phases introduced above. More recent research with bulky surfactant molecules has led to the suggestion of "wormlike" or ribbon micelles that may be best described—conceptually, at least—as "super"aggregations of smaller micellar units or twisted hexagonal systems.

The conditions of temperature and concentration that produce the various solution phases in surfactant–solvent systems can be determined (with a great deal of laboratory work) by the construction of phase diagrams. The construction–interpretation of such diagrams is a complex undertaking that is beyond the scope of this work. However, sufficient literature exists to permit certain generalizations that will help in understanding the activity of most reasonably simple surfactant systems.

Most of the published work on the phase behavior of anionic surfactants has dealt with the simple carboxylic acid soaps, with less information available on the sulfates, sulfonates, and similar compounds. It is generally found, however, that certain trends hold over a wide range of products so that one can predict events with some degree of confidence.

For a typical anionic surfactant the micelles remain approximately spherical or ellipsoidal over a substantial concentration range above the cmc, but ultimately they become rodlike as the concentration continues to increase. At concentrations in the range of 20–30% by weight, a new phase normally appears that is birefringent and quite viscous. X-Ray diffraction studies show that this phase consists of many long, parallel, rodlike aggregates arranged as illustrated in Figure 5.3b. The aggregate interiors are apparently rather fluid, resembling liquid hydrocarbon in many respects. This phase is what is classically referred to as a "liquid crystal" in that it possesses a substantial degree of order in at least one dimension while not being truly crystalline. As indicated previously, the usual terminology for such a structure is the normal hexagonal or simply the hexagonal phase. In the soap industry it is traditionally referred to as the "middle phase."

As the surfactant concentration continues to increase, it may become energetically favorable for the surfactant molecules to arrange themselves into a bilayer structure, the lamellar phase (the "neat" phase to the soapmaker) (Figure 5.3c). Continuation of the process of increasing the surfactant concentration may lead to the formation of the reversed or inverted hexagonal phase, which again involves an array of rodlike aggregates, but now with the rod interiors as the aqueous phase (Figure 5.3d). It is in the region of transition between the normal and the reversed hexagonal phases that the cubic bicontinuous phase appears.

In terms of surfactant structure, one can expect that a more hydrophilic head group will tend to delay the formation of the hexagonal and subsequent mesophases as a result of charge repulsion among adjacent molecules in the micelle. While the effect of changes in counterion will usually be small, if a significant degree of ion binding occurs for a given head group, thereby reducing its hydrophilicity, one can expect the appearance of the hexagonal phase at lower surfactant concentration. Within an homologous series, it is generally found that the appearance of the liquid crystalline phases will occur at lower concentrations for higher members of the series, paralleling the normally observed decrease in cmc.

Much less work has been reported on the phase behavior of cationic surfactants. In general, the phase diagrams for simple quaternary ammonium halides closely resemble those of anionic surfactants. In some cases, it has been found that the phase behavior of such materials is much more sensitive to counterion effects than is that of anionic species.

Dodecyltrimethylammonium chloride in water, for example, shows the presence of two distinct viscous isotropic cubic phases at room temperature. Similar results are found for the decyl and tetradecyl analogs, while the hexadecyl and octadecyl members of the series do not exhibit these phases. The cubic phases are apparently absent in the corresponding bromides of all members of the series.

For a nonionic surfactant of the polyoxyethylene class, the situation is more complex. In a relatively pure sample of such a surfactant one may encounter not only the usual hexagonal and lamellar phases but also one or more isotropic liquid phases. It may be that such phases contain disk-shaped micelles resulting from disruption of the extended lamellar phase by the large steric requirements of the hydrated POE head group. Similar effects have been reported for anionic surfactants at sufficiently high electrolyte concentrations or in the presence of oil-soluble alcohols. In those cases, some form of association phenomena (e.g., ion binding and/or some form of molecular complex formation) might be involved.

In general, the phase behavior of a POE nonionic surfactant (Figure 5.4) is more sensitive to surfactant structure than in the ionic case. Since the vast majority of such materials are, in fact, mixtures of POE chains of various lengths, phase diagrams lose a great deal of their theoretical utility, even though they may still be useful from a practical standpoint. While reproducible results can be obtained for a given sample of surfactant, another material of nominally the same structure may produce different results due to differences in POE chain distributions. As a result, it is not always a safe practice to extrapolate results for one sample to another of nominally the same material, even that provided by the same manufacturer.

Not surprisingly, in nonionic systems temperature can have a much greater effect on phase behavior than in ionic materials. As the temperature of a solution of the nonionic surfactant is increased, the concentration at which the hexagonal phase appears decreases, as does the concentration at which transformation from hexagonal to lamellar occurs. Because the POE surfactant is solubilized in water by hydration of the POE links, the higher temperature reduces the degree of hydration at a

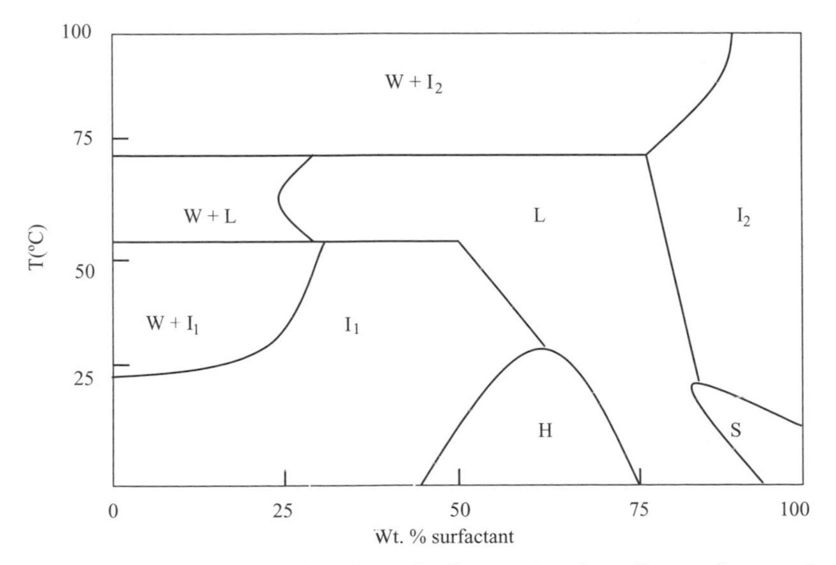

Figure 5.4. Schematic representation of a typically complex phase diagram for a nonionic POE surfactant (W—water; H—hexagonal; L—lamellar; I1,I2—isotropic phases; S—solid), in which phase I1 would normally be a micellar solution.

given concentration, reducing the effective size of the POE group, and facilitating the formation of the lamellar aggregate. The cloud point for the nonionic surfactant will lie at the upper boundary of the liquid isotropic phase for a given surfactant concentration.

It may be noted that the changes in a surfactant system that lead to liquid crystal transformation appear to go hand in hand with changes that affect the size or hydrophilicity of the head group: namely, changes in hydration, changes in the degree of counterion binding, electrical charge screening, and so on. It will be remembered from Chapter 4 that the same factors are found to effect changes in simple micellar solutions. In summary, changes that reduce the effective area of the head group cause the system to be oriented away from spherical aggregates and toward cylinders, lamellar structures, and reversed micelles.

5.3. TEMPERATURE AND ADDITIVE EFFECTS ON PHASE BEHAVIOR

The particular aggregate phase of an amphiphilic material present at a particular concentration will depend on molecular structure, temperature, ionic strength of the aqueous phase, and the presence or absence of other solutes, especially other weakly surface-active materials such as esters, amides, and alcohols. For a typical anionic surfactant, an increase in temperature will usually increase the surfactant concentration at which the hexagonal phase appears. This reflects the increased solubility of the monomeric surfactant, the higher cmc, and the greater thermal mobility and solubility of the hydrophobic tail. An increase in ionic strength of the solution will generally have the opposite effect; that is, the hexagonal phase will appear at lower concentration.

A simplified explanation for those observations is that the increased ionic strength reduces the charge repulsion between adjacent head groups, facilitating the closer packing of molecules into the rodlike aggregates. In effect, the surface area requirements of the head group are reduced, leading to the effect predicted by geometric considerations discussed in Chapter 4. A similar explanation has been invoked for the effect of low concentrations of short-chain alcohols. The nonionic species can pack between the larger surfactant molecules, decreasing the effective dielectric constant of solution in the area of the charged groups and serving to insulate somewhat the neighboring charges, permitting higher packing densities and the transition to aggregate structures with a smaller radius of curvature.

More hydrophobic additives such as free fatty acids and their esters and amides, long-chain monohydric alcohols, and nitriles may have an even more dramatic effect on the phase behavior of a surfactant due to solubilization phenomena. The general subject of solubilization in micellar systems is discussed in Chapter 6. For now, we will focus on the effects that the presence of solubilized materials may have on liquid crystal phases.

Four main classes of solubilizates can be defined on the basis of the overall nature of the additive. These are the completely hydrophobic materials, such as hydrocarbons and halogenated hydrocarbons; weakly surface-active materials, such as nitriles, methyl esters, ketones, and aldehydes; more surface-active monohydric alcohols; and fatty acids. As one might expect, each class has a distinct effect on the phase transitions in a ternary surfactant system.

For hydrophobic additives, one generally finds that an increase in the relative concentration of the additive results in an increase in the range over which one encounters liquid isotropic and middle phases, leading ultimately to a viscous isotropic mesophase. When the additive possesses a slightly polar group (ester, amide, etc.), the solubilizing capacity of both the liquid isotropic and middle phases may be increased significantly, resulting in the formation of a lamellar neat phase at high additive concentrations.

When the additive is a monohydric alcohol, a much more complex system may be encountered, often with the formation of the lamellar neat phase at water concentrations much higher than normal. It is also common to find a second liquid isotropic phase in which the alcohol becomes a secondary solvent. Additional complex mesophases of indefinite structure may arise, leading ultimately to a reversed middle phase. In the presence of fatty acids, the phase diagram will resemble that of the monohydric alcohols except that the liquid isotropic solution in the acid will usually be found to incorporate more water.

Obviously, the phase behavior of a surfactant is a complex matter that may significantly affect its activity in a given application. While the discussion above is highly abbreviated, it should serve to illustrate again the great importance that surfactant structure and environment can have in complex (and sometimes simple) systems.

5.4. SOME CURRENT THEORETICAL ANALYSES OF NOVEL MESOPHASES

The attractive, simple picture of amphiphilic aggregation structures is, as already noted, blurred by reality. The classical method for studying the phase behavior of surfactants has been through the construction of phase diagrams—a delicate and laborious process that requires care and dedication (read: "many ready and willing graduate or undergraduate hands"). Even then, the interpretation and application of the results to predictive theory requires care, knowledge, intuition, and significant mathematical ability (and computer power). In practice, it is still not possible to accurately predict a phase diagram from the molecular composition and structure of an amphiphile alone. That goal is thus far beyond our reach, in part because we still do not understand the finer points of such factors as specific ion effects governing the delicate balance of water–hydrophobe or electrostatic and dispersion force interactions at the molecular level. Nor do we have a really good grasp of membrane rheology, (e.g., bending elastic moduli and undulation forces) or entropic contributions that predominate in the self-assembly process. The basic problem remains the impossibility of deducing a complete theoretical phase diagram from first principles, even for a simple model amphiphile. The problem lies in the impossibility of writing down a complete Boltzmann-type distribution of all possible membrane geometries and topologies, appropriately weighted by their energies and entropies.

An important question in the analysis of mesophase structures is that of the relative stabilities of intermediate versus bicontinuous mesophases. Most analyses of the available data suggest that intermediate mesophases of unusual topologies and geometries fall into three topological classes: noncubic bicontinuous sponges, branched bilayers, and punctured bilayers. A fourth possible class, "ribbon" mesophases, consist of geometric distortions of columnar micelles in hexagonal mesophases. Experimental studies of surfactant–water phase diagrams suggest that intermediate structures form in place of bicontinuous cubic mesophases once the surfactant chains exceed a certain length or their rigidity is enhanced, such as by replacing hydrocarbon with fluorocarbon chains.

5.5. VESICLES AND BILAYER MEMBRANES

The association of surfactants into relatively simple aggregate structures such as spheres, ellipses, and disks allows for a reasonably straightforward analysis of the fundamental aspects of their structure, including their kinetics, thermodynamics, and geometric considerations. The simplest extension of the simple micellar structures, namely, the rodlike micelles often encountered in systems of ionic surfactants in solutions of high salt content, presents a number of theoretical difficulties. Such structures are large (relative to spherical systems) and polydisperse, with no theoretical limit on the length that can be attained. Their average aggregation number is also very sensitive to the total surfactant concentration, so that the properties of the system do not always lend themselves to easy analysis. In general, their unusual properties result from the large dissymmetry in the dimensions of the structural unit and the effects of the ends of the rods, where the associated molecules are forced to pack into hemispherical caps.

As predicted by the geometric approach to aggregation, amphiphilic materials that cannot readily pack into neat, closed structures such as simple micelles are exactly those that are found to produce larger units such as vesicles and extended bilayers. Such materials will have relatively small head groups, or, as is more common, their hydrophobic groups will be too bulky to be packed in a manner necessary for normal micelle formation. Such a state of affairs is particularly common for molecules having more than one hydrocarbon chain, very highly branched chains, or structural units that produce molecular geometries incompatible with effective packing into highly curved structures.

Although extended planar bilayers are a thermodynamically favorable option for the association of some bulky surfactants in aqueous solution, under certain conditions it is more favorable to form closed bilayer systems, leading to the existence of membranes and vesicles. Such a situation arises from two basic causes: (1) even large, highly extended planar bilayers possess edges along which the hydrocarbon core of the structure must be exposed to an aqueous environment, resulting in an unfavorable energetic situation; and (2) the formation of an infinitely extended structure is unfavorable from an entropic standpoint. The formation of spherical closed vesicles, then, addresses both those factors: the edge effect is removed by the formation of a closed system, and the formation of structures of finite size overcomes much of the entropic barrier. As long as the curvature of the vesicle is gentle enough to allow the packed molecules to maintain close to their optimum surface area, vesicles will represent viable structures for the association of surfactants and related materials.

Over the years it has been confirmed that geometric factors control the packing of surfactants and lipids into association structures. The concept has already been introduced, but warrants repetition in the current context for clarity. The packing propensity of a given amphiphilic structure can be conveniently given by the critical packing parameter, denoted here as P_c, and given by

$$P_c = \frac{v}{a_o l_c} \tag{5.1}$$

where v is the volume of the hydrophobic portion of the molecule, a_o is the optimum head group area, and l_c is the critical length of the hydrophobic tail, effectively the maximum extent to which the chain can be stretched out, subject to the restrictions of bond lengths and bond angles. The value of P_c will determine the type of association structure formed in each case. The structures to be expected from molecules falling into various "critical packing" categories are summarized in Table 5.1.

Examples of materials that do not fit neatly into such a scheme may be found, but the general concepts are usually found to be valid. For surfactants and

TABLE 5.1. Expected Aggregate Characteristics of Amphiphiles as Determined by Their Molecular Structure and Packing Parameter P_c

General Surfactant Type	P_c	Expected Structure
Simple surfactants with single chains and relatively large head groups	<0.33	Spherical or ellipsoidal micelles
Simple surfactants with relatively small head groups, or ionic materials in the presence of large amounts of electrolyte	0.33–0.5	Relatively large cylindrical or rod-shaped micelles
Double-chain surfactants with large head groups and flexible chains	0.5–1.0	Vesicles and flexible bilayer structures
Double-chain surfactants with small head groups or rigid, immobile chains	1.0	Planar extended bilayers and cubic bicontinuous phases
Double-chain surfactants with small head groups, very large and bulky hydrophobic groups	>1.0	Inverted micelles

other amphiphiles that form bilayer structures, several generalizations have been found useful that make it easier to understand the geometric consequences of the structure of the amphiphile:

1. Molecules with relatively small head groups, and therefore large values for P_c, will normally form extended bilayers, large (low-curvature) vesicles, or inverted micellar structures. Such results can also be brought about in many "normal" anionic systems by changes in pH, high salt concentrations, or the addition of multivalent cations.
2. Molecules containing unsaturation, especially multiple cis double bonds, will have smaller values for l_c and thus will tend toward the formation of larger vesicles or inverted structures.
3. Multichained molecules held above the melting temperature of the hydrocarbon chain may undergo increased chain motion, allowing trans–gauche chain isomerization, reducing the effective value of l_c and resulting in changes in aggregate structures.

These generalizations on assemblies of surfactants and other amphiphilic molecules offer a broad view of the types of structure that may be formed as a result of the self-assembly process. They consider only the fundamental relationships between structure and the geometric characteristics of the molecules involved. Not considered are any effects on the systems that may exist as an indirect result of curvature or other distortions of the molecular packing. The interested reader can obtain more in-depth discussions in the works cited in the Bibliography. The unique characteristics of the bilayer and vesicle assemblies have attracted the attention of scientists in many disciplines for both theoretical and practical reasons. The following brief discussion only skims the surface of what is sure to become an even more interesting and important area of surfactant-related surface science.

5.5.1. Vesicles

Many naturally occurring and synthetic surfactants and phospholipids cannot undergo simple aggregation to form micelles because of the structural characteristics outlined above. When dispersed in water, they will spontaneously form closed layered structures referred to as *liposomes* or *vesicles*. Such structures are composed of alternating layers of lipid or surfactant bilayers separated by aqueous layers or compartments arranged in approximately concentric circles (Figure 5.5a). If the spontaneously formed multilayer vesicles are subjected to ultrasound or other vigorous agitation, the complex multilayer structure may be disrupted to produce a single bilayer assembly consisting of a unilamellar vesicle in which a portion of the aqueous phase is encapsulated within the single-bilayer membranes (Figure 5.5b). In essence, an assembly resembling a biological cell is produced, although the cell wall is composed of the amphiphilic material without all of the addenda present in biological systems. Typically, a vesicle so produced will have a diameter of 30–100 nm, falling within the size range of classical colloidal systems.

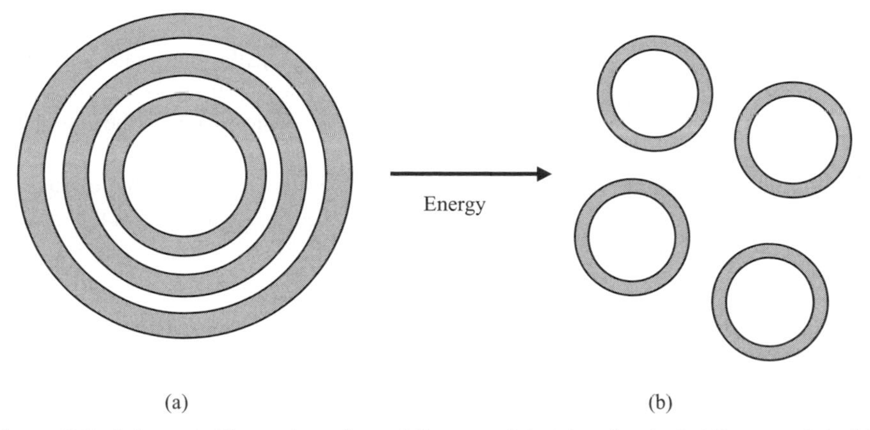

Figure 5.5. Schematic illustration of a multilayer vesicle (a) and a single bilayer vesicle (b).

Like biological systems, however, the vesicle is by nature compartmentalized, which makes it amenable to the inclusion of additives in the three phases present: the external aqueous phase, the hydrophobic interior of the bilayer, and the aqueous internal phase. That availability of "carrying capacity" has made vesicles natural candidates as delivery systems in pharmaceutical, cosmetic, and various other industrial applications. Who hasn't heard of the magical "liposomes" that make the wrinkles of aging disappear? As will be seen below, other "newer" mesophases, such as the cubic bicontinuous phase or "cubosomes," are now beginning to join the corps of amphiphile aggregate workhorses.

Natural and synthetic amphiphiles that form vesicles are inherently of limited solubility in aqueous systems, so that the exchange of individual molecules from the bilayer is often relatively slow. In addition, the bilayer structure has a significant degree of internal stability so that vesicles, once formed, can maintain their original structures for a significant amount of time. Lifetimes of from a few days to several months are common, as would be required of systems designed for drug or cosmetics delivery systems. After a period of time, dictated primarily by the composition of the system, the unilamellar vesicles will usually begin to fuse to produce the more complex aggregate structures of the original systems. The rate of that fusion process can be controlled in several ways, including selection of the appropriate amphiphile structure, the use of mixed amphiphiles, the addition of nonamphiphilic materials that incorporate into the vesicle wall to provide stability, or the use of amphiphiles with unsaturated hydrocarbon chains that can be subsequently cross-linked or polymerized to "fix" the structure.

As mentioned, one of the interesting and useful characteristics of vesicles is their ability to entrap within the assembly a portion of the aqueous phase present at the time of their formation, along with any solute present at the time. It is then possible to alter the composition of the external aqueous phase after vesicle formation by dialysis, diafiltration, or other related purification techniques. Vesicles, therefore, represent a unique microencapsulating technique. Oil-soluble materials can also be incorporated into vesicle systems, although they would naturally be located inside the hydrophobic portion of the vesicle membrane, like materials solubilized in conventional surfactant micelles (see Chapter 6). The potential for the incorporation of both aqueous and nonaqueous additives into vesicles poses the interesting possibility of producing a system containing two or more active components: (1) a component soluble in the internal water phase, the oil-compatible component in the hydrophobic portion of the bilayer; and (2) another component in the external aqueous phase, for simultaneous delivery.

Other interesting and potentially useful physical characteristics of conventional vesicles include their potential activity as microosmotic membranes, their ability to undergo phase transitions from liquid crystalline to a more fluid state, their permeability to many small molecules and ions, especially protons and hydroxide, and their potential applicability as models for biological membranes.

5.5.2. Polymerized Vesicles

Because of their unique properties, vesicles and related lamellar structures are seen as potentially useful vehicles for various controlled delivery and encapsulation processes and as potential pathways to extremely high-resolution lithographic systems for microelectronics applications, as well as serving as good models for naturally occurring membrane systems. A number of reviews and books published more recently cover the current state of the art for such applications in great detail. Major barriers to the use of conventional vesicles in such applications include the inherent long-term instability of the systems, their potential for interaction with enzymes and blood lipoproteins, and their susceptibility to the actions of other surface-active materials. The latter effects especially limit the use of vesicles for oral drug delivery, since the bile acids in the digestive tract may lead to a rapid degradation of the amphiphiles, resulting in an undesirably rapid drug release. For such critical applications as controlled-release drug delivery, even the most stable systems with a nominal shelf life of several months do not begin to approach the usual requirements of pharmaceutical applications.

As a result of the potential utility and relatively low cost of vesicles, a great deal of effort has been applied to the development of polymerized surfactant and phospholipid systems. Covalently crosslinking the vesicle membrane after the encapsulation process should produce a system in which the basic nature of the vesicle as an encapsulating medium is retained while the structural integrity and increased stability of a crosslinked polymeric structure are added.

The general approach used to attain such structures has been the synthesis of conventional vesicle-forming amphiphilic materials containing easily polymerizable functionalities in the molecule. After vesicle formation, subsequent polymerization, preferably by some nonintrusive means such as irradiation, produces the final encapsulated system. The polymerizable functionality can be located at the end of the hydrophobic tail, centrally within the tail, or in association with the ionic or polar head group (Figure 5.6). The choice of a preferred structure will be determined by the requirements of the system and the synthetic availability of

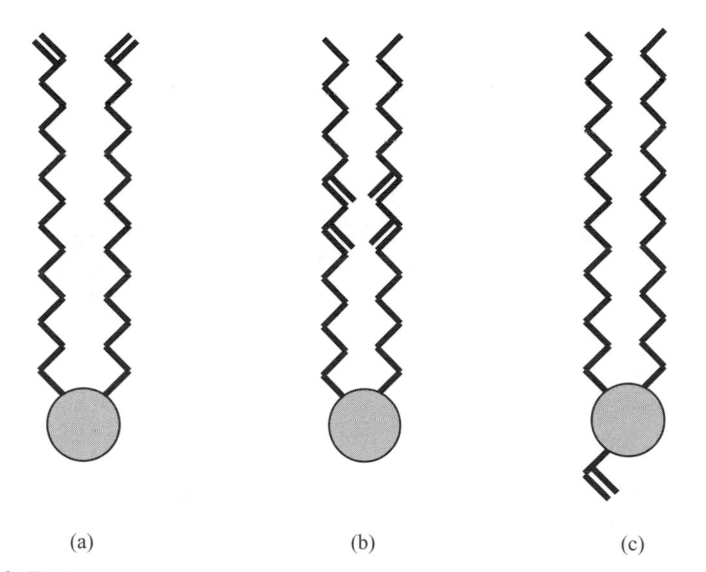

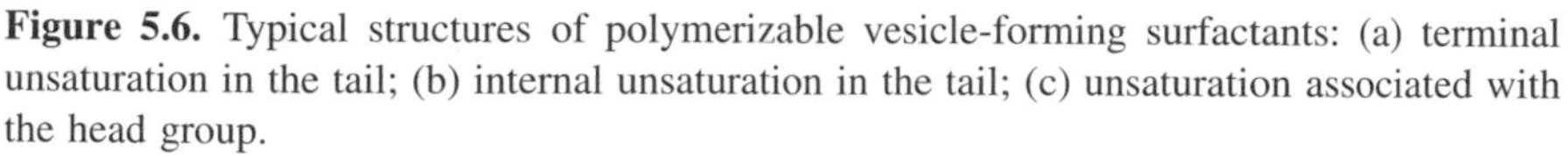

Figure 5.6. Typical structures of polymerizable vesicle-forming surfactants: (a) terminal unsaturation in the tail; (b) internal unsaturation in the tail; (c) unsaturation associated with the head group.

the desired materials. Drug delivery after polymerization would be regulated by the diffusion rate of the active component through the vesicle membrane or by the rate of membrane breakdown by physiological processes.

Considering the impact of amphiphilic species and aggregated structures in biological systems, it has long been a goal of workers in many allied fields to develop a well-characterized synthetic model of biological membranes. The development of new knowledge and techniques in the areas of vesicles, bilayer and multilayer membranes, cubic bicontimuous phases, and their polymerizable analogs provides new opportunities for research in those areas. Research using model systems may be able to provide answers to questions related to natural processes that have so far eluded the research community because of the complexity and intractability of the natural systems. It may also provide new approaches to the development of new "silver bullet" drugs for the treatment of old problems such as cancer, AIDS (autoimmune deficiency syndrome), and the like. With time, the skills of human science may finally begin to approach those of nature in producing the systems and processes capable of mimicking and actively modifying biological reality to the benefit of humankind.

5.6. BIOLOGICAL MEMBRANES

There has been a dramatic increase in interest in the molecular structure of biological membranes. While model systems composed of artificially prepared (or isolated) amphiphilic materials and associated colloids serve a very useful purpose, a

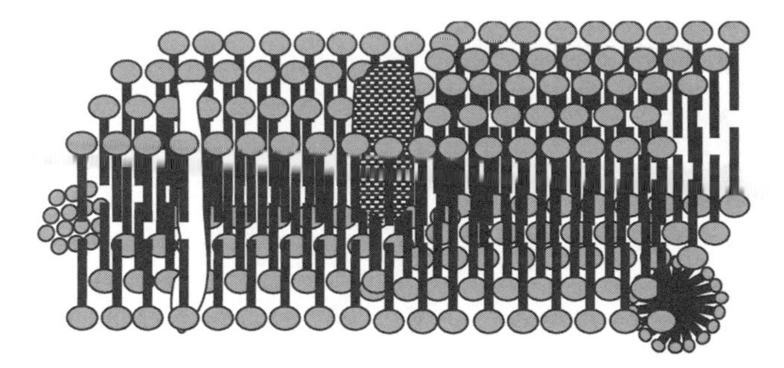

Figure 5.7. A schematic representation of a typical biological membrane including some of the many complex components generally found in such aggregate structures.

better understanding of the reality of biological systems would be invaluable in many areas such as biochemistry, medicine, and pharmaceuticals. Although it is reasonably easy to determine the constituents of a biological membrane, elucidating just how the various components are put together, how they interact, and their exact function within the membrane represents a decidedly more difficult task (Figure 5.7).

New techniques for characterizing the nature of aggregate systems at the colloidal level have opened a wider crack in the door to understanding the finer workings of biological cells and other related structures in life processes. Knowledge gained over the years in terms of cell membrane composition has given way to knowledge of exactly how the various components function as a unit to carry out necessary biological functions.

Aggregated amphiphilic structures in biological systems have been found to include mesh, bicontinuous, and columnar or hexagonal mesophases. Reviews of older data on mesh and bicontinuous mesophases, insights provided by the newer techniques into their relationship to simpler discrete micellar, hexagonal, ribbon and lamellar phases, and analysis of geometries and topologies have significantly clarified our picture of the how's and why's of biological cell function. The standard picture that has developed proposes a gradual evolution of mean curvature in both type 1 and type 2 mesophases within the cell wall that results in the production of a workable system. Those curvatures depend on local conditions such as solvent and solute concentration, specific-ion effects (e.g., Na^+ vs. Ca^{2+}), temperature to a greater or lesser extent, and the presence of specific "activating" chemicals in the vicinity of the wall membrane.

5.6.1. Some Biological Implications of Mesophases

The biological importance of mesh and other intermediate mesophase structures goes far beyond the finer details of phase studies. Most studies of biological membranes have historically focused on membrane proteins and their influence on biological function. Until relatively recently, that emphasis has been justified by the availability of techniques for the study of protein structure. Newer techniques, however, have made it possible to investigate in depth the equally—or perhaps more—important role of amphiphiles in the control of membrane topology and function. Some significant experimental results on membrane fusion and dissolution have appeared that reinforce suggestions about the importance of lipids in determining membrane form and function.

Direct observations of model physiological lipid membranes using atomic force microscopy (AFM) have shown that carefully prepared bilayers deposited on silicon substrates using Langmuir–Blodgett techniques spontaneously form punctures, tunnels, or channels passing through the structure. Bilayers formed by sequential monolayer deposition (A followed by B), for example, were found to form approximately circular bilayer "punctures," channels, or defects occupying 10–20% of the membrane area. The spontaneous formation of such "spontaneous" openings traversing the membrane, if they are indeed spontaneously formed, would seem to indicate a natural propensity for the existence of such defects and remind one of the similarity to channels in natural biological membranes used for the transfer of ions and small molecules into and out of the cell interior. The potential relevance of these "defects" to biological membranes is therefore hard to deny.

Low-temperature TEM images of biological membranes also suggest the importance of unusual "intermediate" membrane topologies in biological processes and the possibility of topological and geometric control of membranes by biochemical variations of the bilayer composition. Experiments have been carried out that explore changes in membrane morphology on the addition of surfactants with small shape parameters ($P_c < 1$) to standard bilayer-forming lipids in which $P_c \geq 1$.

More recent work on biologically relevant systems demonstrates the potential importance of studies of amphiphile self-assembly processes. That importance lies in its potential for increasing our understanding of biological functions and their obvious implications. Assuming that the conditions used in the studies do not unduly influence the bilayer structure and do not affect the activity of the membrane itself, it is becoming clear that membrane-forming lipids found in biological systems can form unusual intermediate (mesh) structures under biologically reasonable constraints. The data suggest a conclusion of major biological importance—punctures or tunnels can form spontaneously and reversibly in lipid bilayer membranes as a result of subtle environmental changes such as the control of water content alone, partially explaining some mechanisms of cellular processes and potentially pointing to new tools and access points for cell manipulation.

The suspicion that biological membrane activity may be controlled by lipids in addition to the conventional protein-governed mechanisms is beginning to be confirmed. That progress in understanding creates a new motivation for the continuation of careful experimental and theoretical study of mesophases in amphiphilic systems. The full class picture of possible mesophases and structures available for self-assembly processes remains to be taken. Further work is required to clarify the balance between protein and lipid activities as the principal functional controls for biological systems. The classic picture of lipids as passive bricks in the membrane wall, supporting the real biomachinery in the form of proteins and glycoprotiens, will need to be retaken, with a better appreciation of the active role of lipids and other membrane-bound agents.

5.6.2. Membrane Surfactants and Lipids

The surface-active components of biological membranes are referred to as *lipids*, with the majority consisting of double-chain phospholipids or glycolipids. The hydrophobic tails normally contain chains of 16–18 carbons, one of which is generally unsaturated. Those structural features immediately indicate that such amphiphiles will have values of $P_c \cong 1$. Those factors guarantee that the lipids will have significant surface activity and will spontaneously form self-assembled bilayer membranes that can encapsulate or isolate different regions and functions in biological systems (e.g., as vesicles), or influence the curvature and conformation of membranes when incorporated into the overall structure. In addition, the long chain lengths ensure that such lipids will have relatively low solubility in water (as the monomer) and a low cmc, and therefore their assemblies will be stable and remain intact while contacting surrounding fluids. The presence of unsaturation in the hydrocarbon chains also helps guarantee that the structures they form will remain relatively fluid and flexible over a wide, biologically relevant temperature range. In that way, their chemical composition helps ensure the functional viability of the biological structure and the organism of which it is a part under varied environmental conditions.

Size, structure, and fluidity of membrane lipids are also important characteristics because those aspects of the amphiphilic molecules make it possible for them to efficiently pack into a variety of bilayer membrane structures with various degrees of curvature and flexibility. That flexibility makes possible the inclusion of other important components of the cell wall, including proteins, glycoproteins, and cholesterol.

In terms of molecular geometry, one can visualize a mixed amphiphilic system in which one class of lipid having a $P_c < 1$ that will produce a truncated cone shape (Figure 5.8a), while another will have $P_c > 1$ for an inverted truncated cone (Figure 5.8b). Combinations of the two can then accommodate the inclusion of, for example, proteins and cholesterol, while maintaining an overall planar structure (or a given degree of curvature), or increase curvature to produce a smaller associated unit. The situation is shown schematically in Figure 5.8c.

Biological membranes, like micelles and vesicles, are theoretically dynamic structures in which the component lipids and proteins can move about and undulate relatively freely. Nevertheless, the exchange of individual molecules with the surrounding solution will be significantly slower than in micelles, so that the structure as a whole remains intact. It simply wouldn't do to have biological cells falling apart too often. To carry out its biological function, the cell membrane will also have heterogeneous regions of lipids, proteins, or other materials, which may serve as specific binding sites, transport "channels," and similar structures. The components of the entire structure, however, must all have one thing in common—they must be able to associate spontaneously to form the necessary stable assembly of

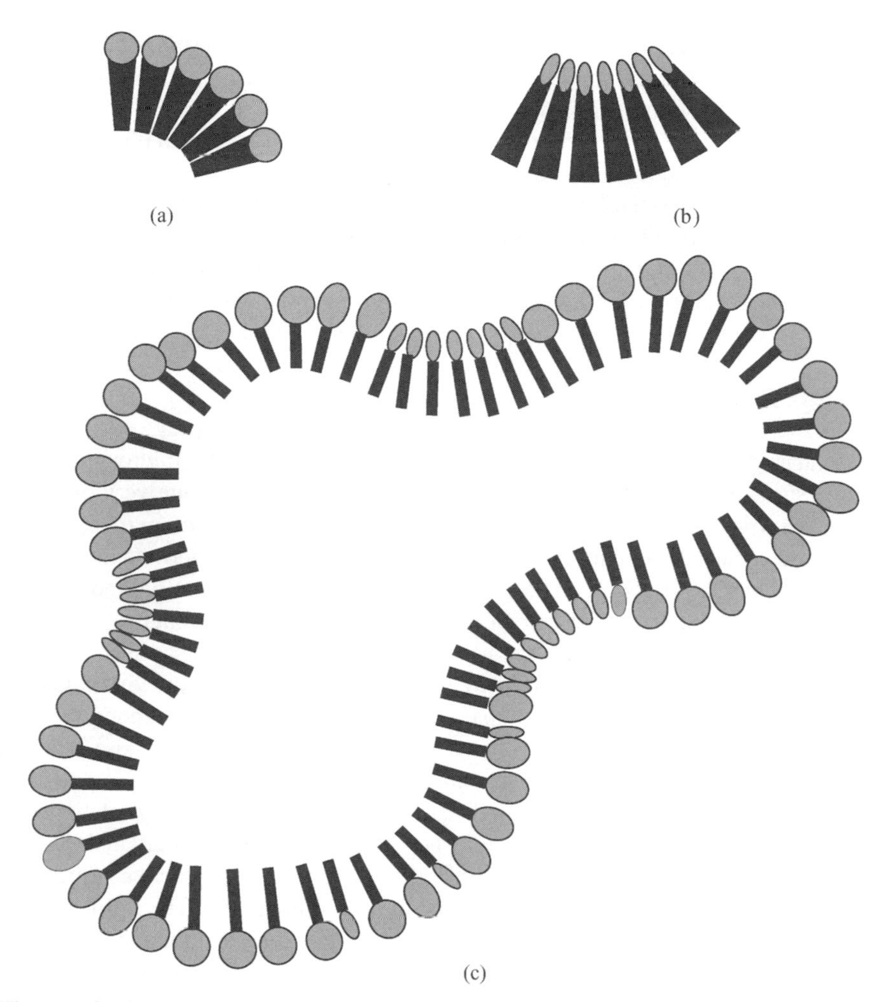

Figure 5.8. Illustration of the role of surfactant molecular geometry in the packing of lipids to form micelles, membranes, and complex cellular structures: (a) a spherical micelle; (b) a reversed micelle; (c) mixed packing to form planar and cellular structures.

molecules to do the job, even when some of the components (e.g., cholesterol and lipids) will not form suitable structures alone. It appears that an organism can "sense" the specific lipid structures needed in a given situation to produce the required membrane structure with the necessary fluidity, surface characteristics, ionic sensibilities, and other properties to carry out its function.

When conditions such as temperature change, the organism often synthesizes new molecules (e.g., more or less saturated fatty acid chains in a lipid) to fit the new conditions. Clearly, the creation and functioning of biological membranes cannot be a haphazard process of trial and error in selecting the proper lipids for a given cell structure. There must be some feedback mechanism through which the organism "knows" what material characteristics are needed under a given set of conditions so that it can be provided when and where called for.

Other aspects of the interactions of lipids and bilayer structures in biological systems can be understood in the context of molecular geometry, association phenomena, and general interfacial interactions. Unfortunately, those topics are too broad to be included here. It will be interesting to see how future research in molecular biology is able to incorporate the fundamentals of surface and colloid science into a better understanding of the function of membranes, cells, and entire organisms.

5.7. MICROEMULSIONS

As indicated above, the status of systems commonly called "microemulsions" among surface and colloid chemists is somewhat uncertain, despite very extensive more recent investigations and discussions. Various experimental approaches have been used in an attempt to ascertain all the details of their thermodynamic and structural characteristics. As a result, new theories of the formation and stability of these interesting but quite complex systems are appearing. Hand in hand with understanding have come ideas for new potentially useful applications. Although a great deal is known about microemulsions, there is much more to be learned about the requirements for their preparation and the relationships among the chemical structure of the oil phase, the composition of the aqueous phase, and the structures of the surfactant and the cosurfactant, where needed. As new data become available, it is clear that any discussion of structure–property relationships between surfactants and microemulsions becomes just as rapidly outdated. Several excellent books and reviews that address the theoretical and practical aspects of microemulsion theory and practice are cited in the Bibliography. The following discussion is limited to the presentation of comparisons among microemulsions and related systems (e.g., swollen micelles and macroemulsions) and some general relationships that have been developed between surfactant structures and microemulsion formation.

The distinction between microemulsions and conventional emulsions is fairly clear. Although emulsions may be kinetically stable for long periods of time, they must all, in the end, suffer the same fate: phase separation to attain a minimum in interfacial free energy. The actions of surfactants, polymers, and other stabilizing aids may shift the rate of droplet coalescence to extremely long times through decreased kinetic rate constants, but the thermodynamic driving force to minimize interfacial area of contact between immiscible phases remains unchanged. Microemulsions, on the other hand, appear to be thermodynamically stable compositions with essentially infinite lifetimes, assuming no change in such factors as composition, temperature, and pressure.

In addition to the thermodynamic distinction usually drawn between macro- and microemulsions, the two classes of colloids differ in several other more tangible characteristics, including the size of droplets formed and the mechanical requirements for their preparation. As far as droplet size is concerned, emulsions generally have relatively large particle diameters, meaning that such systems are usually quite turbid or opaque. Microemulsions, however, normally have droplet diameters of 100 nm or less, many of which are only slightly larger than those in micellar systems. Because those particles are much smaller than the wavelength of visible light, they are normally transparent or slightly bluish.

The energy requirements for the formation of emulsions can be quite substantial. The formation of small droplets requires that the system overcome both the adverse positive interfacial free energy between the two immiscible phases working toward drop coalescence and bulk properties of the dispersed phase such as viscosity. Microemulsions, on the other hand, form spontaneously with little or no agitation required when the proper composition of the system is reached.

When one compares microemulsions and micelles, the demarcation line can become quite blurred and, in some cases, does not exist. As noted earlier, there is some controversy as to the true definition of clear, isotropic solutions of oil, water, and surfactant (and cosurfactant, if needed) as microemulsions rather than swollen micelles. Although the differences between the two systems may appear to be more semantic than physical, several arguments can be put forward that strongly support a differentiation of the two systems. Although there is no need to overemphasize the philosophical aspects of the subject, a synopsis of the current situation may be useful in helping us understand the impact of the molecular structures of the components on the characteristics of the final system.

If one constructs a "spectrum" of the possible situations for the dispersion of one liquid phase in another (e.g., oil in water), the possible sizes of the dispersed phase units range from the molecularly dispersed solution where "droplet" sizes are on the order of a few nanometers, to emulsions with droplet diameters of hundreds or thousands of nanometers (Figure 5.9). Lying between the extremes are

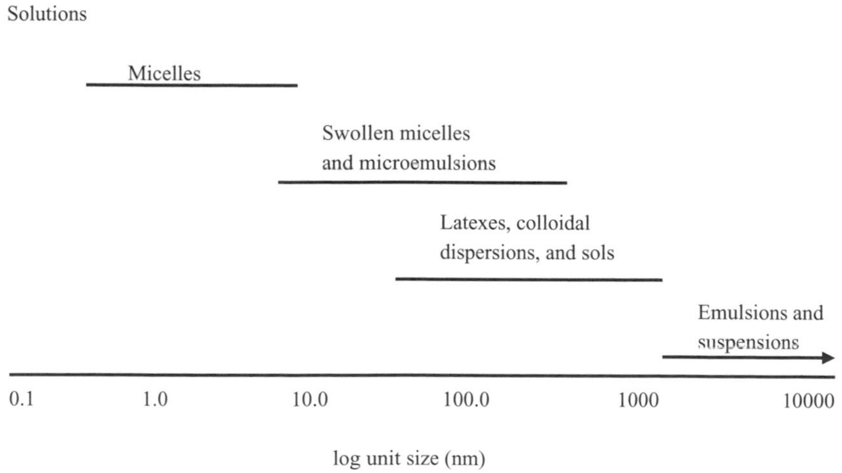

Figure 5.9. A "spectrum" of approximate size ranges for surfactant aggregates and dispersions.

micelles (a few tens of nanometers), macromolecular solutions (tens to hundreds of nanometers), and colloids of several hundred to several thousand nanometers. The systems typically referred to as "microemulsions" will normally have particle sizes between 5 and 200 nm, generally well beyond the range of normal micelles in dilute solution. Classifications such as those illustrated in Figure 5.9 are certainly arbitrary in their ranges and some overlap inevitably occurs; however, the physical differences encountered among most of the different groups are sufficient to obviate any controversy as to their general definitions.

The problem of the thermodynamic differentiation between micelles and microemulsions is less amenable to easy solution. While it is undoubtedly true that, in the smaller size ranges especially, many systems classed as microemulsions are almost indistinguishable from swollen micelles, it is equally true that the larger microemulsion systems far exceed the solubilizing capacity of micelles. Micelles will form under many circumstances, although the specifics of cmc, aggregation number, and other parameters may change with the environmental conditions. The formation of microemulsions, on the other hand, has been shown to have very specific compositional requirements. It is primarily because of those specific demands on the composition of the system and the chemical structures of the various components that the nomenclature for this separate class of dispersed species has developed.

In the literature on the solubilization of hydrocarbons, dyes, and similar compounds in micellar solutions, the ratio of solubilized molecules to surfactant molecules very rarely exceeds, or even approaches, 2. Many microemulsion systems, on the other hand, have been described in which the ratio of dispersed phase to surfactant (and cosurfactant) exceeds 100. Because of the relatively low ratios of additive to surfactant obtainable in micellar systems, it is clear that there can exist no oil phase that can be considered separate from the body of the micelle; that is, the solubilized oil phase is present as individual molecules intimately associated with the micelle structure. In many microemulsions, however, the size of the droplet and the high additive : surfactant ratio require that there be a core of dispersed material that will be essentially equivalent to a bulk phase of the additive. The seemingly obvious conclusion is that the microemulsion systems possess an interfacial region composed primarily of surfactant and cosurfactant, analogous to that encountered in emulsions, but with the difference that the interfacial region is thermodynamically stable.

Extensive work on microemulsions has led to the postulation that the driving force for the spontaneous formation of such dispersed systems is the existence of a transiently negative interfacial tension between the oil and water phases, resulting in a rapid transfer of one of the two phases through the interface, producing the optimum droplet size for the given composition. It must be emphasized that the negative interfacial tension is a transient phenomenon, and at equilibrium must be zero or slightly positive.

As mentioned above, the spontaneous dispersion of one liquid phase in another can occur only if the interfacial tension between the two phases is so low that entropy effects due to the dispersion process can dominate the total energy of the system. Although many surfactants can lower the interfacial tension between oils and water substantially, such factors as micelle formation, solubility limits, and interfacial saturation normally prevent the attainment of the required low values for microemulsion formation. For ionic surfactants in particular, the nature of the head group will usually lead to the formation of rather rigid interfacial films, which limit surfactant mobility in the film and the curvature that may be obtained in the interfacial region. The addition of a cosurfactant, usually an alcohol or amine of short or medium chain length, can serve to reduce the rigidity of the interfacial film and increase adsorption at the interface.

According to the Gibbs equation, the surface or interfacial tension of a system, σ, is related (approximately) to the amounts of surface-active materials preferentially adsorbed at the interface by

$$d\sigma = -\Gamma_i RT\, d(\ln C_i) \tag{5.2}$$

where Γ_i is the surface excess of component i at the interface and C_i is its concentration in the bulk solution. From the equation it is clear that the positive adsorption of any material at the interface will result in a lowering of the interfacial tension. The maximum value of Γ_i attainable for a single-component surfactant system is usually limited by solubility, cmc, or geometric restrictions, so that very few such systems can produce the low values of σ required for spontaneous dispersion. Two notable exceptions to that rule are sodium di-2-ethylhexylsulfosuccinate and some POE nonionic surfactants at temperatures near their cloud point.

If increased adsorption at the interface is needed, then the addition of a material that can circumvent or overcome the negative aspects of the single-component surfactant system may achieve the desired result. From Chapter 3 it can be recalled that the addition of many alcohols of short to medium chain length, referred to above as "cosurfactants," will increase the cmc of ionic surfactants and increase their solubility in the aqueous phase. Both effects work in favor of microemulsion formation as postulated above, especially from the standpoint of the solution properties of the surfactant.

If, in addition to improving the solution properties of the surfactant, the added alcohol can be preferentially adsorbed at the oil–water interface, the third barrier preventing the attainment of very low interfacial tensions can be attacked. Because of the relatively large differences in size between the surfactant and cosurfactant, the alcohol molecules, having a cross-sectional area of only a few square angstroms, can efficiently pack themselves between the larger surfactant chains at the interface. The smaller size and lower hydrophilicity of the hydroxyl group can also moderate the electrostatic and steric interactions among the primary surfactant head groups. The net result is a more densely packed interfacial layer (a much larger positive value of Γ_i), which makes possible very low and transiently negative interfacial tensions. In addition, the mobility of the interfacial layer is increased by the plasticizing (by analogy to polymeric systems) effect of the smaller cosurfactant molecules. This situation is depicted schematically in Figure 5.10.

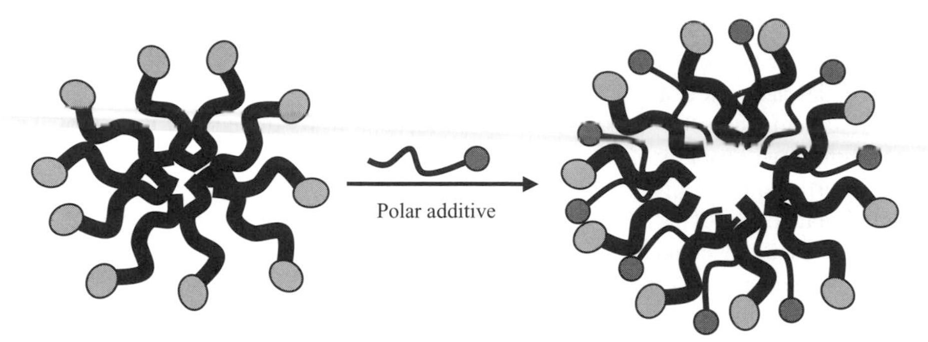

Figure 5.10. Schematic representation of the enhanced packing efficiency of surfactants in the presence of polar additives.

It should be pointed out that although the preceding discussion was concerned with the use of alcohols as cosurfactants in microemulsion formation, many other types of material can also be used to the same end. Especially important are primary amines (commonly used with cationic surfactants) and thiols.

5.7.1. Surfactants, Cosurfactants, and Microemulsion Formation

Microemulsions are composed of two mutually immiscible liquid phases, one spontaneously dispersed in the other with the assistance of one or more surfactants and cosurfactants. While microemulsions of two nonaqueous liquids are theoretically possible (e.g., fluorocarbon/hydrocarbon systems), almost all the reported work is concerned with at least one aqueous phase. The systems may be water-continuous (O/W) or oil-continuous (W/O), as illustrated in Figure 5.11; the result is determined by variables such as the surfactant/cosurfactant system employed, temperature, electrolyte levels, the chemical nature of the oil phase, and the relative ratios of the components.

From Figure 5.11, it is clear that in O/W microemulsions the surfactant tails are most constrained or crowded, while in the W/O case it is the head group that is space-limited. In the case of ionic surfactants the head group crowding will obviously introduce significant electrostatic repulsion among the charges and significantly affect the character of the monolayer. The crowding of the surfactant tails in the O/W case is much less of a problem since such crowding is actually favored, up to a point, by attractive hydrophobic interactions. Such qualitative evaluations have been found to be useful for predicting the "sense" of microemulsions formed, especially for borderline compositions. The use of three- and four-component phase diagrams makes it possible to determine the relationships among the various components with a fair degree of precision and thereby predict the character of the microemulsion to be expected for a given composition.

The character of a microemulsion, or whether one will be formed at all, is critically dependent on the structures of the surfactants and cosurfactants employed

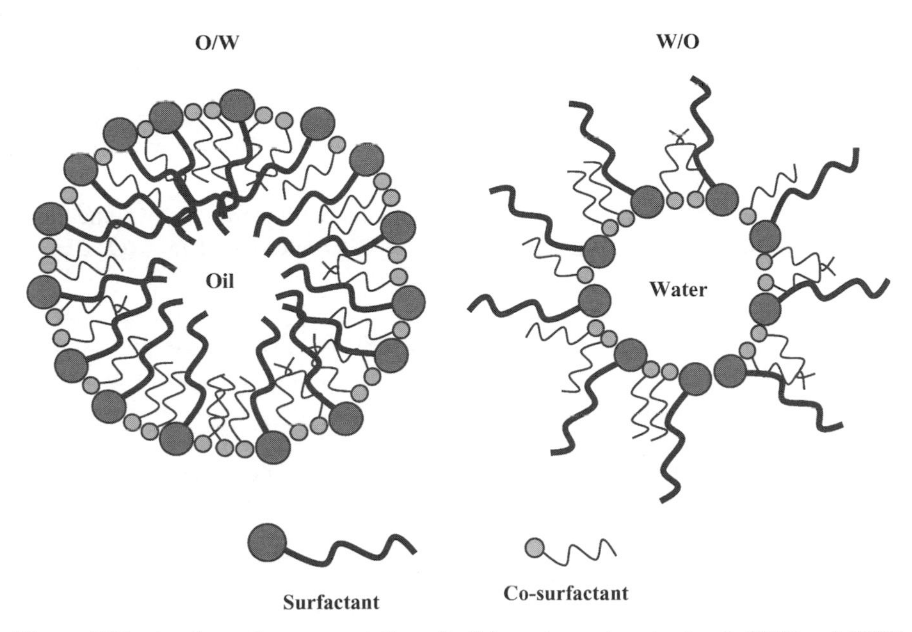

Figure 5.11. A schematic representation of oil-in-water and water-in-oil (O/W and W/O) microemulsions including cosurfactants.

and the relative quantities of each in the system. As mentioned above, most microemulsions, especially those employing an ionic surfactant, will require the addition of a cosurfactant to attain the required interfacial properties to allow for the spontaneous dispersion of one phase in the other. Nonionic surfactants, on the other hand, quite often form microemulsions without the addition of another component.

5.7.1.1. Ionic Surfactant Systems

Using the schematic picture presented in Figure 5.11, it is possible to qualitatively analyze the relationship between surfactant and cosurfactant structures and the most probable microemulsion system formed for ionic surfactants. If the cosurfactant : surfactant ratio is less than 3, the surfactant head groups can approximate a close-packed monolayer for type 1 curvature and the formation of a O/W system is favored. If the ratio is three or greater, close packing of the head groups is not possible and W/O systems are predicted. Longer-chain cosolvents will tend to favor W/O systems, all things being equal, while electrical double layer effects will tend to push for O/W systems.

Although the thermodynamic theory of microemulsions still has some way to go to be more or less complete, a number of generalizations can be made regarding ionic surfactant microemulsions:

1. A cosurfactant is always required to form W/O microemulsions using ionic surfactants, not only to lower the interfacial tension σ but also to reduce head group crowding due to the type 2 curvature.

2. O/W microemulsions require less cosurfactant than W/O systems, all other things being equal (OTBE).
3. Electrical double-layer effects favor O/W systems. As a result, the addition of electrolyte will tend to push the same surfactant/cosurfactant system toward the formation of W/O microemulsions.
4. Increasing the chain length of the cosolvent will tend to move the same surfactant system toward the formation of W/O microemulsions (OTBE).

5.7.1.2. Nonionic Surfactant Systems

Many nonionic surfactants such as the alkylphenol–POE ethers form microemulsions. When compared to systems involving ionic surfactants, a number of important differences are evident:

1. In most case, no cosurfactant is needed, even for pure surfactant samples in which the POE chain has been purified, rather than the normal distribution of chain lengths found in commercial materials.
2. Because of the temperature–solubility relationship for POE nonionic materials, the system temperature becomes an important variable in determining the character of the final microemulsion.
3. Not surprisingly, nonionic microemulsion systems are much less sensitive to electrolytes than are ionic systems, although any effect will be in the same sense as that for ionic systems.
4. For alkylphenol–POE nonionic surfactants of a constant HLB but different alkyl and POE substitutions, an increase in surfactant molecular weight (tail + head group) results in an increase in the amount of oil that can be included in the system before a second phase begins to form. Branching of the hydrocarbon tail, on the other hand, results in a decrease in oil incorporation capacity.

When the length of the POE chain is increased for a given tail, the optimum temperature for solubilization of an oil in aqueous micellar solution is shifted upward. If the temperature of the system for a given POE chain is increased, the hydration of the POE chain by water will decrease, leading in many cases to an inversion in microemulsion type from O/W to W/O.

5.7.2. Applications

The spontaneous formation, almost monodisperse particle size, and thermodynamic stability of microemulsions make them very attractive for applications that involve oil/water mixtures. In many cases, almost transparent systems also provide a degree of a esthetic appeal.

The most important applications of microemulsions to date are probably found in cosmetics and tertiary oil recovery, where aesthetic factors and the attainment of very low interfacial tensions are of prime importance. In the petroleum industry, microemulsions theoretically would allow the difficult to recover residual oil to be pushed efficiently through the rock formation without being retarded by the development of a large pressure drop across the highly curved membranes as given by the La Place equation

$$\Delta p = \frac{2\sigma}{r} \tag{5.3}$$

This equation states that the pressure differential or drop across a curved interface will be directly proportional to the interfacial tension σ and inversely proportional to the radius of curvature r. As the drop size decreases, Δp increases for a given value of σ. Lowering the value of σ helps overcome the increased pressure differential and facilitate the movement of the oil. Although the use of microemulsions in tertiary oil recovery has shown great promise for a number of years, the special conditions of temperature, salinity, and other factors that impact on microemulsion formation and stability place a number of roadblocks in their widespread application.

Other application or potential application areas include cutting oils and special lubricating systems, dry cleaning systems, self-emulsifying oils for plant protection, floor and furniture polishes, leather treatments, cosmetics, and pharmaceuticals.

PROBLEMS

5.1. A very widely used class of surfactants used in the food industry is that of the monoglycerides, especially glycerol monostearate:

$$C_{17}H_{35}COOCH_2CH(OH)CH_2OH$$

The pure (>90% α- or terminally substituted monoglyceride) material is known to have three primary liquid crystalline phases, designated α, β, and β', which differ in the size of the aggregate, its internal structure, and water content. Only one structure, the β' is generally useful in food applications in which a rapid incorporation and activity are necessary. In the absence of any specific knowledge about the three structures, suggest what characteristics you would expect the more useful system to exhibit in order to function well in water-based applications.

5.2. Give three possible technological advantages one might expect a microemulsion to have over a "conventional" emulsion in an intravenous drug application system, assuming that the proper balance of constituents can be found to fit the pharmacological requirements of the system.

5.3. Microemulsions, while often being desirable, can become extremely bothersome when phase separation is the ultimate goal in an industrial process. That can be especially true in some advanced petroleum recovery systems that involve the use of aqueous surfactant systems. Suggest a chemical and a physical process that might be used to "break" a stable crude oil microemulsion resulting from the use of an anionic surfactant/medium chain length alcohol extraction system.

5.4. A newly hired technician was given the job of formulating a simple O/W emulsion for use as a vehicle for a topical antibiotic. After due consideration of the inactive components in the cream, the technician decided to use an alcohol–$(POE)_5$ surfactant that worked well at ambient temperature. However, a sample of the emulsion was left near a radiator and the technician noted a significant change in the viscosity of the warmed emulsion. When cooled down, the sample returned to normal. What might have been happening in the sample to cause the observed changes in the emulsion?

5.5. An O/W emulsion using a soap surfactant was found to invert to a W/O system on the addition of $AlCl_3$. Propose an explanation for the observed change.

5.6. A concentrated surfactant solution containing a hexagonal close-packed mesophase would be appropriate for solubilizing which of the following materials in the discontinuous phase: butyl myrstate, *n*-decanol, or lecithin.

5.7. A system of unilamellar vesicles is formed in an aqueous solution with pH 5. If the average diameter of the vesicles is 25 nm, how many protons will be encapsulated in the average vesicle core.

5.8. A vesicle system prepared with purified soybean lecithin in 1 mM KCl has an inner radius of 15 nm. The system is then diluted with a large excess of distilled water. Estimate the time required for the encapsulated ions to transfer into the outer continuous phase given that the permeability coefficient for both ions is 10^{-11} cm/s and that the dimensions of the vesicles do not change despite the change in osmotic pressure.

5.9. Of the three classes of polymerizable lipids suggested for vesicle work in Figure 5.6, which would probably be most suitable for a polymerization using UV radiation when the internal phase contains a polyunsaturated fatty acid component? Why?

6 Solubilization and Micellar and Phase Transfer Catalysis

In addition to being one of the fundamental thermodynamic consequences of the nature of amphiphilic molecules, micelle formation has a significant practical impact on the application of surfactants in various technological areas. The technologically important phenomena related to micellar solubilization and micellar catalysis deserve special mention in any discussion of surfactant technology, especially in view of the increasing attention they are receiving in both academic and industrial circles. The ability to incorporate inherently insoluble (or only slightly soluble) materials into a solvent system in a stable, reproducible, and readily characterized way has many significant technological implications and applications. Two examples of potentially great economic and social importance include new drug delivery systems and tertiary oil recovery methods. Other areas of application in personal care products, agriculture, medicine, foods, biotechnology, and so on can be conceived of by the innovative scientist.

This chapter focuses on three types of phenomenon that are closely related to the presence of amphiphiles and micelles in solution, and on the roles surfactant structures and other characteristics may play in their application. To exploit the micellar nature of surfactants and to realize their technological potential, it is necessary for the investigator to understand and very carefully control the many variables involved in the various phenomena. It is probable that in the near future we will see a dramatic increase in the use of micellar and related systems to produce better, more effective, more appealing, and (hopefully) cheaper products for the realization of a better world.

From the dawn of science and the "sometimes science" of the alchemists, a major goal has been the emulation of natural processes in bringing about chemical change, in addition to the "classical" desire to change base metals into gold. The quiet prayer of many synthetic chemists has long been that their reaction pathways would someday approach those of nature's enzymes in speed, efficiency, and effectiveness. While modern chemical techniques have made it possible to prepare many unique compounds that do not seem to appear in nature (probably because there is no "natural" need for them), those preparations usually involve such extreme

Surfactant Science and Technology, Third Edition by Drew Myers

measures or produce such low yields that they are in fact little more than scientific curiosities, even if they do add greatly to our basic knowledge of the world about us. On the other hand, a number of natural compounds that can be of great benefit to humans can be obtained synthetically or from biological sources in small quantities and at great expense. We seem to have difficulty reproducing many of the actions of nature's workers, the enzymes. The application of the principles of surface activity and surfactants has allowed science to begin to understand and replicate (roughly) enzymatic processes. The new age of transgenic microbial, plant, or animal production of desired natural products is, of course, one route for circumventing nature. The technological and sociological impact of that new technology remains in question, however.

The following discussions are only brief surveys of the subjects involved and barely begin to address the large amount of theoretical and experimental information available. They are intended to introduce some of the basic concepts involved in each case to enable readers to more readily formulate ideas as to the potential applicability of such phenomena to their own needs. The information presented should serve at least as a good starting point for pursuing those ideas.

6.1. SOLUBILIZATION IN SURFACTANT MICELLES

The increased solubility of organic materials in aqueous surfactant solutions is a phenomenon that has found application in many scientific and technological areas. Only relatively recently has a good understanding of the structural requirements for optimum solubilization begun to develop as a result of extensive experimental and theoretical work. Empiricism is slowly giving way to well-thought-out correlations between the requirements of a system and the chemical structure of surfactant that will provide the necessary environment to promote the solubilization process.

Early work in the twentieth century addressing the mechanisms of micellar solubilization was, unfortunately, often performed with surfactants of questionable purity. As described in Chapter 4, small quantities of impurities and isomeric variations in the structure of the surfactant can have a significant impact on the micellization process and, naturally, on solubilization phenomena related to it. More recently, closer attention has been paid to using the purest or best characterized surfactant systems available, so that more confidence can be placed in the validity and interpretation of experimental results. That is not to say, however, that the pioneering work of the first half of the twentieth century was without merit. To the contrary, modern experimental techniques have done much to confirm the work of that era. Considering the relatively limited resources of the early investigators (compared to the modern chemical laboratory), one can regard their results and interpretations with only the highest respect.

When discussing a subject such as micellar solubilization, it is very important to define exactly what is meant by the term. As is often the case, there is some disagreement within the surfactant literature as to the fine points of the definition of solubilization, particularly as the surfactant : additive ratio decreases and one approaches the nebulous frontier between swollen micellar systems and the microemulsions discussed in Chapter 5 and emulsion regimes to be discussed in Chapter 9. For now, the discussion is limited to systems in which the micelle is clearly the primary vehicle for the observed phenomenon.

For present purposes, solubilization is defined as a spontaneous process leading to a thermodynamically stable, isotropic solution of a substance (the additive) normally insoluble or only slightly soluble in a given solvent produced by the addition of one or more amphiphilic compounds, including polymers, at or above their critical micelle concentration. Using such a definition, we can cover a broad area that includes both dilute and concentrated surfactant solutions, aqueous and nonaqueous solvents, all classes of surfactants and additives, and the effects of complex interactions such as mixed micelle formation. It does not, however, limit the phenomenon to any single mechanism of action.

The history of solubilization research in the first half of the twentieth century has been extensively reviewed, and several pertinent references are listed in the Bibliography. A discussion of some important results can be found in the work of Elworthy, et. al., which also includes a description of many of the experimental techniques that have been developed for investigations into the factors affecting the process.

Although there are many aspects to understanding solubilization phenomena, this discussion is concerned primarily with the correlations that can be made between the molecular structure of a surfactant and its activity and capacity as a solubilizing agent, the related effects of the chemical nature of the additive, and the role of the solution environment. For a specified solvent system, water or aqueous solutions, for example, two variables must be considered in the solubilization process: (1) the molecular nature, purity, and homogeneity of the surfactant and (2) the chemical nature of the additive. From a technological viewpoint, it is important to understand exactly what surfactant structural features serve to maximize the desired solubilizing effect, and the best way to achieve that understanding is through a fundamental knowledge of the molecular and thermodynamic processes involved. In addition, since most technological applications of solubilization (e.g., detergent action) involve complex multicomponent systems, such factors as temperature, electrolyte content, and the presence of polymeric species and other solutes must be examined. Obviously, for such applications as cleaning and detergency, it is not possible to completely specify the system with anticipation; therefore, consideration must always be given to attaining broad solubilizing capabilities, often at the expense of the optimum for a specific "model" set of circumstances.

Before addressing some of the specific aspects of the influence of surfactant structure on solubilization, it will be useful to understand the "geography" of solubilization—that is, the possible positions in (or on) the micelle that can serve as host sites for the additive molecules and the factors that determine exactly where solubilization will occur.

6.1.1. The "Geography" of Solubilization in Micelles

It is well established that the location of a solubilized molecule in a micelle relative to the structural components of the surfactant will be determined primarily by the chemical structure of the additive (Figure 6.1). In aqueous solutions, nonpolar additives such as hydrocarbons are intimately associated with the core of the micelle (Figure 6.1a), while slightly polar materials such as long-chain fatty acids and alcohols, esters, amides, nitriles, and the like are usually located in what is termed the "palisades layer" (Figure 6.1b) lying near the transition zone between the hydrophobic micellar core and the more hydrophilic outer layer of the aqueous micelle. The orientation of such molecules is probably more or less radial, with the hydrocarbon tail remaining closely associated with the micellar core. In some cases, that orientation can potentially have a significant effect on the nature of the system, as

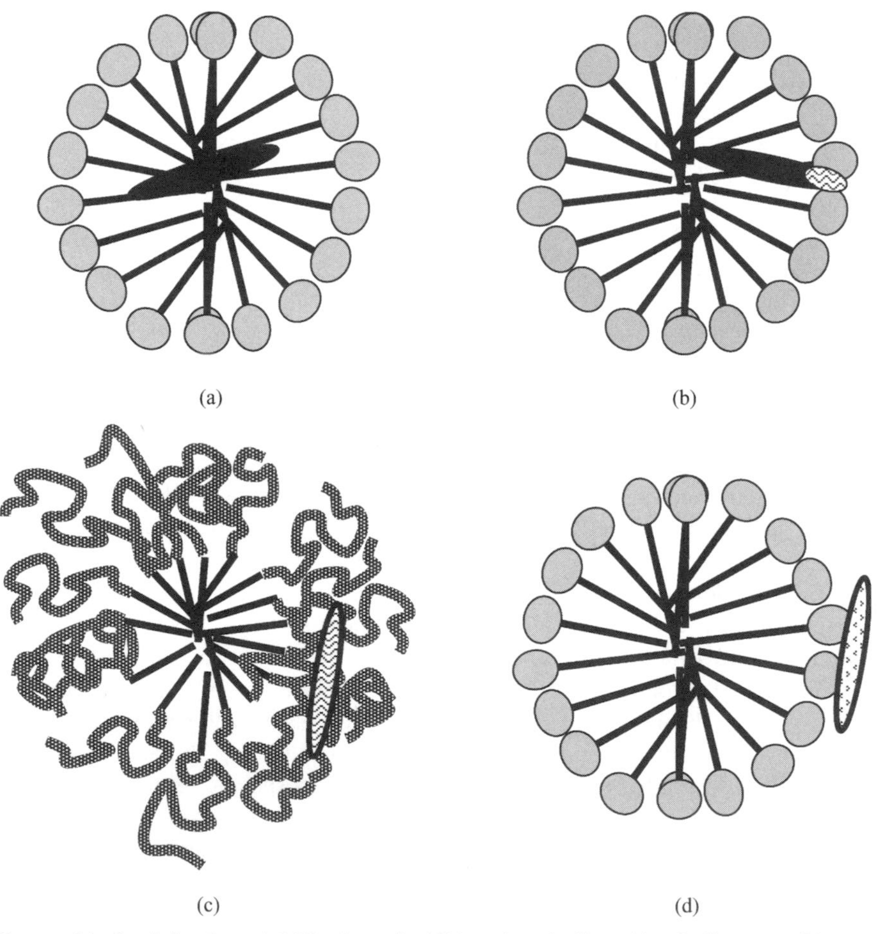

Figure 6.1. Loci for the solubilization of additives in micelles: (a) micelle core; (b) core-palisades interface; (c) surface region (nonionics); (d) on micelle surface (ionics).

discussed in Section 5.7. Other structural factors, such as the charge on the surfactant head group, can significantly affect the locus of solubilization. Materials containing aromatic rings, for example, may be solubilized in or near the core of anionic systems, but in the palisades layer of cationic micelles because of polarization interactions between the aromatic ring and the cationic head group.

In addition to the solubilization of additives in the micellar core and the core–palisades boundary region, they may also be found entirely in the palisades region (Figure 6.1c) and on the micellar surface (Figure 6.1d). The nature of the polar head group of nonionic surfactants, especially the polyoxyethylene derivatives, is such that a relatively large fraction of the micelle volume corresponds to the palisades region. Because of the bulky nature of the POE chain and its attendant solvent molecules, it has been suggested that the hydrophilic chain is arranged in an approximate spiral from the micellar core outward into the solution. As a result, areas of the palisades near the core will be sterically crowded with the POE chains, with relatively little room left for waters of hydration or casual water molecules. As the distance from the core increases, the palisades layer becomes more hydrophilic, acquiring more characteristics of an aqueous solution. The net effect of such a situation is that, deep within the palisades layer, the chemical environment may resemble that of a polyether, so that materials soluble in such solvents will be preferentially located in that region.

Even though chemical structures may dictate the preferred location for the additive, solubilized systems are dynamic, as are the parent micelles, and the location of specific molecules may change over time. It will always be important to remember, then, that while a given region of the micelle may be preferred by an additive on chemical grounds, there is no guarantee that all phenomena related to the system (e.g., catalysis) will be associated with that region.

In surfactant/nonpolar solvent systems where the orientation of the micelle is reversed, the polar interactions of the head groups provide not only a driving force for the aggregation process but also an opportune location for the solubilization of polar additives. Water is, of course, one of the most important potential polar additives to nonaqueous systems, and it is located primarily in the polar core. The nature of such solubilized water is not fixed, however. It has been shown, for example, that in the system benzene/(sodium di-2-ethylhexylsulfosuccinate)/water, the initial water added is tightly associated with the ionic head group of the surfactant (as waters of hydration), while subsequent additions appear to have the character of free bulk water. Other polar additives such as carboxylic acids, which may have some solubility in the organic phase, are probably associated with the micelle in a manner analogous to that for similar materials in aqueous systems.

The effects of solubilized additives on the micellar properties of nonaqueous surfactant systems vary according to the structures of the components. Since such changes are often greater than those found in aqueous solutions, however, care must be exercised in evaluating the effects of even small additions on the aggregation characteristics of surfactants in nonaqueous solvents. Because of the industrial importance of nonaqueous surfactant systems as cutting oils, lubricants, and corrosion inhibitors, a great deal of knowledge about such systems is closely held by the various industrial and governmental organizations involved in the related research. As a result, there are a number of gaps in our understanding of the structural relationships among surfactant, solvent, and additive. Such information it becomes more readily available, might facilitate the extension of current knowledge to new applications.

6.1.2. Surfactant Structure and the Solubilization Process

Earlier chapters introduced some of the wide array of chemical species that exhibit surfactant properties and are potentially useful in solubilization processes. Just as molecular structure is important to such surfactant characteristics as the cmc, aggregation number, and micellar shape, it also controls the ability of a surfactant to solubilize a third component. Conversely, the presence of a third component in a surfactant solution can often affect its aggregation characteristics. It is documented in a number of reports that the presence of a solubilized additive, even though the additive has no inherent surface activity, can change the cmc of a surfactant substantially from that of the pure system. As noted in Chapter 4, the existence of such an effect means that great care must be exercised in the interpretation of experimental data on micellization derived from solubilization results.

Whether micelles formed in the presence of a third component are the same as those formed in its absence is a subject of some controversy. It has been shown that micellar activity may be induced in surfactant solutions below the "normal" cmc in the presence of small amounts of solubilized additives. In some cases such effects have been attributed to additive-induced micellization. In others, effects have been seen at concentrations several orders of magnitude below the cmc, suggesting the presence in solution of submicellar species possessing some properties of the fully aggregated system.

Some researchers have suggested that surfactants in dilute solutions undergo a low level of molecular aggregation at concentrations well below their cmc levels, during which dimers, tetramers, and other "premicellar" aggregates are formed. That may be especially true for surfactants having unusually large or bulky hydrophobic groups, such as the bile acids and tetraalkylammonium halides. Large reaction rate enhancements have been found for such materials when used as phase transfer catalysts (see discussion below), suggesting that they are acting in a micelle-like fashion even though normal micelle formation is precluded by their molecular structure. Some effect due to the formation of tight or solvent-separated ion pair aggregates is usually invoked to explain the observed catalytic results.

Since the cmc's of most surfactants occur at rather low concentrations, evidence of premicellar aggregate formation quite often becomes a question of the interpretation of results lying at the limits of sensitivity and accuracy of many experimental techniques, and, of course, the view of the individual interpreter of those results. There seems to be little doubt that in nonaqueous solvents, the formation of dimer and other lower aggregates occurs readily. Fluorescence and electron spin resonance techniques have also shown the presence of such species in very bulky surfactant systems in water. However, there is presently little unambiguous evidence to confirm the occurrence of premicellar aggregation as a general rule in conventional aqueous surfactant solutions.

The discussions of micelle formation given in Chapter 4 indicated that surfactant properties such as the cmc and aggregation number can be reasonably well correlated with the size and nature of the hydrophobic group. In each case, as the hydrophobic group increases in size for a given head group, the cmc decreases and the aggregation number increases within an homologous series. No comparable relationships have so far been determined that can accurately relate surfactant structure and solubilizing power, mainly because the structure of the additive can play such an important role in the overall aggregation process.

As already mentioned, hydrocarbons and polar organic compounds with low water solubility are usually found to be solubilized in the interior of the micelle or deep within the palisades layer. It has generally been shown that, not surprisingly, the amount of such materials solubilized increases as the size of the micelle increases. As a result, any of the factors discussed in Chapter 4 that cause an increase in micelle size might also be expected to increase the solubilizing power of the system. An increase in the length of the hydrocarbon chain, for example, leads to a lower cmc and larger aggregation number, so that more of a nonpolar additive can be incorporated into the micellar core per mole of surfactant in the system (Table 6.1).

In a study of the solubilization of ethylbenzene in a series of potassium carboxylates ranging from C_8 to C_{16}, it was found that as the concentration of surfactant increased, the amount of ethylbenzene solubilized increased, and that as the length of the carbon chain increased, the quantity of material incorporated per mole of surfactant increased with the carbon chain length. Such results have been criticized because of the assumption that the activity of the monomeric surfactant remained

TABLE 6.1. Effects of Surfactant Hydrocarbon Chain Length on cmc and Solubilization of Ethylbenzene in Potassium Soap Solutions[a]

Surfactant (M)		cmc (M)	Solubilized
$C_7H_{15}COO^-$	(0.48)	0.66	0.141
	(0.83)		0.152
$C_9H_{19}COO^-$	(0.44)	0.17	0.197
	(0.72)		0.233
$C_{11}H_{23}COO^-$	(0.20)	0.046	0.364
	(0.50)		0.407
	(0.86)		0.522
$C_{13}H_{27}COO^-$	(0.24)	0.012	0.745
	(0.50)		0.866
	(0.57)		0.888
$C_{15}H_{31}COO^-$	(0.15)	0.0032	1.15
	(0.29)		1.48

[a]Moles solubilized per mole of soap in micelles.

constant above the cmc so that the amount of surfactant in the micelles could be quantified by subtracting the cmc from the total amount present. It is now known that such an assumption is not strictly valid. Substantially the same results have been found in other studies with different surfactants and additives, however. It was found, for example, that the amount of dimethylaminoazobenzene incorporated by a series of potassium carboxylate soaps appeared to be almost linearly related to the carbon number of the surfactant.

Branching of the hydrocarbon chain of the surfactant usually results in a decrease in the solubilizing power of the system relative to that of the analogous straight-chain material. That is presumably due to geometric and packing constraints, which limit the ability of the micellar core to accommodate the added bulk of the solubilized molecules. The addition of ethylenic unsaturation and aromatic groups also tends to decrease the maximum amount of additive that can be fitted into the core packing arrangement.

In the case of nonionic surfactants, the amount of aliphatic hydrocarbon that can be solubilized generally increases as the length of the hydrophobic tail increases and decreases as that of the POE chain increases. Those results parallel changes in the cmc's and aggregation numbers of the respective materials. Divalent salts of alkyl sulfates quite often exhibit a greater solubilizing capacity than do the corresponding monovalent salts for materials included in the micellar core. That result has been related to the increase in volume of the micellar core of the divalent salts.

The relative solubilizing power of the different types of surfactant with a given hydrophobic tail usually follows the order nonionics $>$ cationics $>$ anionic. The rationale for such a result is usually related to the supposed looser packing of the surfactant molecules in the micelles of the nonionic materials, making available more space for the incorporation of additive molecules.

If one considers additives with a more polar character, which might reasonably reside in or near the palisades layer, fewer generalizations such as those above can be made. The complex interactions among the various components of the system—the surfactant head groups, water (or other solvent) molecules, the exposed micellar core, and the polar group of the additive—appear to be too specifically sensitive to allow for an easy trend analysis.

It was usually found that compounds such as methylisobutyl ether and *n*-octyl alcohol were better solubilized in 0.1 N sodium oleate than in potassium laurate at the same concentration and temperature, contrary to the results for hydrocarbon materials solubilized in the micellar core of the same systems. Octylamine, on the other hand, was incorporated into each to an equal extent. It was also found that the degree of solubilization of 1-*o*-tolyl-azo-2-naphthylamine and related materials in micelles of sodium dodecylpolyoxyethylene sulfates

$$C_{12}H_{25}(OC_2H_4)_xOSO_3^-Na^+$$

where x varied from 1 to 10, increased as the value of x increased. The same material showed no change in solubilization with increase in the POE chain length in the analogous unsulfated nonionic surfactant over the range of $x = 6–20$. That result was attributed to the compensating effects of an increased potential for solubilization due to the increase in the POE chain, and a decrease due to a smaller aggregation number. The solubilization of Yellow OB dye in both the sulfated and unsulfated POE surfactants was greater than that in the corresponding sodium alkyl sulfate.

The addition of a second polar group to a surfactant molecule can either increase or decrease its solubilizing power, depending on the nature of the additive. Studies comparing the solubilizing capacity of the monoesters of maleic acid

$$ROOCCH{=}CHCOO{-}Na^+$$

with those of the disodium salts of the equivalent monoesters of sulfosuccinic acid

$$ROOCCH_2CH(SO_3^-Na^+)COO^-Na^+$$

found that the introduction of the second ionic group decreased the solubilizing capacity for nonpolar additives such as *n*-octane, while that for *n*-octanol was increased. According to the preceding discussion, that result can be explained by the fact that the introduction of the sulfonate group decreases the aggregation number of the micelle, thus limiting its capacity to include the hydrocarbon in the core. At the same time, the bulky ionic groups increased the steric requirements at the micelle surface, increasing the relative volume of the palisades layer available to the more polar *n*-octanol molecules.

The solubilizing power of amphoteric surfactants has not been as widely studied, or at least as widely reported, as that for the simpler ionic and nonionic materials. However, the available data indicate a solubilizing capacity range somewhere between the extremes; the exact results possibly are more sensitive to the nature of the additive than are those for the other classes of surfactants.

6.1.3. Solubilization and the Nature of the Additive

The quantity of a substance that can be solubilized in surfactant micelles will depend on many factors, some of which have already been discussed. From the standpoint of the additive itself, such factors as molecular size and shape, polarity, branching, and the electronegativity of constituent atoms have all been found to be of some significance, depending on the exact system. One of the most extensively explored factors relating the chemical structure of the additive to its solubilization has centered around the relationship between the molar volume of the additive and the maximum amount of material that can be incorporated in a given surfactant solution. Investigations into the solubilization of the hydrocarbons hexane, heptane, and octane, and the aromatics benzene, toluene, ethylbenzene, propylbenzene, and butylbenzene, showed that there existed an inverse relationship between the molecular volume of the additive and the amount of material solubilized. A similar study of polycyclic aromatics in sodium laurate revealed similar results. In each study, linear relationships between the logarithm of the volume of the material solubilized and its molar volume were obtained, although the slopes of the plots differed for the different classes of compounds studied.

In general, the chemical nature of the additive can be classified as either *non-polar*, such as simple hydrocarbons, or *polar*, such as the long-chain alcohols, esters, amides, and nitriles. As is usually found when discussing surfactants, however, such simple classifications are never so nicely clear-cut in reality. In the same way that the hydrophobic character of a surfactant can be varied almost continuously over a wide range, the polar character of a potential additive can be made to span a rather broad spectrum. The aromatic unsaturation of the benzene ring, for example, is sufficiently polarizable that its presence can cause a material to perform like a polar additive, even in the absence of any truly polar functionality, especially in conjunction with cationic surfactant systems.

There does not seem to be a single, simple relationship that correlates a property of a potential additive with its tendency for solubilization in a given surfactant system. Generalizations based on chemical structures are limited and vary a great deal with the characteristics of the complete system—solvent, surfactant, and additive. Examples of such varied results can be seen in Tables 6.2 and 6.3, which list the maximum additive concentrations (MAC) achieved under given conditions of surfactant concentration and temperature. For the case of nonionic surfactants (Table 6.2), it can be seen that as the polarity of the additive increases, the ability of the micelle to accommodate the material increases. Such a result is indicative of solubilization occurring in the palisades layer of the micelle. From the results in Table 6.3, in which various surfactant types can be compared, it is clear that for nonpolar additives, the cationic surfactants are superior solubilizing agents to the two anionic materials tried, and that the unsaturated oleate is better than the related saturated compound on a mole-for-mole basis. For the polarizable aromatic additives, the cationic surfactants lose their advantage, indicating that such materials are prone to polarization by the cation, which results in solubilization outside the core

TABLE 6.2. Maximum Additive Concentrations (MACs) Solubilized in 1% Aqueous Solutions of $C_{10}H_{21}(OE)_{10}CH_3$ Nonionic Surfactant at 27°C

	MAC		
Additive	g/L	mM/L	mol/mol of Surfactant
n-Octane	0.9	7.9	0.48
n-Decane	0.39	2.7	0.17
n-Dodecane	0.16	0.9	0.06
n-Decylchloride	0.45	2.6	0.16
n-Octanol	3.12	24.0	1.47
n-Decanol	2.38	15.1	0.93
n-Dodecanol	2.07	11.1	0.68
n-Decylamine	3.78	24.1	1.48
n-Decanoic acid	2.30	13.4	0.82

TABLE 6.3. MAC Solubilized by Typical Ionic Surfactants Compared to a C_{10} Nonionic Surfactant

	MAC (mol/mol Surfactant)			
Additive	$C_{12}NH_3^+\ Cl^-$	$C_{12}COO^-\ Na^+$	$C_{18}COO^-\ Na^+$	$C_{10}(OE)_{10}CH_3$
n-Hexane	0.75	0.18	0.46	—
n-Heptane	0.54	0.12	0.34	—
n-Octane	0.29	0.08	0.18	0.48
n-Decane	0.13	0.03	0.05	0.17
n-Dodecane	0.06	0.005	0.009	0.06
Cyclohexane	0.87	0.23	0.56	—
Benzene	0.65	0.29	0.76	—
Toluene	0.49	0.13	0.51	—
Ethylbenzene	0.38	0.20	0.40	—
n-Octanol	0.18	0.29	0.59	1.47
2–Ethylhexanol	0.36	0.06	0.47	—

in the palisades region. With truly polar additives, there is a much less clear correlation between surfactant type and the nature of the solubilized material, and the degree of additive incorporation.

It has been suggested that, in general, increasing the chain length of an *n*-alkane or *n*-alkyl-substituted benzene reduces its solubilization in a given surfactant solution. While the presence of unsaturation or cyclic structures tends to increase solubilization, branching appears to have little or no effect. Although such observations can be useful for the simple systems considered, more complicated additive structures fail to behave in such an orderly fashion. The addition of one benzene ring, for example, tends to increase solubility, while a second, fused ring, such as in naphthalene derivatives, produces the opposite effect.

In summary, the relationship between the chemical structure of the additive and its ability to be incorporated into a surfactant solution is quite complex and has so far not lent itself to simple analysis and structural correlation. Perhaps, as our understanding of the geometric packing requirements of molecules in the micellar core and palisades layer improves, and the importance of molecular interactions among the various constituents is better understood, a more rational scheme for predicting solubilization results will emerge.

6.1.4. The Effect of Temperature on Solubilization Phenomena

When considering the effects of temperature changes on the solubilization process, two areas of concern must be addressed:

1. As has been stated often, the ability of a given surfactant to solubilize an additive is intimately related to the characteristics of the micelle (size, shape,

TABLE 6.4. Effects of Temperature on MAC of DMAB in Several Surfactant Systems

	MAC (g/mol Surfactant)		
Surfactant	At 30°C	At 50°C	MAC_{50}/MAC_{30}
$C_9H_{19}COO^-Na^+$	0.64	1.19	1.86
$C_{11}H_{23}COO^-Na^+$	1.50	2.43	1.62
$C_{13}H_{27}COO^-Na^+$	2.71	4.15	1.53
$C_{12}H_{25}NH_3^+Cl^-$	4.32	5.63	1.30

ionic nature, etc.). Since changes in temperature are known to affect some of those characteristics, it should not be surprising to find alterations in the solubilizing properties of surfactants as a result of modifications in micellar structure.

2. Changes in temperature can affect the intermolecular interactions between solvent and solutes (e.g., hydrogen bonding), so that the overall solvent properties of the liquid for surfactant and additive may be significantly altered.

Data illustrating the effect of temperature changes on the ability of several alkali soaps to solubilize *N,N*-dimethyl-aminoazobenzene (DMAB) are listed in Table 6.4. Interestingly, if one considers the relative change (MAC_{50}/MAC_{30}) in the amounts solubilized at 30°C and 50°C, the greatest increase occurs for the system having the lowest initial solubilizing power at the lower temperature. In a relative sense, the effect of the temperature increase can be viewed as the poor getting richer, and the rich not doing too badly.

In a study of the effects of temperature changes on the solubilizing power of the nonionic surfactant $CH_3(CH_2)_9(OCH_2CH_2)_{12}OCH_3$ containing fixed amounts of *n*-decane and *n*-decanol, it was found that in the case of each additive, as the temperature was increased, the apparent micellar aggregation number increased, as expected from results in the absence of additives (Table 6.5), and that the number of additive molecules incorporated per micelle increased. However, examination of the data shows that the ratio of surfactant to additive molecules in each micelle remains constant throughout the temperature range, with values of 10 for the *n*-decane solubilized in the micellar core and 2.5 for the *n*-decanol located in the palisades region.

The effect of temperature changes on the micellization of ionic surfactants is not as simple a relationship as that found for most nonionic materials, and it is to be expected that the effects on solubilization will be correspondingly more complex. It has been reported that micellar solutions of dodecylamine hydrochloride saturated with xylene passed from a clear, isotropic solution to a turbid dispersion as the temperature was increased. It was noted in Chapter 4 that many ionic surfactants pass through a minimum in cmc near room temperature; it would be interesting to know whether a maximum in solubilizing power is attained in the same temperature region as the minimum in cmc.

TABLE 6.5. Effects of Temperature on micelle Size and Aggregation Number of $CH_3(CH_2)_9(OCH_2CH_2)_{12}OCH_3$ Containing Fixed Amounts of *n*-Decane and *n*-Decanol

Temperature (°C)	cmc (mM)	Aggregation Number	Molecules Additive/Micelle
		n-Decane (1.86 wt%)	
10	2.36	65	5.9
30	1.50	67	6.1
50	1.11	71	6.5
60	1.0	85	7.8
69	0.89	110	10.1
		n-Decanol (9.17 wt%)	
10	2.07	73	30
30	1.26	83	33
43	1.09	110	44
50	1.00	140	57
55	0.94	186	76
61	0.89	404	163

6.1.5. The Effects of Nonelectrolyte Solutes

Nonelectrolyte solutes that are not part of the primary solubilized system (solvent–surfactant-additive) can have a significant effect on the solubilizing power of micellar solutions as a result of their effects on cmc's and aggregation numbers. It has become especially obvious that the addition of polar solutes such as phenols and long-chain alcohols and amines can greatly increase the solubility of nonpolar additives in ionic surfactant solutions. A suggested mechanism for such results based on theories of intra- and intermolecular interactions involves the presumed insertion of polar additive molecules between adjacent surfactant molecules in the micelle (Figure 6.2). As a result of that "spacing" of the ionic groups, repulsive interactions among head groups and unfavorable contact between the aqueous phase and exposed hydrocarbon in the core can be reduced. Those two modifications of the micellar surface would allow a decrease in surface curvature of the micelle and a subsequent increase in the capacity of the core to accommodate solubilized nonpolar additives.

Unlike polar solutes with relatively large hydrophobic tails, short-chain alcohols such as ethanol can significantly reduce the solubilizing power of a surfactant. In the discussion of the effects of such materials on the micellization process from Chapter 4, it was shown that the addition of significant quantities of short-chain alcohols, acetone, dioxane, and related compounds could result in profound changes in the cmc and aggregation number of surfactants, even to the point of completely inhibiting micelle formation. It is understandable, then, that such solutes could also adversely affect the solubilization capacity of a surfactant solution.

From the observations above, it seems clear that the effects of an added nonelectrolyte on the solubilizing capacity of a given surfactant system may be quite

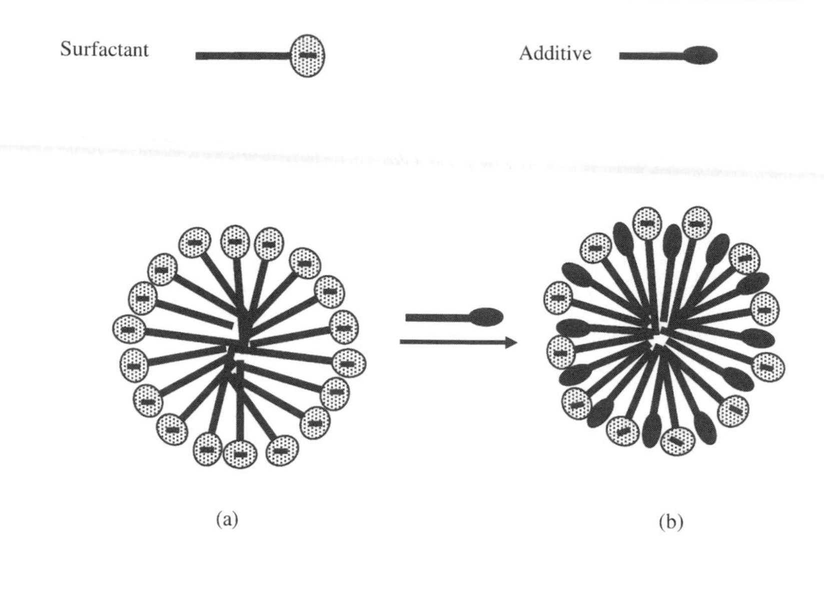

(a) Head group repulsion in ionic surfactants limits micelle size and solubilizing capacity by increasing the effective area per molecule.
(b) The insertion of polar additives between adjacent head groups "insulates" the repulsive interactions between charge group sallowing the micelle to attain a larger radius of curvature and increase its capacity for solubilization.

Figure 6.2. Proposed mechanism (the "Windsor" mechanism) for the role of nonpolar additives in increased solubilization in ionic micellar systems.

complex and may not lend themselves to easy analysis. It can be assumed, however, that the fundamental relationships that exist between the solute and the micellization characteristics of the surfactant, in the absence of the solubilized additive, can be used to good advantage in predicting what may reasonably be expected in the four-component system.

6.1.6. The Effects of Added Electrolyte

The effects of added electrolytes on a micellar system were discussed in Chapter 4. For the case of ionic micelles, the effect of such addition is to decrease the cmc and increase the aggregation number. Such changes are predictable in micellar systems and might be expected to produce parallel effects on solubilization. In fact, however, the results are not always so easily analyzed. At surfactant concentrations near the cmc, it is usually found that the solubilizing power of a system will increase with the addition of electrolyte, as a result of the greater number of micelles available in the system. At surfactant concentrations well above the cmc, however, the simplicity of the relationship between electrolyte concentration, cmc, and solubilization seems to disappear.

A study of the solubilization of decanol in solutions of sodium octanoate showed that at low surfactant concentrations the solubilization of the additive increased rapidly after the cmc was exceeded, and continued to do so for some time as the concentration of sodium chloride was increased. At higher surfactant concentrations, however, it was found that there was an initial increase in decanol incorporation, which reached a maximum and then began to decrease as the salt level continued to increase. When the octanoate concentration well exceeded the cmc, the addition of salt resulted in an immediate decrease in the ability of the system to incorporate the additive. Such complex interactions have been attributed to alterations in the thermodynamics of mixed micelle formation for the decanol and carboxylate salt. Similar results may be seen in systems where the increased electrolyte content produces a change in the character of the micellar system: a sphere-to-rod micellar transformation or the development of a mesophase, for example.

Variations in results for the effect of added electrolytes on solubilization by ionic surfactants might also be related to the nature of the additive and its potential location in the micelle. For nonpolar additives or those lying deep in the palisades layer of the micelle, it seems reasonable to expect that the increased volume of the micellar core produced by electrolytes would lead to a greater capacity for solubilization. For more polar materials, which would be expected to be incorporated less deeply in the micelle, added electrolyte results in a closer packing of ionic head groups, which could reduce the available space for solubilized molecules. Changes in micelle shape, from spheres to rods, for example, would also cause less surface volume to be available in the palisades layer as a result of closer packing of the head groups or, put another way, less curvature in the layer.

In the case of nonionic surfactants, the effects of added electrolytes seem to parallel their effects on the micellization process. When such addition produces an increase in micellar aggregation number, an increase in solubilizing capacity for hydrocarbon additives is also found. The results for the solubilization of polar materials is, again, less clear-cut. A similar trend is generally found for cationic and amphoteric surfactants.

6.1.7. Miscellaneous Factors Affecting Solubilization

Other factors that can affect the ability of a particular surfactant system to solubilize materials include pH and pressure. The effects of such factors, however, have not been as extensively reported in the literature as the factors discussed above, and they are often very specific to each surfactant system. Obviously, surfactants that show special sensitivity to pH such as the carboxylate salts will also be expected to exhibit significant changes in solubilization with changes in that factor. In addition, changes in pH can affect the nature of the additive itself, producing dramatic changes in its interactions with the micelle, including the locus of solubilization. Such effects can be especially important in many applications of solubilization, especially in the pharmaceutical field.

The effects of such a variable as pressure on micelle formation and solubilization is a relatively new field of investigation. It can be assumed that significant effects will be observed once sufficient pressures have been attained. However, such levels lie outside the normally available range of experimental conditions and are of little practical concern except in studies of surfactant activity in supercritical fluid processes.

6.2. MICELLAR CATALYSIS

It is well recognized in all branches of chemistry that the rate of a chemical reaction can be very sensitive to the nature of the reaction environment. Reactions involving polar or ionic transition states can be especially sensitive to the polarity of the reaction medium. It should not be too surprising, then, that many chemical reactions, especially those in which one reactant may be soluble in water and the other in oil, can exhibit a significant enhancement in rate when carried out in the presence of surfactant micelles. The presence of the micellar species can provide a beneficial effect through two possible mechanisms:

1. The palisades region of the micelle represents a transition zone between a polar aqueous environment, which may be either the bulk phase or the micellar core, and a nonpolar hydrophobic region. Such a gradient in polarity can serve as a convenient area of intermediate polarity suitable for increased reactant interaction or for optimizing the energy of transition state formation.
2. The potential for the micelle to solubilize a reactant that would not normally have significant solubility in the reaction medium means that it can serve as a ready reservoir of reactant, in effect increasing the available concentrations of reactants (Figure 6.3). Rate enhancements as high as 10^5 have been reported, which makes such systems very attractive for potential practical applications.

6.2.1. Micellar Catalysis in Aqueous Solution

In aqueous media, a micellar system can serve as a catalyst for organic reactions, but it is also possible for it to retard such reactions. Two possible mechanisms for catalytic action were suggested above; inhibitory actions may arise from unfavorable electrostatic interactions between reactants and changes in the distribution of reactants between the bulk and micellar phases. In the case of electrostatic inhibition, the presence of a charge on the micelle surface can have two effects on a reaction involving a charged species. In the base hydrolysis of water-insoluble esters, for example, if the micelle charge is negative, the transport of hydroxide ion into or through the palisades layer will be retarded by charge repulsion. If the micelle is positively charged, the inclusion of the oppositely charged species will be facilitated. For nonionic and zwitterionic surfactants, there will be little or no effect as a result of electrostatic interactions. Although such a model of electrostatic effects is simple, it has generally been supported by experiment. The basic hydrolysis of

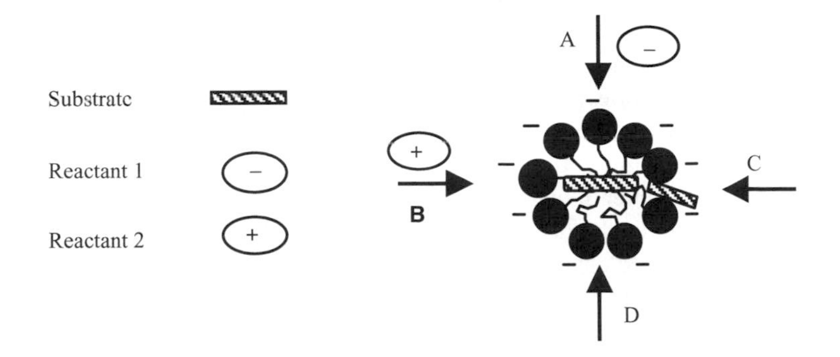

A - Reactant of same sign impeded approach – slow reaction

B - Reactant of opposite sign – accelerated reaction

C - Solubilization in palisades improves solvent character

D - Increased substrate availability due to solubilization

Figure 6.3. Proposed mechanisms for the catalytic (and retarding) effects of micelles in organic reactions.

esters, for example, is catalyzed by cationic and inhibited by anionic surfactants, while the opposite is true for the acid hydrolysis of orthoesters. Nonionic and zwitterionic surfactants can also produce significant rate enhancement, indicating that hydrophobic effects can, in some instances, predominate over electrostatics.

The ability of a micellar system to solubilize a reactant may affect its action as a catalyst or inhibitor in a reaction. When a surfactant system serves as a reservoir for increasing the availability of one reactant, any change that increases the solubilizing capacity of the micelle should also increase its effectiveness as a catalysis. If, on the other hand, the reaction must occur in the bulk phase, increased solubilizing power may remove reactant from the reaction medium and therefore decrease catalytic or increase inhibitor efficiency.

In aqueous solution, the effectiveness of micellar systems as catalysts is quite often found to increase with the length of the alkyl chain. It has been found, for example, that the rate of hydrolysis of methyl orthobenzoate in the presence of sodium alkyl sulfates increased in the order octyl < decyl < dodecyl < tetradecyl < hexadecyl. Such a result may be attributed to either electrostatic or solubilizing effects. Since, as a general rule, the aggregation number of aqueous micelles increases with the chain length in a homologous series, there must be a parallel increase in the surface charge density at the micelle surface. It might be expected, then, that any effects due to electrostatic interactions would also increase. If the hydrolysis is acid-catalyzed, an electrostatic enhancement of the reaction rate would be expected; the base catalyzed reaction would be expected to be slower. Alternatively (or additionally), the larger aggregation number results in an increase in the solubilizing power of the system with a resulting rate increase due to increased substrate availability. The importance of each mechanism will depend on the specifics of the reaction.

In addition to the effects noted for increases in the charge density on the micelle, the charge density on the individual surfactant molecules can be important. It has been found, for example, that cationic surfactants containing two charge groups were significantly better at increasing the rate of nucleophilic aromatic substitutions than analogous singly charged materials. Similar results have been noted for singly versus doubly charged anionic surfactants.

As might be expected, the structure of the reactive substrate can have as much influence on micelle-assisted rate enhancement as that of the surfactant. Since the catalytic effectiveness of the micelle can be related to the location and orientation of the substrate in the micellar structure, the more hydrophobic the substrate (and the surfactant), the more significant may be the catalytic effect.

When nonsurfactant cosolutes are added to the micellar reaction mixture, the results can be quite unpredictable. Studies have found that the presence of excess surfactant counterions or common ions retards the catalytic activity of the micelle, with larger ions more effective in that respect. In contrast, the addition of neutral electrolyte has been found to enhance micellar catalysis in some instances while showing little or no effect in others. It has been proposed that the retardation effect of excess common counterions is due to a competition between the excess ions and the reactive substrate most closely associated with the micelle for the available positions or "binding sites" on or in the micelle. The enhancing effect, however, has been attributed to the more general effects of added electrolyte on the properties of micelles, that is, lowering of the cmc, increasing the aggregation number, and other variations, all of which often tend to increase catalytic activity.

As new experimental techniques produce more detailed information on the location of the various components in micellar systems, and the thermodynamics of substrate–micelle interactions become better understood, our ability to expand the applicability of such systems on the basis of good science and good judgment will be greatly enhanced.

6.2.2. Micellar Catalysis in Nonaqueous Solvents

As noted previously, interactions between polar head groups in nonaqueous solvents provide the primary driving force for the formation of micellar aggregates in such media. Such reversed micellar cores inherently provide a unique location for the solubilization of polar substrates. While keeping in mind the potentially dramatic effects of additives on the properties of micellar solutions, it is obvious that such nonaqueous systems hold great potential from a catalytic standpoint.

One of the first reported instances of catalysis by reversed micelles in the early 1970s concerned the hydrolysis of *p*-nitrophenyldodecanoate in hexanol–water systems containing cetyltrimethylammonium bromide (CTAB). Since that time, nonaqueous systems have gained greater attention as models that mimic the catalytic activity of natural enzymatic reactions. The fundamental principles controlling activity in nonaqueous systems are basically the same as those for aqueous solutions, except that the specificity of the micellar core for the solubilization of polar substrates is much greater for the nonaqueous situation. The popularity of reversed micelles as models for enzyme catalysis stems from the fact that the micellar core is capable of binding substrates in concentrations and orientations that can be very specific to certain functionalities, much as an enzyme would do. As a result, reaction rate enhancements can be obtained comparable to those of the natural systems and far in excess of what can be explained on the basis of the partitioning or availability of substrate.

Work in the area of micellar catalysis in both aqueous and nonaqueous solvent systems is certain to continue to grow in importance as a tool for better understanding the chemistry and mechanics of enzymatic catalysis, as a probe for studying the mechanistic aspects of many reactions, and as a route to improved yields in reactions of academic interest. Of more practical significance, however, may be the expanding use of micellar catalysis in industrial applications as a method for obtaining maximum production with minimum input of time, energy, and materials.

6.3. PHASE TRANSFER CATALYSIS

Modern synthetic organic chemistry has become the principal source of chemical products for practical applications in such important fields as pharmaceuticals, cosmetics and personal care products, agricultural and plant protection agents, dyes, photographic and other imaging chemicals, monomer and polymer fabrication, energy production, and new electronic technologies. The transformation of simple or basic starting materials into more complex final products often requires a number of chemical operations in which expensive and exotic reagents, catalysts, solvents, and other agents are used. Thus, in the course of synthesis, besides the desired products, many waste materials may be produced because such transformations of reactants into products are very seldom quantitative and selective processes. Wastes must be regenerated, recovered, destroyed, or disposed of—all processes that consume a great deal of time and energy, and often represent a significant environmental burden, adding significantly to the economic cost of the final product without adding to its inherent value. It is therefore important to try to develop synthetic methods that minimize these problems.

Phase transfer catalysis (PTC) is a field of research in organic synthesis that began to gain proponents around 1975 because of its apparent potential for enhancing the rates of many two-phase organic reactions involving anionic reactants, particularly bases and nucleophiles. The major attractiveness of the concept was that it appeared that reactions could be carried out in simple, two-phase systems that, under "normal" circumstances, required extreme conditions such as strictly anhydrous, aprotic solvents such as dimethylsulfoxide (DMSO) and dimethylforamide (DMF) and very strong and potentially dangerous basic reactants such as sodium hydride (NaH), sodium amide ($NaNH_2$), potassium *t*-butoxide, and the like, all of which require the manipulation of metallic lithium, sodium, or potassium at some point. Conventional synthetic processes also usually involve the manipulation of significant amounts of potentially dangerous or environmentally damaging waste byproducts. Such conditions imply high costs in terms of raw materials, security, energy, and environmental control. In industry, as well as in the research laboratory, the possibility of avoiding hazards and complications is a powerful inducement to at least try out new synthetic technology.

Phase transfer catalysis is perhaps one of the most general and efficient reaction procedures that fulfill these requirements. PCT technology is applicable to a variety of reactions in which inorganic and organic anions and reactive intermediates such as carbenes react with organic substrates. It involves the use of heterogeneous two-phase systems, one of which is a reservoir or source of reacting anions or base for generation of organic anions, while the other is an organic phase containing oil-soluble reactants and the catalysts, the source of hydrophobic cations that serve as the transport "mules" that make the process work.

Reactions that have been found to be responsive to PTC conditions can be divided into two major categories: (1) reactions involving anions that are readily available as salts, such as sodium cyanide, sodium azide, and sodium acetate, and (2) reactions of anions that must be generated in situ in the organic phase using strongly basic catalysis such as alkoxides, phenolates, *N*-anions of amides or heterocycles, carbenes, and a variety of carbanions. In the former case, the salts are normally present as aqueous solutions or in the form of pellets or powdered solids, whereas the organic phase may contain the organic reactant neat, when liquid, or in an appropriate unreactive solvent. A partial list of standard organic synthetic reactions that have been shown to take place under PCT conditions is given in Table 6.6.

While PTC does not involve solubilization or micellar catalysis in the sense discussed above and seldom involves what are generally considered to be surfactants, it does require the use of catalysts that are amphiphilic in nature. Its basic mechanism, as described below, also brings up some potentially interesting questions related to interfaces, the activity of amphiphilic molecules in multiphase systems, and the transport of ionic species from one phase to the other across such interfaces. Because of those loose, but interesting connections between micellar catalysis and phase transfer catalysis, the following introduction to the theme has been included.

6.3.1. Cross-phase Reactions

The generally accepted sequence of events or physical mechanism for phase transfer catalysis involves the use of special amphiphilic cations as a transport "mules" for moving reactive anionic species through the interface between an aqueous solution or a solid reactant phase and an organic reaction medium, as shown schematically in Figure 6.4.

The overall process illustrated in Figure 6.4 involves the following basic steps, where subscripts org, aq, int, denote organic, aqueous, and interfacial regions, respectively. In the first step, the catalyst cationic amphiphile Q^+ enters or makes contact with the aqueous phase or solid reactant anion source to form the ion pair Q^+X^-. The ion pair is then transported across the interfacial region into the organic

TABLE 6.6. Organic Reactions Shown to Be Effectively Catalyzed Using PTC

O-Alkylation (etherification)
N-Alkylation
C-Alkylation
S-Alkylation (thiolation)
Dehydrohalogenation
Esterification
Transesterification
O-/*N*-/*S*-Acylation using acetic anhydride, benzoyl chloride, PCl, and other water-sensitive reactants
Displacement or substitution reactions using cyanide, fluoride, bromide, iodide (under limited conditions), azide, thiocyanate/cyanate, sulfide/sulfite, nitrite/nitrate, hydroxide/hydrolysis
Thiophosphorylation
Other nucleophilic aliphatic and aromatic substitutions
Other strong base reactions such as Michael additions, aldol condensations, Wittig reactions, Darzens condensations, carbene reactions
Oxidations using hypochlorite, hydrogen peroxide, oxygen, permanganate
Epoxidations
Reductions with borohydride and hydrogenation
Carbonylation
HCl/HBr additions
Transition metal cocatalysis
Other reactions involving anions

phase where the anion, X^-, reacts with the organic substrate, R—Y, to produce the organic product, R—X, and the ion pair product, Q^+Y^-. The product ion pair then returns across the interfacial region and leaves the product anion, Y^-, in the aqueous phase, picks up another reactant anion, and repeats the process until the reaction is complete. That description is, of course, simplified and many PTC reactions involve several more steps, some of which will be mentioned below.

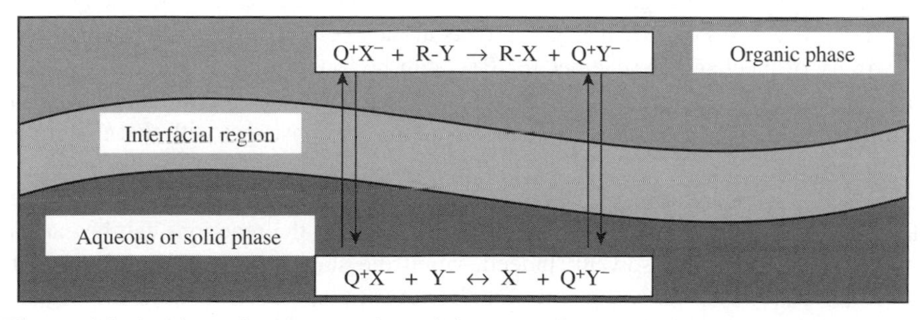

Figure 6.4. A schematic representation of the generally accepted extraction/solubilization mechanism of phase transfer catalysis.

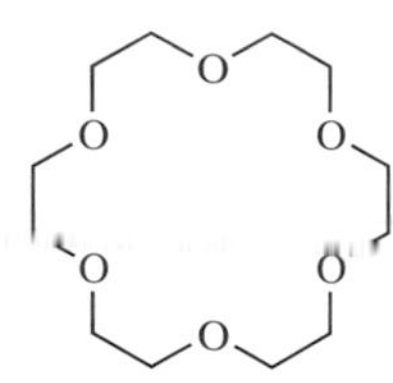

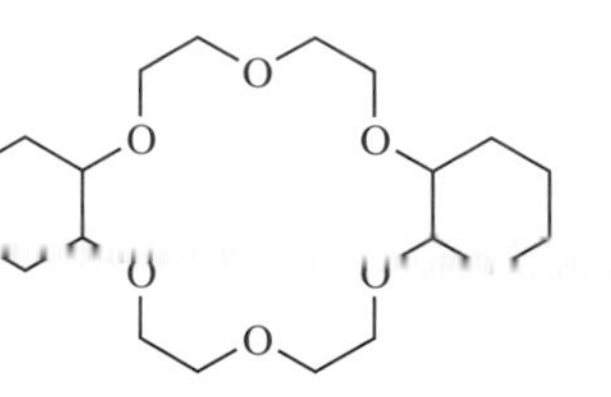

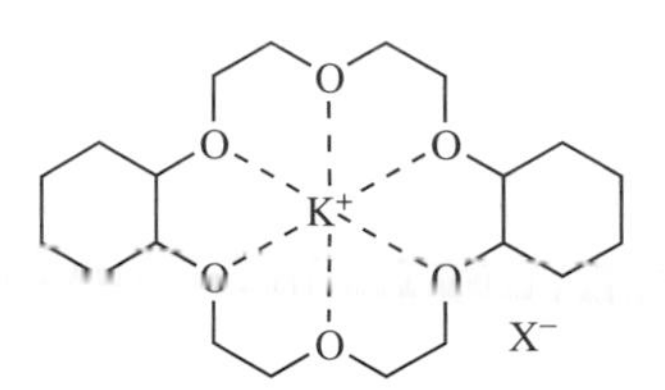

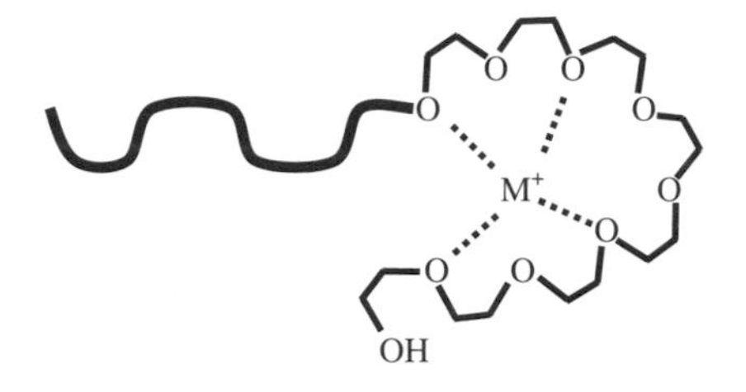

Figure 6.5. Molecular structures of typical crown ether and POE nonionic surfactant catalysts and complexes.

There are two types of PTC processes mentioned in the literature. If two liquid phases are involved, the reaction process is termed *liquid–liquid phase transfer catalysis* (ll-PTC). If the nonorganic phase is a solid, the process is termed solid–liquid phase transfer catalysis (sl-PTC). Since the catalyst, routinely referred to in the literature as the "quat," is the vehicle for the transfer of anions between the two liquid phases, the mechanistic scheme for ll-PCT can be considered an extraction process. For sl-PCT, it may be considered a solubilization process, since the normally organic insoluble reactive anions are solubilized by the formation of more hydrophobic ion pairs at the interface. Typical ll-PTC catalysts are quaternary ammonium salts ($R_4N^+ X^-$) and phosphonium salts ($R_4P^+ X^-$). Crown ethers, illustrated in Figure 6.5, and other cation complexing agents, and some nonionic, straight-chain polyoxyethylene materials have also been found to be efficient catalysts for sl-PTC processes. Polymer bound catalysts are also effective in some applications. A variety of mechanistic schemes for PTC can be found in the literature, but most are variations on the extraction or solubilization mechanisms. Nevertheless, the fundamental principles governing these mechanisms are consistent with that illustrated in Figure 6.4.

Since the concentration of the reacting anions in the organic phase in PTC reactions cannot exceed that of the catalyst, in many cases the reactions can be carried out without an organic solvent. Indeed, when the starting materials and products are liquid, they can act as solvents for the reacting species and form the organic phase when used neat; the reacting anions are introduced directly by the catalyst. Since the catalyst is usually present at approximately 1 mol%, the reactions proceed as if in a dilute solution in terms of the reacting anions. Since anions associated with tetraalkylammonium cations exhibit a high degree of activity, and since the concentration of the substrate reacting with the anions (when it is used neat) is high, the reactions proceed rapidly. Because of the specificity of the PTC mechanism, reactions generally proceed in high yields and high selectivity, while side products are held to a minimum.

In general, for PTC processes involving relatively diffuse or polarizable (termed "soft" in much of the literature) anionic reagents (cyanide, acetate, etc.), the most effective quats are those that contain large hydrophobic alkyl or aryl substituents that substantially separate the cationic center (the positively charged nitrogen) from the anionic center of the nucleophilic reagent. In contrast, for PTC processes involving "hard," poorly polarizable anions such as OH^-, quats that have significant hydrophobic character but that have at least one alkyl substitutent that allows the anion to approach close to the center of positive (e.g., $-CH_3$) charge are usually the most effective. These are termed "accessible quats." For hydroxide-promoted reactions, care must be taken to avoid the degradation of the quat catalyst by Hofmann elimination pathways.

6.3.2. Some Examples of PTC Applications

The use of PTC technology instead of traditional processes for industrial synthesis can potentially provide substantial cost and environmental benefits. Major potential advantages of PTC in industrial applications include

1. Elimination of organic solvents
2. Elimination of dangerous, inconvenient, and expensive reactants (NaOH, KOH, K_2CO_3, etc. instead of NaH, $NaNH_2$, *t*-BuOK, R_2NLi, etc.)
3. High reactivity and selectivity of the active species
4. High yields and purity of products
5. Simplicity of the procedure
6. Low investment cost
7. Low energy consumption
8. Possibility of mimicking countercurrent process
9. Minimization of industrial wastes

Some of these benefits can be illustrated with some simple examples.

6.3.2.1. Alkylnitrile Synthesis

The synthesis of alkylnitriles from alkyl chlorides, shown in the scheme below, is an important industrial pathway to organic nitriles and subsequent derivatives:

$$R\text{—}Cl + NaCN \rightarrow R\text{—}CN + NaCl$$

Since the reactants for such transformations are mutually immiscible, conventional synthetic procedures make it is necessary to use some suitable solvent to produce a homogeneous reaction medium, or at least obtain sufficient concentration of the two reactants to bring about a reasonable reaction rate. The process is usually carried out in mixtures of a low-molecular-weight alcohols and water, where, due to strong solvation of the anions, the reaction is rather slow and requires prolonged heating. Isolation of the product involves separation from the solvent, which should be recovered for both economic and environmental reasons, and waste byproducts, which are produced in substantial quantities, must be disposed of or destroyed. The reaction proceeds much faster in polar aprotic solvents such as DMF or DMSO, but isolation of the product and recovery of the solvents is even more troublesome in those cases. The significant initial cost of such solvents is also an important factor.

Using PTC technology, neat alkyl chloride containing 1 mol% catalyst is stirred at room temperature with a saturated aqueous solution of NaCN. On completion of the reaction, the organic phase, which is often the pure product, can be separated and the product purified or used "as is" in subsequent reactions. The reaction product in the aqueous phase, NaCl, can be removed fairly easily, fresh NaCN added, and further reaction with R—Cl carried out. In the end, the only reaction waste product is NaCl.

A second example is the alkylation of arylacetonitriles to produce -arylalkylnitriles, which is an important process in the pharmaceutical industry. It proceeds in general according to the reaction scheme

$$ArCH_2CN + R{-}X + B^- \rightarrow ArRCHCN + X^- + BH$$

Traditional industrial processes for this reaction involve multistep reactions. First, the starting nitrile, $ArCH_2CN$, is dissolved in an anhydrous solvent such as toluene and then treated with a strong base such as $NaNH_2$ or NaH to produce the intermediate carbanion in the form of the sodium salt. The intermediate salt—maintained under strictly anhydrous conditions—is then treated with the appropriate alkyl halide to form the final product. Isolation of the product requires the treatment of the mixture with water and removal of the solvent. Aside from its inherent toxicity, the solvent must be recovered for economic and environmental reasons. The overall process requires a large capital investment because of safety requirements for working with $NaNH_2$ or NaH and the necessity of keeping the system strictly anhydrous, which requires a great deal of energy and produces large quantities of waste products.

The same process using PTC technology consists of vigorously stirring neat $ArCH_2CN$ and R—X with 50% aqueous NaOH and about 1 mol% catalyst. The reaction is mildly exothermic, and on completion, the product is isolated by dilution with water and separation of the phases. PCT technology requires much lower capital investment, consumes much less energy, and produces much less wastes, particularly considering that the high selectivity of the process gives higher yields of the final product than traditional technology. Other acidic —CH groups, alcohols, amines, amides, and similar compounds can be effectively alkylated in a similar way.

6.3.2.2. Dihalocyclopropanes

A standard dichlorocarbene reaction in organic synthesis would involve the treatment of chloroform ($CHCl_3$) with a very strong and water-sensitive base such as potassium *tert*-butoxide [$K^+\ ^-OC(CH_3)_3$] in an exhaustively dried solvent under a dry inert-gas atmosphere (Figure 6.6a). The carbene produced (Cl_2C:) would then ideally react rapidly with an ethylene group present in the reaction medium. Using PCT (Figure 6.6b), a solution of the substrate RHC=CRH in chloroform is vigorously stirred in contact with a concentrated solution of NaOH, or solid NaOH pellets, in the presence of up to 5 mol% quat catalyst, at room temperature for an hour or two to produce the desired cyclopropyl derivative in near-quantitative yield with no need for the extreme reaction conditions required for the "normal" synthetic process.

PTC has also found application in organometallic chemistry, for example, in metal-catalyzed carbonylation of alkyl or aryl halides. Interestingly, some reactions are found to occur in PTC systems that do not work under "normal" synthetic reaction conditions.

The above mentioned simple examples of the application of PTC technology illustrate its substantial potential advantages in comparison with traditional processes. There are hundreds other possible industrial applications of PTC for a

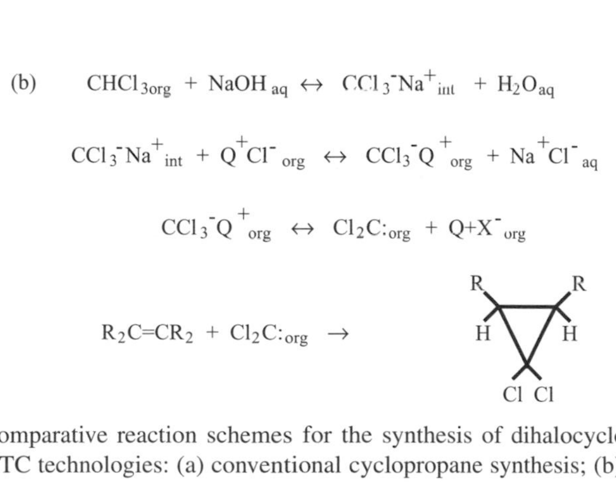

(a) $K^{+}\ ^{-}OC(CH_3)_3 + CHCl_3 \rightarrow Cl_2C: + HOC(CH_3)_3 + KCl$

$Cl_2C: + RHC{=}CRH \rightarrow$ (cyclopropane ring bearing R, H / R, H and Cl Cl)

(b) $CHCl_{3\,org} + NaOH_{aq} \leftrightarrow CCl_3^-Na^+_{int} + H_2O_{aq}$

$CCl_3^-Na^+_{int} + Q^+Cl^-_{org} \leftrightarrow CCl_3^-Q^+_{org} + Na^+Cl^-_{aq}$

$CCl_3^-Q^+_{org} \leftrightarrow Cl_2C:_{org} + Q{+}X^-_{org}$

$R_2C{=}CR_2 + Cl_2C:_{org} \rightarrow$ (cyclopropane ring bearing R, H / R, H and Cl Cl)

Figure 6.6. Comparative reaction schemes for the synthesis of dihalocyclopropanes using standard and PTC technologies: (a) conventional cyclopropane synthesis; (b) PTC synthesis.

variety of processes of organic synthesis. Those applications will almost always require less capital investment, consume less energy, and generate fewer industrial wastes compared to the traditional processes. For both economic and environmental reasons, it is obvious that any measures that can save energy and money offer substantial direct and secondary benefits for the environment.

6.3.3. Some Notes on the Use of PTC

PTC reactions involving organic anions produced in situ in the organic phase are mechanistically more complicated because of the more hydrophobic nature of such species. In such cases, the aqueous phase normally contains a base such as aqueous concentrated or solid NaOH or KOH or solid K_2CO_3, and the organic phase contains the anion precursor (an alcohol, amine, or other "acid"), an electrophilic reactant, and in some cases, an aprotic solvent. Alkylation of phenylacetonitrile by reaction of its carbanion with alkyl halide is an example of such an application of PTC and helps illustrate how the process functions.

Mechanistic studies strongly indicate that the generation of carbanions, alkoxides, or other organic anions does not occur via direct ion exchange between $Q^+_{org}\ X^-$ and $NaOH_{aq}$ followed by deprotonation of the carbanion precursor in the organic phase. This is assumed because the ion exchange equilibria for $X^- = Cl^-$ and Br^- lie strongly to the left of the reaction equation; therefore the concentration of Q^+OH^- in the organic phase will always be very small. All indications are that deprotonation of the acidic precursor takes place at the interface.

6.3.4. Some Requirements for a Successful PTC Reaction

In order for phase transfer catalysis to take place, the catalyst must effectively transport the reactant anion or nucleophile from the aqueous phase through the interface into the organic phase. The successful transport of the anion, however, does not mean that catalysis will automatically take place. Once in the organic phase, the anion must be in a sufficiently reactive state to rapidly react with the target or electrophilic organic reagent. Finally, for successful PTC, the catalyst must transport the product anion or leaving group from the organic phase through the interface back into the aqueous phase so that the catalytic cycle can begin again. If this final stage does not take place, the reaction will stop; if the backtransfer of the product anion does not occur, the reaction is said to be "poisoned." This most often occurs when the leaving group is significantly hydrophobic (e.g., iodide, *p*-toluenesulfonate) and strong ion pairs with the catalyst remain in the organic phase, thus occupying all of the available "mules" and shutting down the catalytic cycle. The efficiency of phase transfer catalysis is influenced by the bulkiness of the groups attached to the phase transfer catalyst and its hydrophobicity, as well as that of its counterion.

The phase transfer catalytic process involves at least three important steps: (1) the *reactant transfer step*, in which the reactant anion is transferred from the aqueous or solid phase into the organic phase; (2) the *reaction step*, in which the product is formed in the organic phase, or in some cases, at the interface; and (3) the *product transfer step*, in which the product anion is transferred back from the organic phase into the aqueous or solid phase. The optimization of each of these steps is a function of several variables, the most important of which is the structure of the phase transfer catalyst.

Since Murphy's laws rule in organic synthesis as well as in the rest of the world, it is important to take into consideration as much previous knowledge as possible when making a first attempt at using new or unfamiliar technology such as PTC. With PTC, it is possible to minimize disappointments by considering a few essential points:

1. PTC reactions can be tricky, so patience and perseverance are important ingredients in the stew. It is important not to expect an immediate boom (speaking figuratively and hopefully, not literally). Optimizing trials will almost certainly be needed to follow up preliminary reactions.
2. In both ll-PTC and sl-PTC inert, polar, nonprotic solvents are recommended. In addition, in ll-PTC the solvent should not be water-miscible.
3. Work with relatively concentrated solutions, using one of the reactants as a solvent if practical.
4. A catalyst concentration of 5 mol% relative to the amount of substrate should be used on the first try. In order to ensure an important point of reference, the proposed reaction should be attempted under PTC conditions, but without the catalyst. In that way, one can be sure that PTC is actually occurring. If so, the catalyst concentration can probably be reduced in subsequent experiments.
5. Agitation of the reaction medium should be as rapid and turbulent as possible to maximize the interfacial contact area between phases.
6. For initial trials, it is convenient to try the tetra-*n*-butyl ammonium, $(C_4H_9)_4N^+$, cation as catalyst. It is readily available with a variety of anions, it is relatively inexpensive, and it will work to some degree in almost every PTC reaction, although other catalysts may give better results as determined in follow-up trials. For reactive anions of low polarizability such as HO^- and F^-, "accessible" quats such as methyltributylammonium, $(C_4H_9)_3CH_3N^+$; benzyltriethylammonium, $C_6H_5CH_2(C_2H_5)_3N^+$; or methyltrioctylammonium, $(C_8H_{17})_3CH_3N^+$ are good second choices, while for highly polarizable anions such as CN^- or CH_3COO^-, more hydrophobic quats such as tetraoctylammonium, $(C_8H_{17})_4N^+$, are good second choices. Alkylations with concentrated NaOH are usually found to proceed well with $(C_2H_5)_3C_6H_5CH_2N^+$ as the catalyst. For reactions conducted at high temperatures (well over 100°C), crown ethers such 18-crown-6 or dicyclohexyl-18-crown-6 are recommended.
7. Hydrophilic, highly solvated catalyst anions (X^-) such as sulfate, hydrogen sulfate, or chloride are preferred because they are less likely to produce "poisoning" problems as the reaction proceeds. Bromide is acceptable under many conditions, but iodide should be avoided because of its greater solubility in organic solvents.

8. Where possible, the leaving group (Y^-) should be as hydrophilic as possible; good candidates are Cl^- and mesylate. Anionic leaving groups that are particularly hydrophobic such as iodide and *p*-toluenesulfonate, tend to form tighter ion pairs with the quat cation and slow down or stop the PTC process.

Chemical processes related to or involving interfacial interactions such as those introduced in this chapter present great opportunities for new and better pathways for the transformation of raw materials into useful chemical products. They have been utilized in nature since the origin of life, without understanding by humankind for thousands of years, and by design for about a hundred. Economic, environmental, and intellectual driving forces will undoubtedly lead to their greater use in the years to come.

PROBLEMS

6.1. Given the following chemical structures, predict the probable location of each of the following compounds if solubilized in aqueous micellar solutions of (a) sodium dodecyl sulfate, (b) *n*-hexadecyltrimethyl ammonium bromide, (c) dodecylphenol$(POE)_7$:

n-Hexadecane	*n*-Octanoic acid n-butyl lactate	Chlorobenzene
Methyl oleate	2,5-Di-*t*-butylhydroquinone	Di-*n*-butyl ether
Naphthalene	*N,N*-dimethylhexadecyl amine	Tetrahydrofuran

6.2. Which of the following surfactants would be expected to be most efficient at solubilizing hexadecane: sodium *n*-nonylbenzene sulfonate, sodium hexadecylsulfate, benzyltrimethylammonium acetate, or SDS? For solubilizing cholesterol?

6.3. A system of aqueous micelles of a nonionic POE surfactant is found to solubilize an average of 2 molecules of a material per micelle at 25°C. If the temperature of the system is raised to 50°C, would you expect the solubilizing capacity per micelle to increase or decrease? Based on the information provided, for the same total surfactant concentration, what can you say about the total solubilizing capacity of the system at the two temperatures?

6.4. If an aqueous micellar solution of sodium tetradecylsulfate is employed to solubilize a polar dye, would you expect the addition of dodecyl alcohol to increase, decrease, or not affect the capacity of the system?

6.5. For the solubilization of high-molecular-weight hydrocarbons in anionic micelles, would the addition of electrolyte be expected to increase, decrease, or not affect the capacity of the system?

6.6. The base-catalyzed hydrolysis of esters is accelerated by the presence of 10^{-4} M nonylphenyl–POE_9 ether surfactant, unaffected by 10^{-4} M SDS, and retarded by 10^{-3} M SDS. Explain the observed effects of surfactant class and concentration.

6.7. An amphoteric amine sulfate surfactant is tested as a catalytic system for the base-catalyzed hydrolysis of a triglyceride. What pH range (< 5, 5–8, > 8) would you expect to be most appropriate for maximum rate enhancement? Explain why.

6.8. It is found that the solubilizing capacity of sodium 2-ethylhexylsuccinate for dodecane is greater than that of disodium 2-ethylhexylsulfosuccinate, while the opposite is true for the solubilization of *n*-dodecanol. Suggest an explanation for those results.

6.9. The solubilizing capacity (grams of additive per gram of surfactant) of typical POE nonionic surfactants normally increases with increase in temperature. For nonpolar additives, that result is presumed to be a result of the availability of more room in the larger micelle core. What explanation can you suggest for the same effect with polar additives?

6.10. The reaction of chloroform ($CHCl_3$) with an alkene to form the corresponding dichlorocyclopropyl compound using PTC conditions is rapid and almost quantitative. The same conditions using iodoform (CHI_3) results in the production of almost no cyclopropyl product. Explain that result.

7 Polymeric Surfactants and Surfactant–Polymer Interactions

A glance at ingredients lists of many consumer products will quickly show that surfactants constitute some of the most functionally important ingredients in cosmetic and toiletry products, foods, coatings, pharmaceuticals, and many other systems of wide economic and technological importance. In many, if not most, of those applications, polymeric materials, either natural or synthetic, are present in the final product formulations or in the ultimate targets for their use. Other surfactant applications, especially in the medical and biological fields, also potentially involve the interaction of polymers (including proteins, nucleosides, etc.) with surfactant-containing systems. In addition, many natural and synthetic polymers are themselves amphiphilic, or potentially so, so one must consider not only the surface activity of a "normal" surfactant in a system but also that of any polymeric species present. Nor surprisingly, the situation can become very interesting as a system becomes more complicated in terms of the addition of amphiphilic materials. This chapter presents some of the basic aspects of normal amphiphile–polymer interactions, an introduction to the subject of polymeric surfactants, and some of the potential complications that might arise in the presence of both types of amphiphiles.

7.1. POLYMERIC SURFACTANTS AND AMPHIPHILES

Polymers can generally be categorized into five classifications based on their specific chemical makeup (monomer content) and the exact manner in which the monomers are arranged along the polymer chain. Within each class there are finer subdivisions based on the presence or absence of branching along the chain and the stereochemical relationship among the monomer units. Since polymer chemistry is a vast area of colloid science or materials science, as one prefers, the current discussion will be limited to the primary chemical structure of polymers with the more refined aspects left for other venues.

Surfactant Science and Technology, Third Edition by Drew Myers

The five general classes of polymers can be defined as

1. *Homopolymers*, in which all the monomer units are identical.
2. *Heteropolymers* or *random copolymers*, in which at least two different monomer units are arranged in a more-or-less random manner along the chain.
3. *Block and graft or comb copolymers*, in which different monomer units are linked in homogeneous groups to make up the chain or different monomer units are grafted onto a main chain or backbone to form a comblike structure.
4. *Polyelectrolytes*, which can be members of any of the three classes above, but with the characteristic that they carry a significant number of electrical charges along the chain that impart special characteristics to the class.

Schematic representations of the five classes are shown in Figure 7.1. The representations given are intended to be illustrative and in no way show the true complexity of most polymer chains. The polymers structures illustrated in the figure are generally synthetic, but natural polymers can also fall into the same categories.

Polymers such as cellulose, starches, natural gums, acrylates, acrylamides, polyethyleneimines, and pyrollidones have been staples of the water-soluble polymer industry for many years. In an attempt to expand the basic utility of such materials, and to solve certain technological needs not satisfied by more conventional amphiphilic materials, they have been modified to give special performance additives in

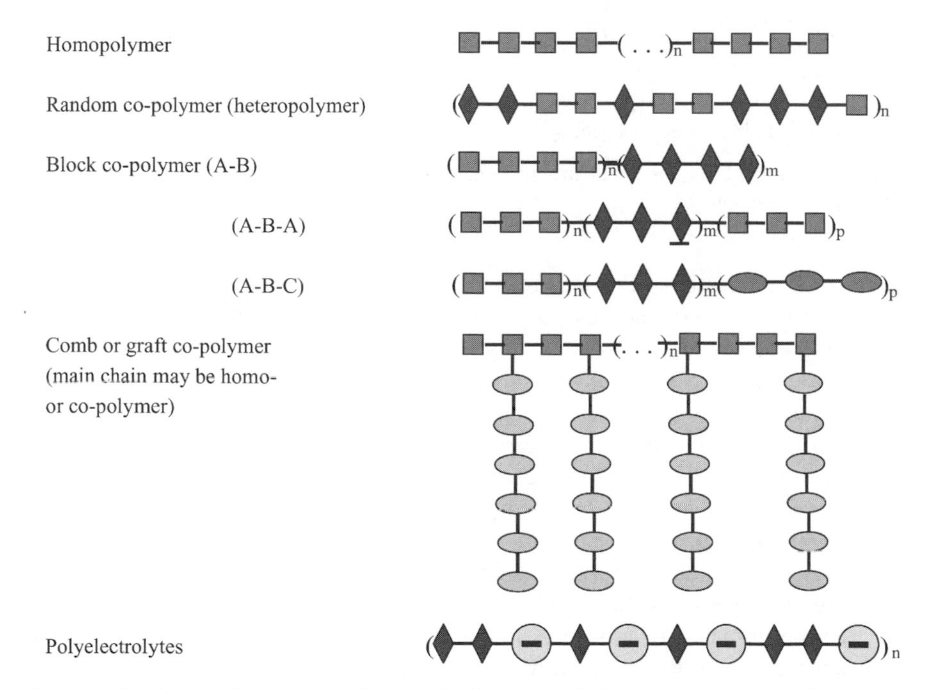

Figure 7.1. Schematic representations of the five general classes of polymers, where A,B,C represent different monomer units.

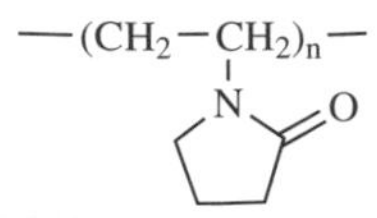

Polyvinyl pyrrolidone (PVP)

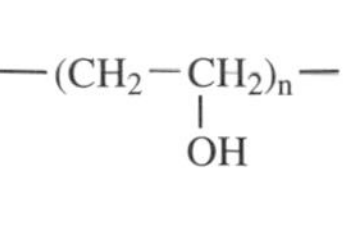

Polyvinyl alcohol (PVA)

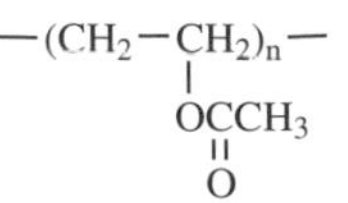

Polyvinyl acetate (PVAc)

$HO-(CH_2CH_2O)_n-OH$

Polyoxyethylene (POE) or
Polyethylene glycol

$HO-(CH_2CH(CH_3)-O)_nH$

Polypropylene glycol (PPG)

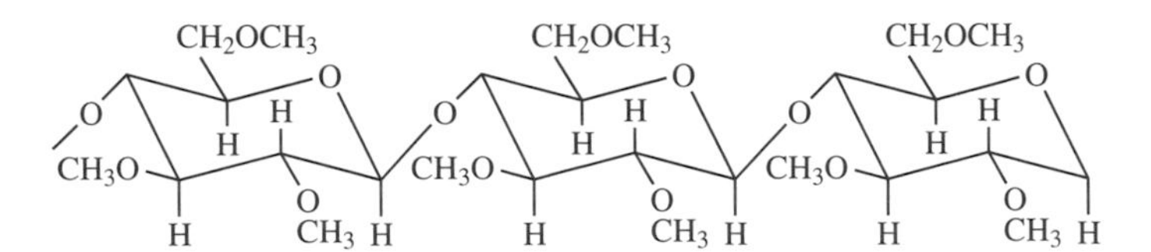

Methyl cellulose (MC)

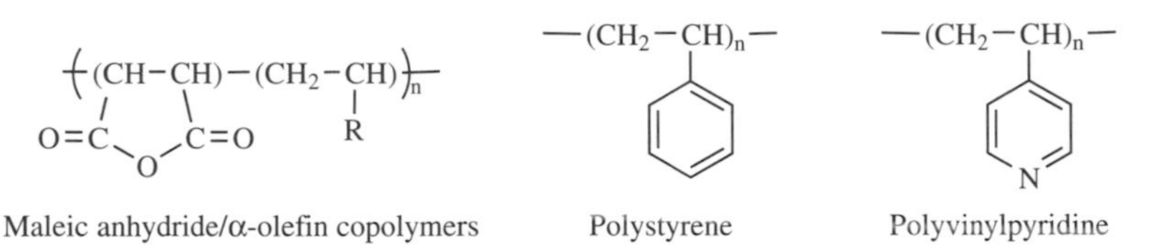

Maleic anhydride/α-olefin copolymers Polystyrene Polyvinylpyridine

Figure 7.2. Some basic chemical structures of common polymers used in the preparation of polymeric surfactants.

commercial applications. Coming from the opposite direction in terms of solubility, hydrophobic polymers such as polystyrene, A–B block copolymers, A–B–A–B alternating copolymers, and random copolymers can be functionalized to increase their hydrophilicity. Chemical structures of some of the more common candidates for functionalization are shown in Figure 7.2.

Polymers of those and other families can be modified to increase their hydrophobic or hydrophilic characteristics and are finding increasing use in the formulation of consumer products and in industrial operations. While their end-use performance depends on the nature and degree of functionalization and the basic nature of the base polymer backbone, the relationship between structural modifications and performance is not yet sufficiently understood to allow sweeping generalizations about structure–performance relationships similar to those generally applied to monomeric surfactants. Many of the same basic concepts are found to apply, however, so that the formulator looking into the use of such materials will not always start out working in the dark.

7.2. SOME BASIC CHEMISTRY OF POLYMERIC SURFACTANT SYNTHESIS

As noted above, many classes of water-soluble and water-insoluble polymers can be functionalized to increase their hydrophobicity or hydrophilicity, as needed. In some cases the resulting materials will be statistically random modifications, while in other cases the distribution of the relevant functional modifications will be more regular or uniform.

7.2.1. Modification of Natural Cellulosics, Gums, and Proteins

Natural product polymers such as cellulose, starches, gums, and proteins have been employed in a wide variety of applications for centuries. In many cases, particularly in the cases of gums and some proteins, the natural product is amphiphilic as isolated and can be used directly. While such "natural" amphiphilic materials have found wide use in the food, cosmetic, and pharmaceutical industries, as well as in industrial process applications, more stringent environmental, regulatory, and end-use requirements of modern products often require new modified products that, in some minds at least, take them out of the "natural" category. Semantic nit picking aside, modified natural polymers have a number of inherent advantages, including their generally lower environmental load in manufacture, their enhanced biodegradability because of their sources, and the theoretically renewable nature of their sources.

The modification of natural polymers to introduce or modify their amphiphilic character can take two general approaches: (1) degradation processes such as hydrolysis to reduce molecular weights or break crosslinking bonds to increase water solubility in normally insoluble materials such as cellulose; (2) chemical modification through the addition of molecular components that alter the nature of the molecules, usually involving esterification, amination, oxidation, etherification, or similar reactions; and (3) enzymatic modifications.

A classical example of such functionalization is the carboxymethylation of cellulose to produce carboxymethylcellulose (CMC), used in a wide variety of applications as a thickener, stabilizer, binder, or other agent. The functional characteristics of such materials can be controlled with some exactness by controlling the degree of functionalization and the molecular weight of the base polymer. Similar classes of materials can be prepared by the hydrolysis of proteins or the use of various reactions well known to the organic chemist.

7.2.2. Synthetic Polymeric Surfactants

Synthetic polymers can be equally useful as bases for the preparation of polymeric amphiphilc materials. A common example of the functionalization of a water-insoluble material is the sulfonation of polystyrene

$$\mathrm{P{-}C_6H_5 + SO_3 \rightarrow P{-}C_6H_4SO_3^-}$$

where P represents a polymer chain. Depending on the degree of sulfonation attained, such materials can remain water-insoluble, but with sufficient electrical charge to make then useful as binding reagents, or they can attain such a high charge that they dissolve or at least swell substantially in aqueous solution. In the form of crosslinked particles, sulfonated polystyrenes produce swollen microgel particles that find wide application as ion exchange resins. Other derivatives of polystyrene and other aromatic polymers can be prepared by any number of standard organic reaction schemes to produce other anionic or cationic polymers. Special reactive units can also be attached to the polymer sidechain that will, for example, bind specific antibodies for special applications. The sulfonation and derivitization of most water-insoluble polymers are essentially random processes, so that the exact distribution of charges or other modifications along the polymer chain will be somewhat variable. Close process control can, however, ensure a functionally reproducible product.

Polyacrylic acid and polyacrylamide are typical water-soluble polymers that can be functionalized by the addition of hydrophobic sidechains that turn them into amphiphilic materials. The usual processes would include esterification and amidation of the carboxyl groups. Similar results can be obtained by preparing copolymers of acrylic and methacrylic acid with the desired preformed esters. A third alternative is the formation of polyacrylate or polymethacrylate esters followed by controlled saponification or hydrolysis to a desired degree of free carboxyl groups. The derivitization of acrylics is also an essentially random process, so that the exact characteristics of the final product may be somewhat variable. Close process control can, again, ensure a functionally reproducible product.

$$\underset{\text{(esterification)}}{\text{P–COOH} + \text{ROH} \rightarrow \text{P–COOR}} \qquad \underset{\text{(amidation)}}{\text{PCOOH} + \text{RNH}_2 \rightarrow \text{PCONHR}}$$

$$\underset{\text{(saponification or hydrolysis)}}{\text{P–COOR} + \text{OH}^- \rightarrow \text{P–COO}^- + \text{ROH}}$$

An interesting class of polymers that can be functionalized in a more regular manner are alternating A–B polymers prepared by the reaction of maleic anhydride with terminal or α-alkenes discussed below.

Perhaps the most common and well-studied regular polymeric surfactant family is that of the block copolymers. Denoting one monomer type as A and the other as B, typical basic compositions would be A–B, A–B–A, and B–A–B. The chemistry and technology for preparing such polymers are well developed. The most common addition process essentially involves a multistep reaction sequence in which one monomer (e.g., A) is reacted with an acid or base catalyst to produce a homopolymer with a reactive terminal unit (A_n^-). When monomer A is used up or reacted to the desired extent, the reaction is treated with monomer B, which then begins to polymerize on the active terminal of A_n^-. After a suitable reaction time, the polymerization can be terminated to produce the A_n–B_m copolymer, or a second addition of A to $A_nB_m^-$ will produce the A–B–A family. The amphiphilic properties of the resulting product will depend on the relative amounts of A and B added and, to some extent, the order of addition. The process is illustrated schematically for a base catalyzed addition process as follows:

$$AH + B^- \rightarrow BH + A^- \quad \text{(initiation step)} \qquad A^- + nA \rightarrow A_n^- \quad \text{(propagation step 1)} \qquad A_n^- + mB \rightarrow A_nB_m^- \quad \text{(propagation step 2)}$$

$$A_nB_m^- + H^+ \rightarrow A_nB_mH \quad \text{(termination step)} \quad \text{or} \quad A_nB_m^- + qA \rightarrow A_nB_mA_q^- + H^+ \rightarrow A_nB_mA_qH \quad \text{(propagation step 3 and termination)}$$

The description here is, of course, very simplified. In reality, the process requires a great deal of technological skill; it is, nevertheless, quite achievable with a little work. In a few cases in which the reactivity of two monomers are very different, block copolymers can be prepared by a more simple free-radical polymerization process.

In this class of materials, the careful control of monomer feed and reaction conditions allows the preparation of surfactants in which such characteristics as the hydrophile–lipophile balance (HLB), solubility, wetting, and foaming properties can be closely and reproducibly controlled. The classification of these surfactants is based primarily on the nature of the initiator employed in the formation of the initial polymer block, with subclasses determined by the compositions of the various blocks. In fact, there is no need for the initiator in these materials to be particularly hydrophobic; the hydrophobicity is derived from one of the two polymeric blocks. Typical initiators would be monohydric alcohols such as butanol and dihydric materials (glycol, glycerol and higher polyols, ethylene diamine, etc.). A generic alkylene oxide molecule can be represented by

$$H_2C{-}CH{-}R \quad (\text{epoxide ring, bridged by } O)$$

where $R = H$ (ethylene oxide), $R = CH_3$ (propylene oxide), and so on. It is theoretically possible to use di-, tri-, and even tetra-substituted oxides, either symmetrically or asymmetrically substituted, although they are less common because of higher costs and poorer reactivity.

It is possible to envisage four subclasses of surfactants in this group based on the nature of the polymer blocks in the molecule. The first and simplest is that in which each block is homogeneous; that is, a single alkylene oxide is used in the monomer feed during each step in the preparation to give a species of the form

$$(A{-}A{-}A{-}A)_n{-}(B{-}B{-}B{-}B)_m$$

where A and B designate different alkylene oxides. Such materials will be referred to as all-block (AB) surfactants. A second subclass is termed the block–heteric

(BH) or heteric–block (HB) nonionics

$$(\text{AAAA})_n\text{–}(\text{B–B–C–B–B–B–C–C–B–B})_m$$

(BH)

or

$$(\text{A–C–C–A–A–A–A–C–C–A})_n\text{–}(\text{B–B–B–B})_m$$

(HB)

in which one portion of the molecule is composed of a single alkylene oxide while the other is a mixture of two or more such materials, one of which (C) may be the same as that of the homogeneous block portion of the molecule. In the preparation of such materials, the hetero portion of the molecule will be totally random. The properties of BH and HB surfactants will almost certainly differ from those of the AB surfactants. The fourth subclass will be that in which both steps in the preparation involve the addition of mixtures of alkylene oxides to give mixed heteric–block (MHB) materials with the schematic structure

$$(\text{A–A–C–A–C–C–C–A–C–A})_n\text{–}(\text{B–B–D–B–D–D–D–B–B–D})_m$$

(MHB)

In the preparation of the AB surfactants, a monofunctional initiator (i.e., one having a single acidic hydrogen) such as a monohydric alcohol (ROH), an acid (RCOOH), a mercaptan (RSH), a secondary amine (RR′NH), or an *N*-substituted amide (RCONR′H) is employed as the initiator. A controlled feed of the A oxide is added to obtain the desired average degree of polymerization, *m*, after which the feed of the B oxide is started. Such materials would generally be given by the formula

$$\text{I–A}_m\text{–B}_n$$

where I is the initiator molecule. For purposes of discussion, the A portion is an alkylene oxide unit in which at least one hydrogen of the alkylene oxide has been replaced by an alkyl or aryl group, and *m* is the degree of polymerization, usually greater than 6. The B linkage would then be a hydrophilic group such as polyoxyethylene, where *n* is again the average degree of polymerization. The most common commercially available members of this family are the polyoxypropylene–polyoxyethylene block copolymers. A wide range of surfactant properties can be achieved by the proper control of *m* and *n*.

The other subclasses of block polymer surfactants, BH, HB, and MHB, are prepared in essentially the same way. The major difference is that the monomer feed for the alkylene oxide in each step will be

For BH materials: pure A followed by B + C
For HB materials: A + C followed by pure B
For MHB materials: A + C followed by B + D

The H blocks will be random copolymers reflecting the composition of the oxide feed with its solubility characteristics determined by the relative ratios of potentially water-soluble and water-insoluble materials. These materials have a potential manufacturing advantage in that it is possible to vary the monomer feed composition during the reaction to continuously change the composition of the growing polymer chain. Specifically, as the reaction is being carried out, the number of each type of alkylene oxide available for addition to the growing chain changes and the chain composition changes accordingly. In this way, the resultant material may possess a high proportion of hydrophobic (or hydrophilic) units near the initiator, with a smooth transition to hydrophilic (or hydrophobic) toward the terminus of the chain.

It is also possible to employ multifunctional initiators to prepare AB surfactants of the form

$$[B_n\text{–}A_m]_x\text{–I–}[A_m\text{–}B_n]_y$$

where x and y may be 1, 2, and so on. Multifunctional initiators can also be employed in materials of BH, HB, and MHB types, although the chemistry and engineering for the commercial production of such materials could get a bit involved.

Probably the most studied and used members of the block copolymer surfactants are members of the polyoxypropylene–polyoxyethylene (POP–POE) family. A wide range of materials have been developed industrially that range from oil-soluble to highly water-soluble materials. The surfactant characteristics of each member of the family have been well documented so that the formulator can usually determine relatively quickly which product, if any, best suits her needs.

It is also possible to use functionalities other than alkylene oxides in the preparation of the polymer block surfactants (e.g., polyamines); however, there has been much less published about activity in that area.

Polymeric nonionic surfactants can also be prepared from prepolymerized hydrocarbon backbones. An interesting class of materials already mentioned, alternating copolymers, can be synthesized from polymers of maleic anhydride and styrene, α-olefins, and other interesting molecules. Maleic anhydride has the interesting (and useful) property that it tends to copolymerize with most other olefins in a regularly alternating manner: $MA + O \rightarrow (MA\text{–}O)_n$, where MA represents the maleic anhydride unit and O, the olefin component. An example of such a syntnesis would be the reaction of maleic anhydride with 1-hexadecene. The resulting polyanhydride can be further reacted with an alcohol, amine, and so on (RXH) to produce a "comb" polymer. If the "teeth" of the comb are, for example, short to medium-length POE chains, the resulting material is a potentially interesting nonionic comb polymer surfactant:

$$\underset{\text{O=C}\diagdown_{\text{O}}\diagup\text{C=O}}{\text{CH=CH}} + CH_2{=}CH{-}(CH_3)_{13}CH_3 \longrightarrow \underset{\text{O=C}\diagdown_{\text{O}}\diagup\text{C=O}}{(CH_2{-}CH_2{-}CH_2{-}CH{-}(CH_2)_{13}CH_3)_n}$$

$$\downarrow \text{RXH}$$

$$(CH_2{-}\underset{\substack{|\\ \text{O=C}\\ |\\ \text{X}\\ |\\ \text{R}}}{CH_2}{-}\underset{\substack{|\\ \text{C=O}\\ |\\ \text{X}\\ |\\ \text{H}}}{CH_2}{-}CH{-}(CH_2)_{13}CH_3)_n$$

The production of the half-ester of the maleic acid group (only one RXX group added per anhydride) leaves a free carboxylic acid just waiting to be reacted or neutralized to produce a new member of the family. As is often the case, the variety of surfactants attainable within a given group is sometimes limited only by the imagination of the synthetic organic chemist.

Another interesting possibility is the reaction of maleic anhydride with butadiene to give an unsaturated polymer backbone that can be further derivatized through the remaining ethylenic linkage:

$$\underset{\text{O=C}\diagdown_{\text{O}}\diagup\text{C=O}}{\text{CH=CH}} + CH_2{=}CH{-}CH{=}CH_2 \longrightarrow \underset{\text{O=C}\diagdown_{\text{O}}\diagup\text{C=O}}{(CH{-}CH{-}CH_2{-}CH{=}CH{-}CH_2)_n}$$

It is also possible to functionalize random polymers to produce interesting surfactant species, but they do not seem quite so elegant.

The discussion so far has been confined to hydrocarbon–based initiators and block groups. There is no fundamental reason for not including materials of other types in the construction of such surfactants. Fluorine-substituted carbon chains, polysiloxanes, and fluorinated polysiloxanes have become of increasing interest technologically because of their exceptional surface activity and utility in a wide variety of solvents and chemical environments. General interest in such materials is limited by the lack of reactive starting materials, difficulties in working with such materials (toxicity, etc.), and costs.

Polymeric biosurfactants have begun to receive attention because of their renewable source and potential natural compatibility for use in medical and pharmaceutical applications. They may be produced and extracted from a number of microbiological sources and have indicated great potential utility in specific applications. Since they are generally produced as products of fermentation processes, their cost, for now, at least, may preclude their wide application in general industrial or commercial products. It has also been found that the exact structure of the amphiphilic material produced and the overall yield of useful product can be highly sensitive to the characteristics of the source and the nature of the nutrients employed in the fermentation. The complexity of their chemical structures also make biosurfactants hard to study at the level of basic interfacial chemistry.

7.3. POLYMERIC SURFACTANTS AT INTERFACES: STRUCTURE AND METHODOLOGY

This section will introduce some of the basic factors controlling the behavior and function of polymeric surfactants in solution and in colloidal systems. Polymeric surfactants, in principle, at least, can perform all the same functions of "normal" surfactants; the main differences between the two are the wide variation in molecular mass between the two classes of materials, the higher-order structural conformations found for polymeric materials in solution, and the energetic and kinetic consequences of those conformations and changes to them imparted by changes in the complete functional system.

Certain classes of polymeric surfactants are found to associate in solution to give micelles, while others seem to prefer the solitary life of the lone, coiled polymer chain. Polymeric surfactant micelles can apparently perform most of the functions of normal micelles, including the solubilization of materials such as drugs, or they can associate into higher-level structures to give large changes in viscosity, which are useful for stabilizing colloidal dispersions. Although a great deal of information is not readily available, it may be assumed that at least some polymeric systems can form the equivalents of the mesophases found for normal surfactants. For a typical A–B block copolymer, a single-chain micelle may be pictured as shown in Figure 7.3, where the A units (heavy line) are hydrophobic and the B units (light lines) are hydrophilic. A similar picture can be used for protein micelles or globules where the different portions of the chain are sufficiently different in terms of their hydrophobic/hydrophilic character.

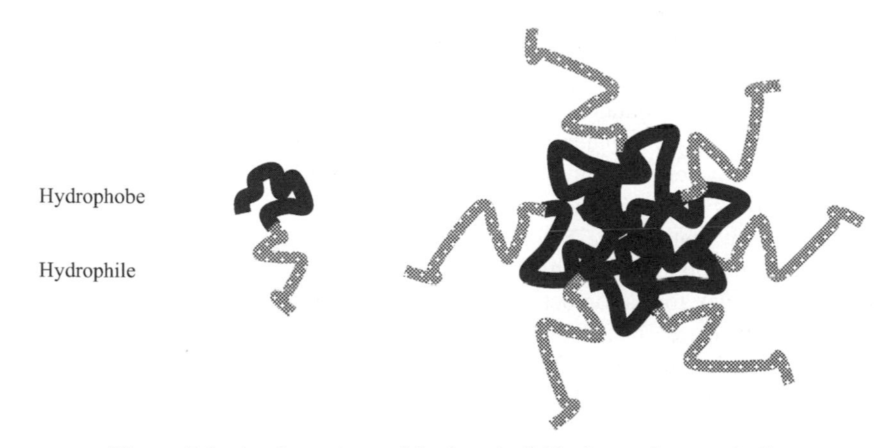

Figure 7.3. A schematic model of an A–B block copolymer micelle.

Polymeric surfactants adsorb at solid–liquid interfaces to give enhanced colloidal stability and at liquid–fluid interfaces to control (increase or inhibit) foaming or emulsion stability. The relatively wide molecular weight distribution of most commercial polymeric surfactants means that competitive adsorption processes can affect and be affected by other components in a formulation, as well as the end-use function of the system. With polymers, it often takes patience to know the final effect they produce.

Polymeric surfactants in general react to the same changes in solution properties as "normal" surfactants. Changing solution conditions such as temperature, electrolyte concentration, and type (i.e., monovalent vs. polyvalent), or the addition of nonelectrolyte solutes will usually result in some alteration of the characteristics of the system related to surfactant properties. That may be especially true of charged materials or polyelectrolytes. Because they are polymers, however, they also undergo complex conformational changes in solution that often respond to the same or different influences, and those conformational changes will almost certainly alter to some extent the surfactant properties of the material. Obviously, the complications of polymer solution characteristics superimposed on surfactant solution properties promise an interesting life to the intrepid voyager into the realm. Nevertheless, nature also usually provides a solution, and we find that materials or systems can usually be designed that are either resistant to many of the potential complications or that can actually take advantage of them to produce novel and useful effects.

7.4. INTERACTIONS OF "NORMAL" SURFACTANTS WITH POLYMERS

Interactions between surfactants and natural and synthetic polymers have been studied for many years because they are vitally important to the success of product formulations in many areas. Although the basic mechanisms of interaction are reasonably well understood, there still exists disagreement on the details of some of the surfactant–polymer interactions at the molecular level. Observations on changes in the interfacial, rheological, spectroscopic, and other physicochemical properties of surfactant systems containing polymers indicate that such interactions, regardless of the exact molecular explanation, can significantly alter the macroscopic characteristics of the system and ultimately its end-use functionality.

It is generally accepted that surfactant–polymer interactions may occur between individual surfactant molecules and the polymer chain (i.e., simple adsorption), or in the form of polymer–surfactant aggregate complexes. In the latter case, there may be a complex formation between the polymer chain and micelles, premicellar or submicellar aggregates, liquid crystals, and bicontinuous phases—that is, with any and all of the various surfactant aggregate structures described in Chapters 4 and 5. Other association mechanisms may result in the direct formation of what are sometimes called "hemimicelles" along the polymer chain. The term *hemimicelle* may be defined, for present purposes, as a surfactant aggregate formed

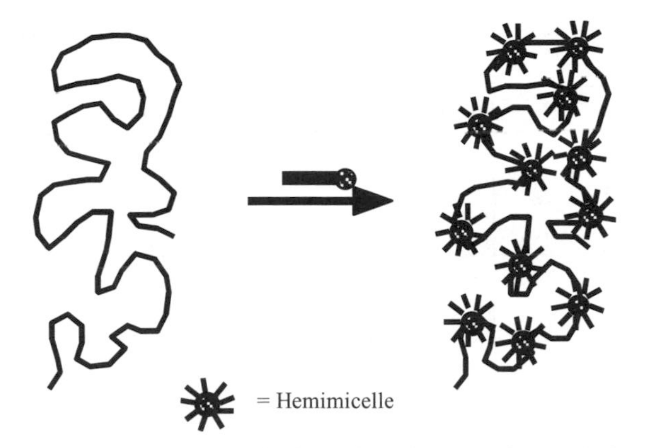

Figure 7.4. Hemimicelle formation along a polymer chain.

in the presence of a polymer chain or solid surface having many of the characteristics of a micelle, but being intimately associated with the locus of formation; hemimicelles obviously cannot exist as such in solution. The formation of such structures in surfactant–polymer systems is often illustrated as resembling a string of pearls or water droplets on a spider's web (Figure 7.4). It could also be hypothesized that hemimicelle structures could form on solid surfaces, especially heterogeneous surfaces that offer variations in the hydrophobic–hydrophilic environments available to interested surfactant molecules. The existence of such aggregates is still speculative, but surface chemistry still has a few ideas to teach us.

The basic forces controlling surfactant interactions with polymers are the same as those involved in other solution or interfacial properties of amphiphilic systems, namely, van der Waals and dispersion forces, the hydrophobic effect, dipolar and acid–base interactions, and electrostatic interactions. The relative importance of each type of interaction will vary with the natures of the polymer and surfactant so that the exact characters of the complexes formed may be almost as varied as the types of material available for study.

Experimental methods for investigating polymer–surfactant interactions vary widely, but they generally fall into two categories: those that measure macroscopic properties of a system such as viscosity, conductivity, and dye solubilization, and those that detect changes in the molecular environment of the interacting species such as nuclear magnetic resonance, optical rotary dispersion, and circular dichroism. A comparison of the experimental results of various studies can be complicated by variations in the sensitivity of experimental techniques and the physical manifestations of the interactions occurring, as well as differences in the purity and characterization of the experimental components. The results of each experimental approach, although useful in understanding the "symptoms" of surfactant–polymer interactions, do not always provide an unequivocal distinction among the possible mechanisms at the molecular level. Newer techniques such as small-angle neutron scattering, which come close to "photographing" the relative relationships among polymer and surfactant units, promise to clarify many

TABLE 7.1. Comparison of cac's and cmc's for Typical Surfactant–Polymer Combinations

Surfactant	Polymer[a]	cac (mM)	cmc (nM)
$C_{12}H_{25}–OSO_3^- Na^+$	PVP	2.5	8.3
$C_{12}H_{25}–OSO_3^- Na^+$	POE	5.7	8.3
$C_{10}H_{21}–OSO_3^- Na^+$	PVP	10	32
$C_{12}H_{25}N(CH_3)_3^+ Br^+$	$PAA^- Na^+$	0.03	16
$C_{12}H_{25}N(CH_3)_3^+ Br^+$	Sodium alginate	0.4	16
$C_{14}H_{29}N(CH_3)_3^+ Br^+$	$PAA^- Na^+$	0.0025	3.8
$C_{14}H_{29}N(CH_3)_3^+ Br^+$	Sodium alginate	0.03	3.8

[a]*Abbreviations:* PVP = polyvinylpyrrolidone, POE = polyoxyethylene, $PAA^- Na^+$ = sodium ployacrylate.

questions now in dispute. Unfortunately, it is not possible to purchase the required equipment out of the usual laboratory supplier's catalogs.

7.4.1. Surfactant–Polymer Complex Formation

When a surfactant is added to a polymer solution, it is often observed that processes such as micellization appear to begin at surfactant concentrations below the cmc of the material in the absence of polymer. In many cases, a complex aggregate structure is formed in association with the polymer coil at a concentration sometimes referred to as the "critical aggregation concentration" (cac), where cac < cmc, and varies with the nature of the amphiphile and the polymer (Table 7.1). It can be seen from the table that the difference between cac and cmc can vary by a factor of 10–1000 in some cases.

It is assumed, on the basis of experimental data, that the complex formed in these cases is between the polymer chain and a micelle or micelle-like structure. The significant differences between the cac's and cmc's is attributed to a stabilization of the aggregate structure by the polymer units. In the case of nonionic polymers such as PVP and POE, it might be assumed that the basic interaction is the hydrophobic interaction of the surfactant tail with the polymer chain.

The model of the aggregate would then be the "string of pearls" already mentioned. In the case of ionic polymers of charge opposite that of the amphiphile, it would be very surprising if the basic interaction were not electrostatic, at least until all the charge sites on the polymer chain are bound with surfactant molecules. In that case one can visualize a more complex structure that would involve an initial "coating" of charged sites along the chain through electrostatic interactions, producing a "hairy worm" complex. With the addition of more surfactant, hemimicelle seed regions might form that could then grow by the association of additional surfactant molecules through normal hydrophobic interactions. A hypothetical picture of such a process is given in Figure 7.5.

In the case of the interaction of SDS with the polar polymer POE, it is generally felt that the lower cac results from a stabilization of micelles by the adsorption of

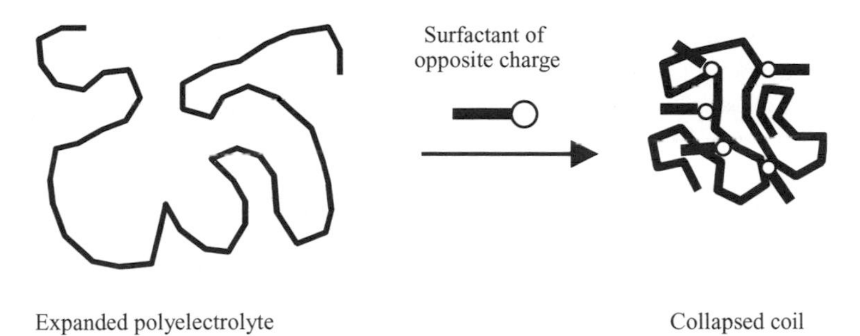

Figure 7.5. A schematic model for the interaction of charged polymer chains with surfactants of opposite charge.

polymer onto the surface of the aggregate (Figure 7.6). The question is whether the micelle exists before polymer adsorption occurs, or whether the polymer acts as a seed for the aggregation process. On the basis of the observed acceleration of the aggregation process, it would seem logical to think that the polymer chain is an active participant in the process (seed) rather than just sitting around to adsorb onto a preformed aggregate.

Figure 7.6. A schematic illustration of the effect of polymers on the aggregation of amphiphiles in solution.

A classical model for surfactant–polymer interaction is based on a stepwise sequence of binding between surfactant monomers (S) and the polymer chain (P), with each step governed by the laws of mass action, and with a unique rate constants k_i controlling each step:

$$\begin{aligned} P + S &\rightarrow PS && (k_1) \\ PS + S &\rightarrow PS_2 && (k_2) \\ PS_2 + S &\rightarrow PS_3 && (k_3) \\ PS_{n-1} + S &\rightarrow PS_n && (k_n) \end{aligned}$$

The question of whether the sequential addition of surfactant molecules to the complex involves an addition to an aggregate unit or simply the adsorption of individual molecules followed by the aggregation process when the coil is "saturated" cannot be easily answered. In either case, a given polymer chain can accommodate only a limited number of surfactant molecules or aggregates. When the surfactant concentration exceeds the capacity of the polymer present to form the complex, additional surfactant appears to continue its "life as usual" until its normal cmc is reached. The values of the various surfactant–polymer interaction rate constants and their dependence on experimental conditions (e.g., temperature, solvent, ionic strength, pH) serve as a basis for formulating feasible descriptions of the molecular processes involved in the interactions. The combination of macroscopic and molecular information can provide valuable insight into the overall process.

In the model described above, it is assumed that the stepwise binding process occurs initially through surfactant monomeric units; that is, there is no significant direct association of micelles or other aggregates with the polymer chain. The existence of such aggregate–polymer complexes is not excluded, however, since they may form on the chain as the total concentration of bound surfactant increases. It is only the stepwise association process that is limited to monomeric surfactant species. If polymer is added to a solution already containing micelles or other aggregate structures, a form of adsorption of polymer onto or into the aggregate cannot be ruled out. The nature of polymers in solution, especially the relatively slow kinetics of adsorption usually encountered, makes it difficult to say exactly what may be happening in the short term after mixing the two ingredients. That picture is also complicated by the differences of adsorption rates for polymer chains of different molecular weights. A smaller polymer chain will adsorb more rapidly, but it will also desorb more easily. A higher-molecular-weight chain will adsorb slowly, but be very slow to desorb. The result is that a given system may exhibit certain solution characteristics (e.g., viscosity) that will change with time as the polymer chains and surfactant accommodate themselves to the most favorable energetic situation; that is, as thermodynamics overtakes kinetics.

Like all surfactant-related phenomena, surfactant–polymer interactions involve a complex balance of factors encouraging and retarding association and are understandable only if those factors can be reasonably estimated. The complications added by the energetics of polymer conformations in solution only add to the potential for confusion. The dominating forces can be broken down into the categories of electrostatic attractions and repulsions, dipolar interactions, including hydrogen bonding or acid–base chemistry, dispersion and van der Waals forces, and the overall hydrophobic effect. In most cases, combinations of those forces are involved, thereby adding to the fun of interpreting the experimental results. While the electrostatic processes are fairly straightforward, involving the interaction of charged species on the polymer with those on the surfactant molecule, the remaining interactions are less easily quantified and can be quite complex. Polymers in particular add their own new twists, since in solution they will have secondary and tertiary structures that may be altered during the binding process in order to accommodate the bound surfactant molecules. The nature of the surfactant–polymer complex may significantly alter the overall energy of the system so that major changes in polymer chain conformation will result. Any and all of those changes may result in major alterations in the microscopic and macroscopic properties of the system.

Forces opposing the association of surfactant molecules with polymer chains include thermal energy, entropic considerations, solvent effects, and repulsive interactions among electrical charges of the same sign. It is clear that the strength and character of surfactant–polymer interactions depend on the properties of both components and the medium in which the interactions occur. However, even in systems where identical mechanisms are active for different surfactant and/or polymer types, the macroscopic symptoms of those interactions may be manifested in such a way that entirely different conclusions could easily be drawn.

As in the case of surfactants, four general types of polymer can be defined with respect to the electronic nature of the species: anionics, cationics, nonionics, and amphoterics. Not surprisingly, each polymer type will exhibit characteristic interactions with each surfactant class, with variations occurring within each group. It is little wonder, then, that surfactant–polymer interactions can produce some very "interesting" effects and become the subject of some lively discussions.

With the understanding that a great deal remains to be learned about the subject as a whole, the following comments will introduce a few of the observed facts about this field of study.

7.4.2. Nonionic Polymers

Probably the largest volume of published work in the field of surfactant–polymer interactions has involved surfactants and nonionic polymers such as polyvinylpyrrolidone (PVP), polyvinyl alcohol (PVA), polyvinyl acetate (PVAc), polypropylene glycol (PPG), methyl cellulose (MC), and polyethylene oxide (POE). The preferred surfactant has been (of course!) the classic—sodium dodecylsulfate (SDS). The results of most studies with SDS and similar surfactants indicate that the more hydrophobic the polymer, the greater is the interaction with anionic surfactants. For a given anionic surfactant, it has been found that adsorption progresses in the order PVP $\geq$ PPG $>$ PVAc $>$ MC $>$ PEG $>$ PVA. In such systems, the primary driving force for surfactant–polymer interaction will be van der Waals forces and

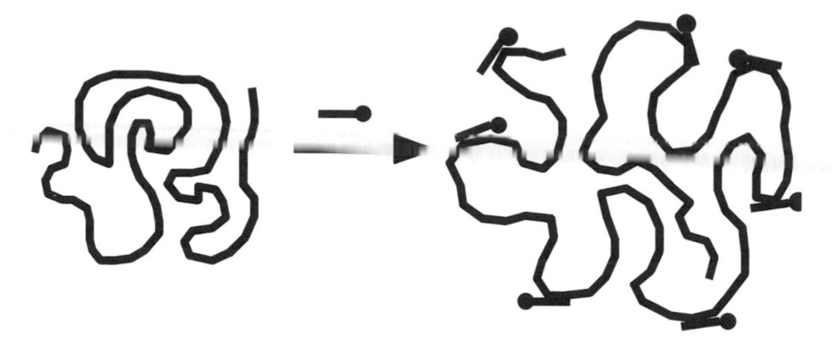

Figure 7.7. A schematic illustration of polymer coil expansion due to dispersion (hydrophobic) surfactant–polymer interaction.

the hydrophobic effect. Dipolar and acid–base interactions may be present, depending on the exact nature of the system. Ionic interactions will be minimal or nonexistent. For the polymer, it is reasonable to infer that the impact of the hydrophobic effect will be related to the ability of the polymer to undergo hydrogen bonding with the solvent (water), as well as the relative availability of nonpolar binding sites along the polymer chain.

If the primary mechanism of ionic surfactant–nonionic polymer interaction is hydrophobic, the adsorption of surfactant molecules will produce changes in the polymer chain conformation that are due to repulsions between the adsorbed ionic surfactant head groups (Figure 7.7). The properties of the solution will be altered as a result of such changes. For example, the solution viscosity may be found to increase substantially since the repulsion will force the polymer chain to uncoil or expand in the solution. If neutral salt is then added to such a system, repulsion between neighboring groups will be screened and the expanded coil will contract or collapse, thus again affecting various macroscopic properties of the solution such as viscosity. Such expansion and collapse of surfactant–polymer complexes as a function of the extent of surfactant adsorption may be seen as being analogous to the solution behavior of polyelectrolytes as a function of the degree of dissociation and electrolyte content. In fact, the surfactant–polymer complex may be viewed as a noncovalent polyelectrolyte.

The bulk of the work on cationic surfactant–nonionic polymer interactions has utilized long-chain alkylammonium surfactants. It has been found that the interactions between such species become stronger as the chain length of the surfactant is increased, reflecting the greater drive to substitute surfactant–polymer for surfactant–water and polymer–water interactions. The nature of the cationic head group seems to have a significant effect on polymer–surfactant interactions. It has been found that the reduced viscosity of aqueous solutions of dodecylpyridinium thiocyanate/PVAc changed very little with variations in the surfactant concentration, whereas solutions of dodecylammonium thiocyanate/PVAc showed considerable viscosity increases with increasing surfactant concentration. Such a result might be interpreted as reflecting a reduced extent of surfactant adsorption onto the polymer chain due to the greater hydrophilicity of the pyridinium ring relative to that of the simple ammonium group. The relative binding strengths between nonionic polymers and cationic or anionic surfactants are difficult to compare. The general trends are that with a given polymer, anionics will exhibit stronger interactions than analogous cationic surfactants, all other things being equal.

The interactions between nonionic surfactants and nonionic polymers has been much less intensively studied than those for ionic surfactants. The limited number of reports available indicate that there is little evidence to indicate extensive direct surfactant–polymer association in such systems. Considering the size of the hydrophilic groups of most nonionic surfactants, their low cmc's, and the absence of significant possibilities for head group–polymer interactions, the apparent absence of substantial interactions is not conceptually difficult to accept. An assertion that binding does not occur under any circumstance, however, would be foolish, given the complexities of polymer and surfactant science in general.

7.4.3. Ionic Polymers and Proteins

In many applications, it is more common for surfactants to encounter charged polymeric species rather than the nonionic examples discussed above. Practically all natural polymers, including proteins, cellulosics, gums, and resins, carry some degree of electrical charge. Many of the most widely used synthetic polymers such as polyacrylate (PAA^-M^+) and polymethacrylate (PMA^-M^+) salts do as well.

Polymers carrying electrical charges are usually termed "polyelectrolytes," although the term may not always be applied to natural polymers or gums that have a small number of charges per chain. When one compares the possibilities for interactions between polyelectrolytes and surfactants with those for nonionic polymers, it is obvious that the presence of discrete electrical charges along the polymer backbone introduces the probability of significant Coulombic interaction, in addition to the nonionic factors mentioned previously. The polymers may be positively or negatively charged, or they may be amphoteric. The presence of charge on a polymer complicates the understanding of the solution properties of the polyelectrolytes. The potential for surfactant–polyelectrolyte interactions does so even more.

Polyelectrolytes, whether natural or synthetic, are of particular interest to surfactant users because of their applications such as viscosity enhancers (thickening agents), dispersing aids, stabilizers, gelling agents, membrane formers, and binders. They are also encountered in fibers and textiles and in natural surfaces and membranes in biological systems. Common synthetic polyelectrolytes include polyacrylic and methacrylic acids and their salts, cellulosic derivatives such as carboxymethylcellulose; polypeptides such as poly-L-lysine; sulfonated polystyrenes and related strong-acid-containing polymers, polymeric polyammonium, and imonium salts; and quaternized polyamines. Commonly encountered natural polyelectrolytes include cellulose, proteins, and various gums such as pectin, arabic, xantham, locust bean, carrageenan, rosin acids, lignins, and keratins. In most cases, the charge on the polymer is fixed as either positive or negative, so that possible interactions with surfactants of a given charge type can be reasonably well defined. While such factors as pH, electrolyte content, and the nature of the polymer counterion will affect the extent of interaction in given systems, the sense of the interaction (anion–anion, anion–cation, etc.) may not change significantly except where protonation or deprotonation of weak acids and bases occurs. Other polymers, proteins in particular, will be amphoteric in nature, with the character of the net charge determined by pH.

Not surprisingly, interactions between surfactants and polymers of similar charge are usually minimal, with electrostatic repulsion serving to inhibit the effectiveness of any nonelectrostatic attractions. This is especially true for polymers having relatively high charge densities along the chain. That would be true for chains in which the charge distribution is relatively uniform. If the charges are, for some reason, clumped in specific regions of the chain, the door is left open for significant hydrophobic interaction in the un-ionized "bare" spaces along the chain.

When opposite charges are present, however, the expected high degree of electrostatic interaction is usually found to occur. In aqueous solution, the result of surfactant binding by electrostatic attraction is normally a reduction in the viscosity of the system, a loss of polymer solubility, at least to the point of charge reversal, and a reduction in the effective concentration of surfactant, as reflected by surface tension increases over what would be measured for that surfactant concentration in the absence of polymer.

Many naturally occurring random coil polyelectrolytes of a single charge type, including some carbohydrates, pectins, and keratins, are anionic and exhibit the same general surfactant interactions as do their synthetic cousins. Proteins, on the other hand, are amphoteric polyelectrolytes, which possess a net charge character (anionic or cationic) that depends on the pH of the aqueous solution. Unlike most synthetic polyelectrolytes, natural polyelectrolytes such as proteins usually have well-defined secondary and tertiary structures in solution that can affect, and be affected by, surfactant binding. When secondary and tertiary structures are present, complications arise because of alterations to those structures during surfactant adsorption. The denaturation of proteins by surfactants is, of course, just such a process of the disruption of higher orders of structure in the dissolved polymer molecule.

The question of exactly how a surfactant interacts with a protein molecule has been the subject of a great deal of discussion over the years. In the case of interactions between bovine serum albumin and sodium dodecylsulfate, the initial binding was found to involve the electrostatic association of oppositely charged species, especially at bound surfactant levels (i.e., the low number of surfactant molecules bound per polymer chain) of less than 10. As such binding occurs, the electronic character of the protein changes, possibly resulting in changes in its secondary and tertiary structures. Such changes may then lead to the exposure of previously inaccessible charge sites for further electrostatic binding or of hydrophobic portions of the molecule previously protected from water contact by the secondary protein structure. Ultimately, as charge neutralization occurs, precipitation of the protein may result.

As the charges on a protein are neutralized by specific surfactant adsorption, association between the hydrophobic tail of the surfactant and similar areas on the polymer becomes more favorable, again changing the net electrical character of the polymer complex. At sufficiently high surfactant–protein ratios, reversal of the native charge of the protein will be the result, as illustrated in the lower portion of Figure 7.8. Macroscopically, the foregoing events may lead to dramatic changes in the viscosity of the system as a result of, first, collapse of the polymer coil, followed by a rapid expansion after charge reversal has taken place. In addition, a minimum in the solubility of the polymer may be encountered, as evidenced by precipitation followed by repeptidization. The effect is similar to that for proteins as the pH of the solution is changed, causing the polymer to pass through its zero point of charge (zpc) at which its water solubility is at a minimum.

When the bound surfactant level is high, exceeding approximately 20 surfactant molecules per chain, evidence supports the view that both the head group and the hydrophobic portion of the molecule become involved in the binding process. Behavior suggesting such complex formation has been found for deionized bone gelatin in the presence of several anionic surfactants. It was found that the extent of interaction as reflected by increases in the viscosity of the system was highly dependent on the length of the hydrocarbon tail of the surfactant. For a series of sodium alkyl sulfates, the effect increased rapidly in the order $C_8 < C_{10} < C_{12} < C_{14}$.

The interactions between cationic and nonionic surfactants and proteins has received substantially less attention than the anionic case. Nonylphenol–POE nonionic surfactants undergo limited binding with proteins, although there is little evidence for sufficient interaction to induce the conformational changes found in the case of anionic materials. The limited number of results published on protein–cationic surfactant systems indicate that little cooperative association occurs in those systems, even if the native protein charge is negative.

Although it is clear that the surfactant binding processes are controlled by the same basic forces as the other solution and surface properties of surfactants, the location of binding sites on the polymer chain, the relative importance of the surfactant tail and head groups, and the exact role of the polymer structure remain to be more accurately defined. In any case, anyone proposing to use a surfactant in aformulation containing polymers, or in an application that entails surfactant–polymer interactions, must always consider the effect of each on the performance of the other.

More specific polymer–surfactant interactions can be obtained if the polymer chain is modified to introduce more hydrophobic groups along the chain. Such modifications provide more opportunity for hydrophobic interactions between the surfactant tail and the polymer. In addition, they may result in increased intra- and interchain polymer interactions, producing more compact chain conformations in aqueous solution or multichain polymer aggregates or micelles (Figure 7.9).

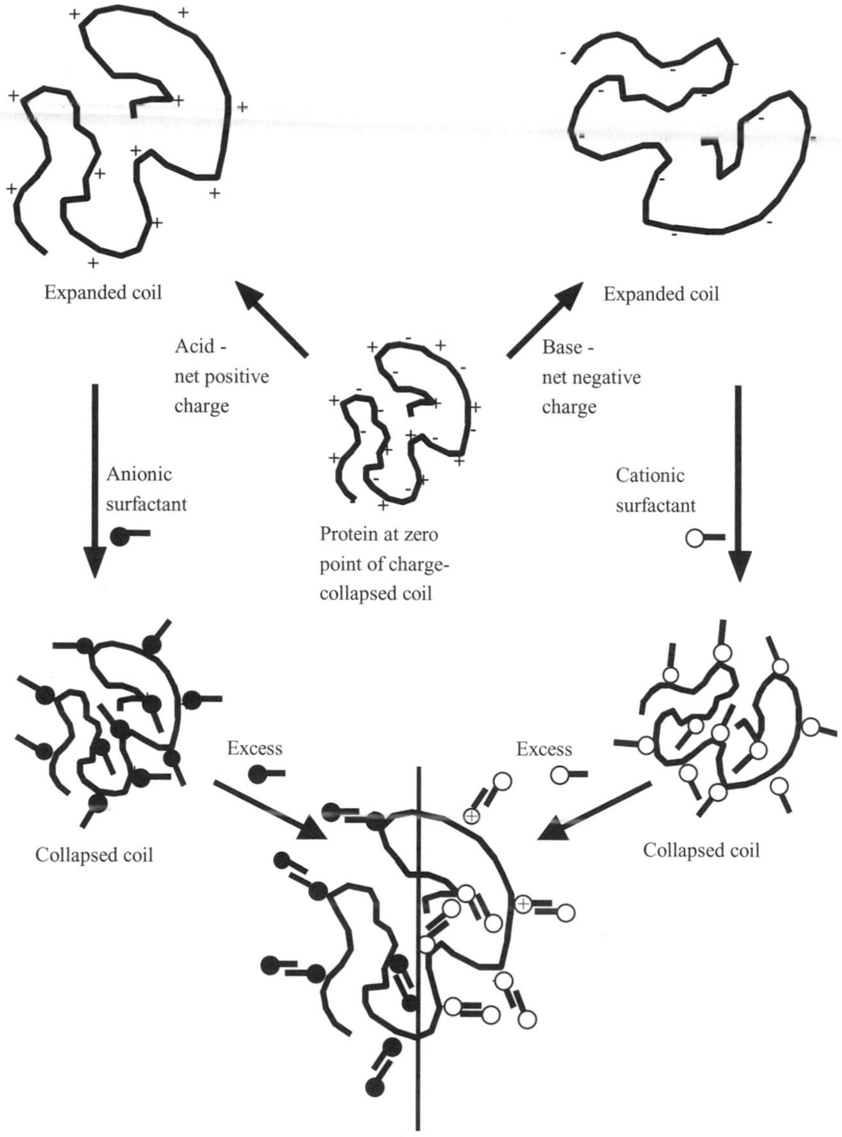

Figure 7.8. A representation of the collapse and charge reversal of proteins with change of pH and adsorption of surfactant of opposite charge.

7.5. POLYMERS, SURFACTANTS, AND SOLUBILIZATION

As discussed in Chapter 6, a useful characteristic of many micellar systems is their ability to solubilize water-insoluble materials such as hydrocarbons, dyes, flavors, and fragrances. Surfactant–polymer complexes, like polymeric surfactants, have

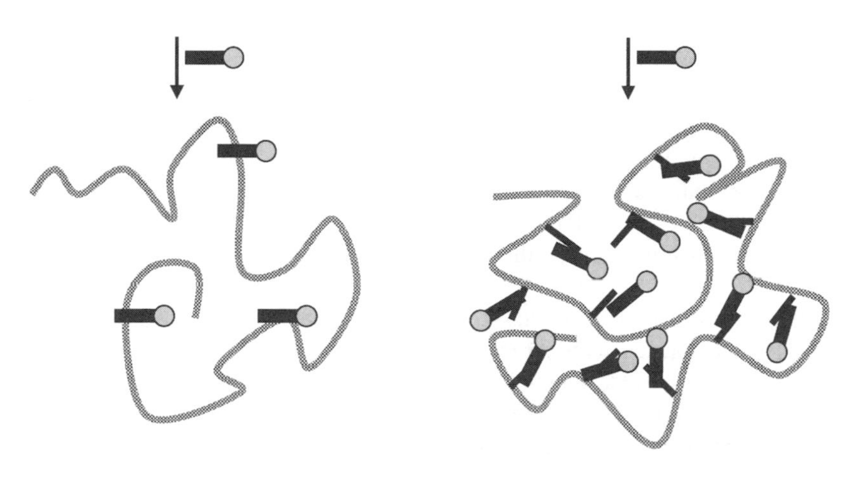

Figure 7.9. Possible solution effects of hydrophobic substitution on polymer chains and their potential affects on polymer–surfactant interactions.

been shown to solubilize materials at surfactant concentrations well below the cmc of the surfactant in the absence of polymer. The effectiveness of such complexes differs quantitatively from that of conventional micelles.

Studies have shown that the amounts of a hydrocarbon such as isooctane taken up by complexes of bovine serum albumin (BSA) and sodium dodecylsulfate (SDS) depended linearly on the number of surfactant molecules bound to the protein molecule. They also showed that a minimum SDS cluster size of about 30 molecules bound to the BSA was required before any solubilization occurred. The solubilization isotherms for BSA/SDS also differed from those of SDS micelles alone. Those results indicate that, for the same isooctane activity, the BSA/SDS complexes had a greater solubilizing power than did the micelles. In a similar way, it has been found that the addition of a polymer to a surfactant solution increased its solubilizing power, although no clear-cut correlations were established between the chemical structure of the polymer and its effect on solubilization. In general, larger effects are observed for aromatic than aliphatic materials. The present state of knowledge in this area is not sufficient to allow quantitative predictions to be made about the solubilizing properties of surfactant–polymer complexes based solely on chemical composition, although it is known that their effectiveness depends on the nature of the polymeric component and the polymer : surfactant ratio.

7.6. SURFACTANT–POLYMER INTERACTIONS IN EMULSION POLYMERIZATION

Surfactant–polymer systems have additional technological significance since surfactants are normally used in the emulsion polymerization of many materials, often involving the solubilization of monomer in micelles prior to polymerization and particle formation. Surfactants have also been shown to increase the solubility of some polymers in aqueous solution. The combined actions of the surfactant as a locus for latex particle formation (the micelle) in some cases, particle stabilization by adsorbed surfactant, and as a solubilizer for monomer permit us to expect quite complex relationships between the nature of the surfactant and that of the resulting latex.

Within a given surfactant class, it is usually found that materials with high cmc's produce latexes with larger particle sizes and broader size distributions. For example, the particle sizes of polyethylacrylate latexes prepared with sodium alkylsulfates ranged from 500 nm for the octyl to 50 nm for the dodecyl surfactant. No similar trend was found for nonionic POE surfactants as a function of POE chain length.

The ability of surfactants to form complexes with polymer chains may also affect the ultimate properties and stability of the resulting polymer, especially when the macromolecule exhibits some affinity for or reactivity with water. Perhaps the best documented case of the effect of surfactant on latex stability is that of polyvinyl acetate. The stability of PVAc latexes has been found to vary significantly depending on the surfactant employed in its preparation. It has also been found that PVAc could be dissolved in concentrated aqueous solutions of SDS and that it did not precipitate on dilution. The results suggest that, in this case at least, solubilization did not occur in the micelle, but that extensive adsorption of surfactant onto the polymer chain was required. They also indicate that a strong, stable PVAc–SDS complex is formed that produces a water-soluble structure that is essentially irreversible, unlike normal micelle formation. Cationic and nonionic surfactants had little or no solubilizing effect under identical conditions, indicating the specific nature of many, if not most, polymer–surfactant interactions.

PVAc is, of course, the precursor for the preparation of polyvinyl alcohol (PVA). The fact that surfactants such as SDS can promote the solubilization of polyvinyl acetate has been used to explain the observed increase in the rate of hydrolysis of PVAc polymers prepared with that surfactant relative to materials prepared with a less strongly interacting POE–sulfate surfactant. The hypothesis is that the SDS adsorbs onto and solubilizes the polymer surface, leading to a partial swelling of the particles, giving greater exposure to the hydrolytic environment, and loss of surfactant available for particle stabilization. As the surface of the PVAc particle is "eaten away" by the solubilization–hydrolysis process, the interior material becomes more exposed and subject to hydrolysis. The nonsolubilizing surfactants, on the other hand, would not facilitate the hydrolysis process, reducing the rate of PVAc exposure to reactants in the aqueous phase. They would also remain adsorbed at the particle surface and continue to perform their function as stabilizers.

In polymers that have very little affinity for water, such as for polystyrene or alkyl acrylates and methacrylates, little effect of surfactant on water solubility would be expected. The action of surfactants in such latex systems would be limited to their action as monomer solubilizers during preparation and as adsorbed stabilizers afterward.

The complex relationships that can develop between polymers and surfactants add a great many question marks to the interpretation of data obtained from such systems. They also open the door to possible new and novel applications of such combinations, however, and will no doubt provide many interesting hours of experimentation and thought for graduate students and industrial researchers in the future.

PROBLEMS

7.1. Other things being equal, the effect of changes in pH (range 2–12) on the viscosity of a dilute solution of a high-molecular-weight sodium carboxylate polymer will (a) be negligible, (b) increase at higher pH, (c) decrease at higher pH, (d) reach a maximum at intermediate pH, (e) reach a minimum at intermediate pH. Explain.

7.2. Adsorption isotherms of polymers on surfaces usually exhibit a "high affinity" character; that is, at low polymer concentration virtually all of the polymer is adsorbed, with very little left in solution (often immeasurable quantities). It is also common to find that the adsorption process is very slow and that adsorbed polymer cannot be readily removed by washing with the same solvent used for adsorption. Explain these observations using logical physical reasoning at the molecular level and, where possible, thermodynamic arguments as support.

7.3. It has been observed that for some solid dispersion–polymer systems, the direct, rapid addition of the dispersion to a polymer solution containing electrolyte results in little or no flocculation, while the addition of the same dispersion to the same solution in small portions results in flocculation of the system. For example, if a particular mineral dispersion is added in one step to a dilute solution of polyvinyl alcohol (PVA) with electrolyte, little change in the characteristics of the dispersion is observed. If the same dispersion is added in two portions—50% followed by the remaining 50%— complete flocculation occurred on the second addition. Explain.

7.4. Increasing the surfactant concentration in an emulsion polymerization usually leads to the formation of more latex particles. If the rest of the reaction mixture is unchanged, what will be the expected effect of the amount of surfactant on the final average particle size, on the rate of polymerization, and on the average degree of polymerization?

7.5. Many natural gums are used as thickeners and stabilizers in food products such as ice cream. It is often found that a mixture of gums and other polymers gives better results in terms of the final characteristics of the product than does any one material alone. Suggest some possible reasons for such an apparently synergistic effect.

7.6. A scientist wants to prepare a stable dispersion of a positively charged mineral in a high-molecular-weight hydrocarbon carrier. One requirement of the system is that it contain no low-molecular-weight components such as normal surfactants. Suggest a relatively accessible synthetic pathway for a polymeric surfactant that may suit her needs beginning with maleic acid and any hydrocarbon materials that may be needed. Specific reaction conditions are not required.

7.7. A series of A–B block copolymers, where A = POE and B = POP chains, were evaluated as solubilizers for a series of dyes. In the series, the molar ratios of the two chains (A/B) were as follows:

Surfactant	A/B
1	0.5
2	1.0
3	1.5

Which surfactant would you expect to perform better for each of the following dye classes: (a) a hydrophobic dye; (b) a polar, uncharged dye; (c) a positively charged dye?

7.8. On the basis of your conclusions in problem 7.7, where would you expect each dye to be "located" in the corresponding micelle?

7.9. What would you expect to happen when the following polymer and surfactant solution are mixed:

(a) Polyvinyl alcohol + SDS
(b) Sodium polyacrylate + SDS
(c) Sodium polyacrylate + $(CH_3)_3C_{12}H_{25}$–N + Cl^-
(d) Carboxymethyl cellulose + glycerol monostearate
(e) Pectin + Ca^{2+} $(C_{17}H_{35}COO^-)_2$
(f) Gelatin (zpc = pH 4.8) at pH =7.0 + $(CH_3)_3C_{12}H_{25}$–N + Cl^-
(g) Gelatin (f) + SDS

8 Foams and Liquid Aerosols

It was pointed out earlier that foams and emulsions are related in that they represent a physical state in which one fluid phase is finely dispersed in a second phase, and that the state of dispersion and the long-term stability (persistence) normally are dependent on the presence of one or more additives that alter the energy of the interface between the two phases. In emulsions, as each phase is a liquid, such factors as the solubility of additives in each phase must be considered. In foams, one phase (the dispersed phase) is a gas, so problems related to transfer of materials from the continuous to the dispersed phase effectively do not exist.

Liquid aerosols are, of course, the inverse of foams—liquid drops dispersed in a gas. However, the "nature of the beast" in the case of aerosols pretty much precludes any significant surfactant effects beyond that of affecting the particle size produced at the atomization stage. Some aspects of liquid aerosol technology are presented below. For the most part, however, this chapter addresses the basic role of surfactants in the formation and stabilization of foams, and gives some leads as to surfactant properties that may be useful for the suppression or elimination of foams where their presence would be considered detrimental.

The presence of foam in a product or process may or may not be desirable. Foams have wide technical importance in such fields as firefighting, in polymeric foams and foam rubbers, in foamed structural materials such as concrete, and, of course, in a myriad of food products. They also have certain aesthetic utility in many detergent and personal care products, although their presence may not add much to the overall effectiveness of the process. Foams also serve useful purposes in industrial processes such as mineral separation (froth flotation) and for environmental reasons for the suppression of liquid and fume emissions in some processes such as electroplating. In the latter case, the presence of a foam blanket over the electroplating solution helps prevent solution splattering and the loss of volatile materials, therefore reducing the costs of maintaining an acceptable working environment.

Unwanted foams may be a significant problem in many technical processes, including sewage treatment, coatings applications, and crude oil processing. By understanding the physical and chemical characteristics of materials that produce and sustain foams, it becomes easier to identify ways to counteract or overcome those foaming tendencies. The following sections cover some of the basic physical principles of foam formation and stabilization and discuss the molecular requirements of surfactants employed in foam formation or foam suppression.

As is so often the case in discussions of surfactants and their applications, the overwhelming bulk of the information available is based on results in water or aqueous solvent systems. The following material, therefore, concentrates on such systems. Foaming in nonaqueous systems is less commonly encountered, although the same physical principles would apply in such cases.

8.1. THE PHYSICAL BASIS FOR FOAM FORMATION

The basic anatomy of a foam is illustrated schematically in Figure 8.1. Because they are encountered in so many important technological areas, foams have been the subject of a significant amount of discussion in the literature. A number of reviews published over the years that cover most aspects of foam formation and stabilization are listed in the Bibliography. While the theoretical aspects of foam stabilization are reasonably well worked out, a great deal remains to be understood concerning the details of surfactant structural relationships to foam formation, persistence, and prevention. The physical nature of foams is quite complicated, and

Isolated foam bubbles

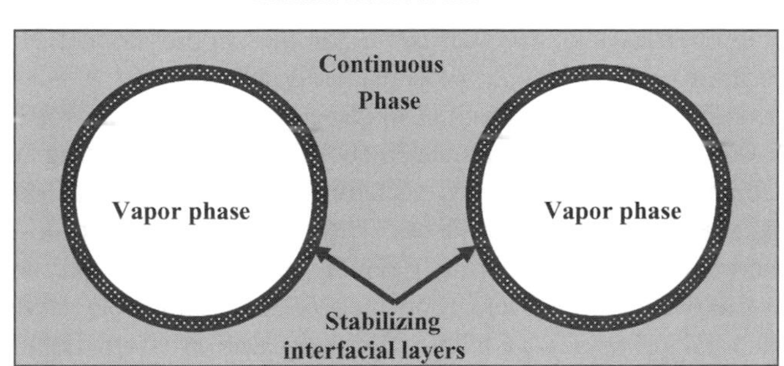

Contacting foam bubbles

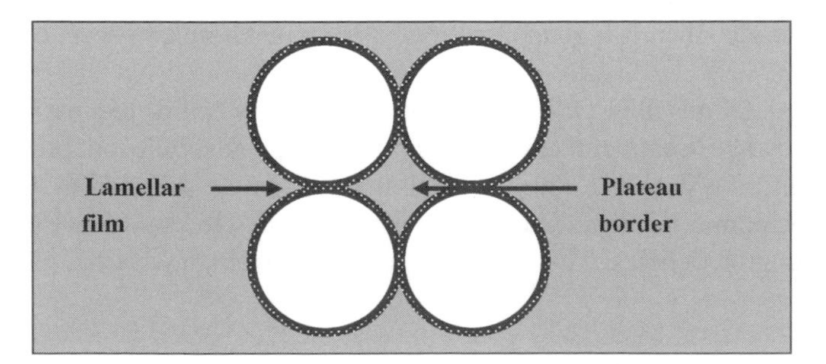

Figure 8.1. The basic anatomy of a foam structure.

conflicting explanations for foam stability found throughout the literature can sometimes be attributed to unwarranted extrapolations of data, over generalization, experimental anomalies, and, of course, honest differences in interpretation. It has been pointed out several times that the banc of the surfactant chemist is quite often the presence in a system of unwanted or unidentified amphiphilic materials. The presence of small amounts of unreacted starting materials, reaction byproducts, chemical homologs, and other factors can thwart even the most careful experimentalist, and the possibility of the existence of such circumstances must always be considered.

Foams, like almost all systems containing two or more immiscible phases, involve thermodynamic conditions in which the primary driving force is to reduce the total interfacial area between the phases—that is, the systems are thermodynamically unstable. The amount of thermodynamically reversible work required to create an interface is given by

$$W = \sigma_i \, \Delta A \tag{8.1}$$

where W is work in mJ, σ_i is surface tension in mJ/m^2, and A is in m^2.

In spite of their tendency to collapse, however, foams can be prepared that have a lifetime or "persistence" of minutes, days, or even months. The reasons for such extended lifetimes can be one or several of the following: (1) a high viscosity in the liquid phase, retarding drainage of the liquid from between the bubble interfaces, as well as providing a cushion effect to absorb shocks resulting from random or induced motion; (2) a high surface viscosity, which also retards liquid loss from between interfaces and dampens film deformation prior to bubble collapse; (3) surface effects such as the Gibbs and Marangoni effects (see text below), which act to "heal" areas of film thinning due to liquid loss; and (4) electrostatic and steric repulsion between adjacent interfaces due the adsorption of ionic and nonionic surfactants, polymers, and other compounds.

In general, one can define three classes of foams: unstable, "metastable," and solid. While all foams containing fluid phases are thermodynamically unstable, their degree of stability or persistence can vary from seconds to weeks. Unstable or low-persistence foams, as the term implies, remain for a very short time and collapse as a result of the overwhelming effects of surface tension and gravitational forces. More-or-less persistent foams can, however, be produced in the presence of extremely small amounts of amphiphilic substances or in the presence of polymers. As little as 5 ppm of saponin, a natural polymeric surfactant extracted from certain trees, in water can produce a foam of finite, though transient, stability.

Metastable foams have a persistence lasting from a few minutes to months. They are stabilized at the liquid–gas interface by the presence of amphiphilic and/or polymeric materials that retard the loss or drainage of liquid from the area between bubbles, or form a somewhat rigid, mechanically strong bilayer that maintains the foam structure. Because the stabilizing structure of the metastable foams is fluid, it can be disrupted by a number of factors such as vibration, dust particles, evaporation, pressure, and other environmental changes. Even the most stable of the metastable foams must eventually collapse, since the diffusion of gas from smaller to larger bubbles (Ostwald ripening) will occur, regardless of the strength of the interfacial film.

Neither of the first two foam types can be considered to be thermodynamically stable. The third class, the solid foams, could be so considered since they possess a mechanically rigid structure formed as a result of a (presumably) irreversible chemical process during or just after foam formation. Although formulations for the production of solid foams contain additives such as surfactants and blowing agents to produce the foam matrix, their action in sustaining the foam structure is negligible. Such foams therefore are not discussed further here.

Foams and emulsions have a great deal in common with regard to the basic physical principles controlling their stability. Some of the equations and concepts presented in this chapter, therefore, will be referred to in the following chapter, although the exact form may change due to the circumstances. The major differences lie in the natures of the dispersed phases (liquid vs. gas) and in the fact that foams will generally involve a much higher volume fraction of dispersed phase than normally is encountered in emulsions. For example, a typical foam (say, angelfood cake or ice cream) may have several factors of 10 more dispersed phase volume than would that of the continuous phase. The ratio of dispersed to continuous phase in an emulsion is unlikely to exceed 3 : 1. The theoretical limit for a monodisperse emulsion of spheres is about 76% dispersed phase, although that can be greater in polydisperse systems. If the dispersed phase is present as deformed spheres or polygons, its content can also exceed the theoretical limit.

When considering the physical and chemical factors involved in the formation and stabilization of foams, it is necessary to consider differences between foaming and nonfoaming systems in general. A foam is produced by the introduction of air or other gas into a liquid phase, during which time the bubbles become encapsulated in a film of the liquid. The thin liquid film separating two or more gas bubbles is referred to as a *lamellar film*, indicating that its nature is related to a layered (laminated) structure that possesses two essentially identical interfaces in close proximity. In the case of a foam of small bubble size, each interface will possess a significant degree of curvature, concave toward the gas phase. The Laplace equation, in the form

$$\Delta p = \sigma\left(\frac{l}{r_1} + \frac{l}{r_2}\right) \tag{8.2}$$

states that there will exist a pressure difference across each interface related to the major radii of curvature of the system, r_1 and r_2, and the interfacial tension σ. When three or more bubbles are in contact, especially when the foam has reached a generally stable honeycombed structure, a region will be developed in which the curvature of the lamellae is much greater than that in the main body of the system. These regions, referred to as "plateau borders" (Figure 8.1), possess a greater pressure difference than exists elsewhere in the foam. Since the gas pressure within the bubble must be the same throughout, the liquid pressure within the plateau borders must be lower than in the more parallel areas. As a result, fluid drainage occurs from the lamellar regions into the plateau borders. Liquid will also be drained from the lamellae because of gravitational forces, and the lamellae will become thinner and thinner; if a critical thickness is reached at which point the system can no longer sustain the pressure, collapse occurs.

The question of foam stability and bubble coalescence requires the consideration of both the static and dynamic aspects of bubble interaction. In the initial stages of film drainage, where relatively thick lamellar films exist between gas bubbles, gravity can make a significant contribution to the drainage of liquid from between foam bubbles. Once the films have thinned to a thickness of a few hundred nanometers, however, gravity effects become negligible and interfacial interactions begin to predominate. When the two sides of the lamellar film are in sufficiently close proximity, interactions can occur involving the dispersion, electrostatic, and steric forces already discussed in various surfactant contexts. Such forces, acting normal to (across) the lamellar film, make up what is normally referred to as the "disjoining pressure" of the system, $\pi(h)$. The net interaction energy between bubbles as a function of distance of separation through the lamellar phase will have a form similar to that in Figure 8.2, where the minima will correspond to metastable states in which $\pi = 0$, and the films will have some degree of equilibrium stability. In the plane parallel regions of the lamellar film, the Laplace (or capillary) pressure given by Eq. (8.2) will be zero. In the plateau border regions, however, that will not be the case and mechanical equilibrium requires that

$$\Delta p = -\pi(h) \tag{8.3}$$

Thus, the internal pressure of the bubbles is just balanced by the interfacial forces acting across the lamellar film.

Dynamically, foams may be subjected to any number of environmental stresses that will act to precipitate bubble coalescence and ultimate foam collapse. Regardless

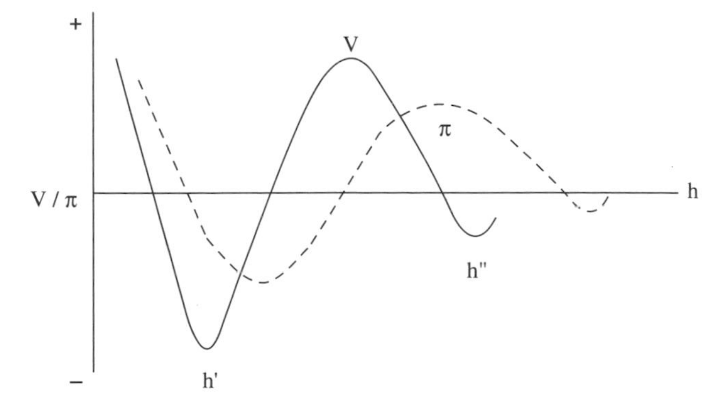

Figure 8.2. Foam bubble interaction energy diagram for total interaction energy (V) between bubbles and their disjoining pressure (π) as a function of separation distance (h).

of the nature of the stress, however, the ultimate cause of bubble coalescence will always be the same—the loss of liquid from the lamellar layer until a critical thickness of 5–15 nm is reached and the liquid film can no longer support the pressure of the gas in the bubble. As pointed out above, the loss or drainage of liquid from the lamellae can be affected by a number of factors, including (1) a high viscosity in the liquid, which will slow the drainage process and, in some cases, have a dissipating or buffering effect on many types of mechanical disturbances; (2) surface rheological effects, which can retard the loss of liquid by a viscous drag type of mechanism; and (3) the presence of repulsive electrical or steric interactions across the lamellae, which can oppose drainage through the effects of the disjoining pressure. The addition of surfactants to a foaming system can alter any or all of these system characteristics and therefore enhance (or reduce) the stability of the foam. Surfactants will also have the effect of lowering the surface tension of the system, thereby reducing the work required for the initial formation of the foam [Eq. (8.1)]. With the exception of bulk rheological effects, each of the phenomena relating surfactants to foam formation and stability is discussed in more detail below.

8.2. THE ROLE OF SURFACTANT IN FOAMS

For a liquid to form a foam, it must be able to form a membrane around the gas bubble possessing a form of elasticity that opposes the thinning of the lamellae as a result of drainage. Foaming does not occur in pure liquids because no such mechanism for the retardation of lamellae drainage or interfacial stabilization exists. When amphiphilic materials or polymers are present, however, their adsorption at the gas–liquid interface serves to retard the loss of liquid from the lamellae and, in some instances, to produce a more mechanically stable system. Theories related to such film formation and persistence, especially film elasticity, derive from a number of experimental observations about the surface tension of liquids:

1. As given by the well-known Gibbs adsorption equation

$$\Gamma_i = \frac{-1}{RT}\left(\frac{\delta\sigma_i}{\delta \ln a}\right)_T \tag{8.4}$$

   the surface tension of a solution will decrease as the concentration of the surface-active material in solution increases (assuming positive adsorption) up to its critical micelle concentration (the Gibbs effect).

2. The instantaneous or dynamic surface tension at a newly formed surface is always higher than the equilibrium value; that is, there is a finite time during which the amphiphilic molecules in the bulk solution diffuse to the interface in order to lower the surface tension (the Marangoni effect). The two effects are complementary, often discussed as the combined Gibbs–Marangoni effect,

and they serve as the basis for describing the mechanism of film elasticity under various conditions.

It is to be expected, as is observed, that the effects of normal surfactants and polymers or polymeric surfactants can be very time-dependent because of the significant differences in the rates of adsorption of the different species at the interface and, in some cases, the time required for specific polymer adsorption phenomena to take place. For example, in a polymer with a range of molecular weight species present, the smaller chains will routinely adsorb first, but they also desorb more readily and are, over time, replaced by the higher-molecular-weight chains. Such an exchange over time can produce a "ripening" effect in the adsorbed film leading to a more viscous or rigid structure. In systems that contain both monomeric amphiphiles and polymers, the original rapid adsorption of surfactant may be followed by a slow displacement of those species by the adsorption of the polymer.

The fundamental impact of surfactant concentration and diffusion rate in lamellar films can be illustrated as shown in Figure 8.3; as the lamellar film between adjacent bubbles is stretched as a result of gravity, agitation, drainage, and other motion, new surface is formed at some locations in the film having a lower instantaneous surfactant concentration, and a local surface tension increase occurs (*a*). A surface tension gradient (*b*) along the film is produced, causing liquid to flow from regions of low σ toward the new stretched surface, thereby opposing film thinning. Additional stabilizing action is thought to result from the fact that the diffusion of new surfactant molecules to the surface must also involve the transport of associated solvent into the surface area, again countering the thinning effect of liquid drainage. The mechanism can be characterized as producing a "healing" effect at the site of thinning.

Even though the Gibbs and Marangoni effects are complementary, they are generally important in different surfactant concentration regimes. The Marangoni

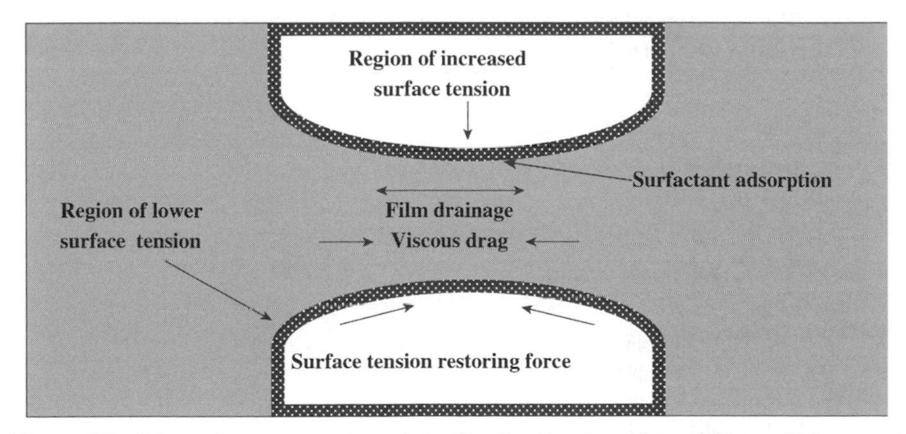

Figure 8.3. Schematic representation of the "healing" action of the Gibbs and Marangoni effects.

effect is usually of importance in fairly dilute surfactant solutions and over a relatively narrow concentration range. In the absence of external agitation, the amount of surfactant adsorbed at a new interface can be estimated by the equation

$$n = 2\left(\frac{D}{\pi}\right)^{1/2} ct^{1/2}\frac{N}{1000} \tag{8.5}$$

where n is the number of molecules per square centimeter, D is the bulk diffusion constant (cm^2/s), c is the bulk concentration of the surfactant (mol/L), t is the time in seconds, and N is Avogadro's number. Using Eq. (8.5), it is possible to estimate the time required for the adsorption of a given amount of surfactant at a new interface compared to the rate of generation of that interface. If the surfactant solution is too dilute, the surface tension of the solution will not differ sufficiently from that of the pure solvent for the restoring force to counteract the effects of thermal and mechanical agitation. As a result, the foam produced will be very transient. In line with the Marangoni theory, there should be an optimum surfactant concentration for producing the maximum amount of foam in a given system, under defined circumstances. Such effects have been verified experimentally (see Table 8.1).

In the case of the Gibbs effect, the increase in surface tension occurring as the film is stretched results from a local depletion of the surfactant concentration in the bulk phase just below the newly formed interface. Obviously, in systems such as foams where the available bulk phase in the narrow lamellae may be small compared to the amount of interface being formed, the effect will be enhanced. As with the Marangoni effect, if the surfactant concentration in the bulk phase is too low, a surface tension gradient of sufficient size to produce the necessary "healing" action will not be produced. Conversely, if the concentration is too large, well above the cmc, the amount of "immediately" available surfactant will be such that no gradient is formed.

TABLE 8.1. Typical Surfactant Concentrations Required to Attain Maximum foam Height, MFH

Surfactant	cmc (mM)	Concentration for MFH (mM)
$C_{12}H_{25}SO_3^-Na^+$	11	13
$C_{12}H_{25}OSO_3^-Na^+$	9	5
$C_{14}H_{29}SO_3^-K^+$	3	3
$C_{14}H_{29}OSO_3^-Na^+$	2.3	3
$C_{16}H_{33}SO_3^-K^+$	0.9	0.8
$C_{16}H_{33}OSO_3^-\ Na^+$	0.7	0.8
p-$C_8H_{17}C_6H_4SO_3^-\ Na^+$	16	13
p-$C_{10}H_{21}C_6H_4SO_3^-\ Na^+$	3	4.5
p-$C_{12}H_{25}C_6H_4SO_3^-\ Na^+$	1.2	4
o-$C_{12}H_{25}C_6H_4SO_3^-\ Na^+$	3	4
$(C_8H_{17})_2CHSO_4^-\ Na^+$	2.3	4

Quantitatively, the Gibbs effect can be described in terms of a coefficient of surface elasticity E, which is defined as the ratio of the surface stress to the strain per unit area

$$E = 2A\frac{\delta\sigma}{\delta A} \tag{8.6}$$

where A is total surface or interfacial area and σ is surface tension. Since the elasticity is the resistance of the film to deformation, the larger the value of E, the greater will be the ability of the film to resist mechanical shocks without rupture. As mentioned earlier, when a film of a pure liquid is stretched, no significant change in surface tension will occur and the elasticity as defined by Eq. (8.6) will be zero. This is the theoretical basis for the observation that pure liquids will not foam.

In addition to the Gibbs and Marangoni effects, foam stability can also be affected by surface and bulk solution transport phenomena. The surface tension gradients induced by the stretching of a film will result in the flow of liquid from a region of low to one of high surface tension. Since the movement of molecules in a liquid is never independent of adjacent molecules, such surface flow will result in the transport of bulk solution beneath the surface in the same direction; that is, underlying solution will be dragged in the same direction as, but to a smaller extent than, the surface layer itself.

The relationship between surface elasticity and surface transport is important since, if a film has a significant value of $E > 0$, stretching the film will produce an increase in the local surface tension and induce flow of bulk liquid into the stretched area, acting to restore the original thickness of the lamellae. Two surfactant solution processes must be considered in conjunction with these foam-stabilizing mechanisms. One is the rate of surface diffusion of surfactant molecules from regions of low to those of high surface tension. The second is the rate of adsorption of surfactant from the underlying bulk phase into the surface. In each case, a too rapid arrival of surfactant molecules at the new surface will destroy the surface tension gradient and prevent the restoring action of the Gibbs–Marangoni "healing" process

8.2.1. Foam Formation and Surfactant Structure

The relationship between the foaming power of a surfactant and its chemical structure is, as is the case for most surfactant applications, quite complex. The correlation of structure and foaming ability is further complicated by the fact that there is not necessarily a direct relationship between the ability of a given structure to produce foam and its ability to sustain that foam. It is generally found that the amount of foam produced by a surfactant under a given set of circumstances will increase with its bulk concentration up to a maximum, which occurs somewhere near the cmc. It appears, then, that surfactant cmc can be used as a guide in predicting the foaming ability of a material, but not necessarily the persistence of such foams. Any structural modification that leads to a lowering of the cmc of a particular class

TABLE 8.2. Foaming Characteristics of Typical Anionic and Nonionic Surfactants in Distilled Water (Ross–Miles Procedure, at 60°C)

Surfactant	Concentrtation (wt%)	Foam Height	
		Initial	After (min)
$C_{12}H_{25}SO_3^- \ Na^+$	0.25	225	205(1)
$C_{12}H_{25}OSO_3^- \ Na^+$	0.25	220	175(5)
$C_{14}H_{29}SO_3^- \ Na^+$	0.11	240	214(1)
$C_{14}H_{29}OSO_3^- \ Na^+$	0.25	231	184(5)
$C_{16}H_{33}SO_3^- \ K^+$	0.033	245	233(1)
$C_{16}H_{33}SO_3^- \ Na^+$	0.25	245	240(5)
$C_{18}H_{37}OSO_3^- \ Na^+$	0.25	227	227(5)
o-$C_8H_{17}C_6H_4SO_3^- \ Na^+$	0.15	148	—
p-$C_8H_{17}C_6H_4SO_3^- \ Na^+$	0.15	134	—
o-$C_{12}H_{25}C_6H_4SO_3^- Na^+$	0.25	208	—
t-$C_9H_{19}C_6H_4O(CH_2CH_2O)_8H$	0.10	55	45(5)
t-$C_9H_{19}C_6H_4O(CH_2CH_2O)_9H$	0.10	80	60(5)
t-$C_9H_{19}C_6H_4O(CH_2CH_2O)_9H$	0.10	110	80(5)
t-$C_9H_{19}C_6H_4O(CH_2CH_2O)_{13}H$	0.10	130	110(5)
t-$C_9H_{19}C_6H_4O(CH_2CH_2O)_{20}H$	0.10	120	110(5)

of surfactants, such as increasing the chain length of an alkyl sulfate, can be expected to increase its efficiency as a foaming agent. Conversely, branching of the hydrophobic chain or moving the hydrophilic group to an internal position, all of which increase the cmc, will usually result in a lower foaming efficiency. Typical foaming characteristics for several anionic and nonionic surfactants are given in Table 8.2, where foaming efficiency and persistence were determined according to the industry-standard Ross–Miles procedure.

The ability of a surfactant to perform as a foaming agent is dependent primarily on its effectiveness at reducing the surface tension of the solution, its diffusion characteristics, its properties with regard to disjoining pressures in thin films, and the elastic properties it imparts to interfaces. The amount of foam that can be produced in a solution under given conditions (i.e., for a set amount of work input) will be related to the product of the surface tension and the new surface area generated during the foaming process [Eq. (8.1)]. Obviously, the lower the surface tension of the solution, the greater will be the surface area that can be expected to be developed by the input of a given amount of work. The amount of foam produced by a surfactant solution is only one part of the foaming story, however. Maintenance of the foam may be as important as original formation.

It is often observed that the amount of foam produced by the members of a homologous series of surfactants will go through a maximum as the chain length of the hydrophobic group increases. This is probably due to the conflicting effects of the structural changes. In one case, a longer-chain hydrophobe will result in a lower cmc and a more rapid lowering of surface tension. However, if the chain length grows too long, low solubility and slow diffusion may become a problem.

It has been found in many instances that surfactants with branched hydrophobic groups will lower the surface tension of a solution more rapidly than will a straight-chain material of equal carbon number. However, since the branching of the chain increases the cmc and reduces the amount of lateral chain interaction, and the associated surface viscosity, the cohesive strength of the adsorbed layer and the film elasticity will be reduced, yielding a system with higher initial foam height but reduced foam stability. Similarly, if the hydrophilic group is moved from a terminal to an internal position along the chain, higher foam heights, but lower persistence, can be expected. In all such cases, comparison of foaming abilities must be made at concentrations above their cmc's.

Ionic surfactants can contribute to foam formation and stabilization as a result of the presence at the interface of an electrical double layer that can interact with the opposing interface in the form of the disjoining pressure. Additional stabilizing effect may be gained because of the requirement of the ionic group for a significant number of bound solvent molecules that will contribute to the steric (or entropic) contribution to the disjoining pressure π. Not surprisingly, it is found that the foaming effectiveness of such surfactants can be related to the nature of the counterion associated with the surfactant. It is found, for example, that the effectiveness of a series of dodecylsulfate surfactants with ammonium counter ions as foam stabilizers decreases in the order $NH_4^+ > (CH_3)_4N^+ > (C_2H_5)_4N^+ > (C_4H_9)_4N^+$. Such an order may reflect changes in the solvation state of the surfactant from highly dissociated (ammonium) to a more tightly ion-paired system (tetrabutylammonium).

Nonionic surfactants generally produce less initial foam and less stable foams than do ionics in aqueous solution. Because such materials must, by nature of their solvation mechanism, have rather large surface areas per molecule, it becomes difficult for the adsorbed molecules to interact laterally to a significant degree, resulting in a lower interfacial elasticity. In addition, the bulky, highly solvated nonionic groups will generally result in lower diffusion rates and less efficient "healing" via the Gibbs–Marangoni effect. POE nonionic surfactants in particular exhibit a strong sensitivity of foaming ability to the length of the hydrophilic chain. At short chain lengths, the material may not have sufficient water solubility to lower the surface tension and produce foam. A chain that is too long, on the other hand, will greatly expand the surface area required to accommodate the adsorbed molecules and will also reduce the interfacial elasticity. This characteristic of POE nonionic surfactants has made it possible to design highly surface-active, yet low-foaming surfactant formulations. Even more dramatic effects can be obtained by the use of "double-ended" surfactants in which both ends of the POE chain are substituted. In many cases, substitution of a methyl group on the end of a surfactant chain will significantly reduce foaming in materials with the same primary hydrophobic group and POE chain length.

If the solubility of a surfactant is highly temperature-dependent, its foaming ability will generally increase in step with its solubility. Nonionic POE surfactants, for example, exhibit a decrease in foam production as the temperature is increased and the cloud point is approached. Long-chain carboxylate salts, on the other hand, which may have limited solubility in water and poor foaming properties at room temperature, will be more soluble and will foam more as the temperature increases.

Quantitatively, the foaming abilities of some surfactants have been correlated with their Hildebrand or Hansen solubility parameters (see Bibliography), which is a semiquantitative, thermodynamically based molecular cohesion parameter that provides a simple method for predicting and correlating the cohesive and adhesive properties of materials on the basis of knowledge of their constituent parts. Although less has been published in relation to foaming in this area than in the areas of polymer solubility and miscibility or emulsions, it remains a potentially interesting approach for investigating structure–property relationships between surfactant structures and their activity in foam systems. Since foaming can be related to the solubility of the surfactant (too high a solubility results in low adsorption; too low, in insufficient availability of surfactant molecules), it is reasonable to expect good correlations between surfactant structure and foaming ability using Hansen parameters or related cohesive energy density approaches.

8.2.2. Amphiphilic Mesophases and Foam Stability

As we have seen, the stability of foams depends on a wide variety of factors involving several aspects of surface science. The potential importance of mesophase formation to the stability of emulsions and foams was briefly mentioned in Chapter 5. Although the phenomenon of mesophase stabilization of aqueous foams has been recognized for some time, although the role of such phases in nonaqueous foaming systems has been less well documented. However, since nonaqueous systems lack the advantages of electrostatic interactions in most aspects of their surface and colloid chemistry, it is not surprising to find that the presence of mesophase can serve as a sufficient condition for the production of stable foams in organic systems.

The role of mesophases in stabilizing a foam can be related to their effects on several mechanisms involved in foam collapse, including film drainage and the mechanical strength of the liquid film. The effect of mesophases on film drainage can be considered to be twofold. In the first place, the more extended and ordered nature of the mesophases impart a higher viscosity to the film than a normal surfactant monolayer. A simple-minded physical picture of the potential extent of mesophase penetration into the liquid lamellar phase would intuitively suggest that they should significantly affect the flow and drainage of liquid from between the two monolayers making up the bilayer film (Figure 8.4). It might also be expected that the sheer physical interaction between neighboring mesophase units would impart mechanical rigidity to the system. In addition, it has been shown that mesophases tend to accumulate in the plateau border areas. Their presence there results in an increase in the size of the areas, a larger radius of curvature, and thus a smaller Laplace pressure. The second stabilizing function of the mesophases can be related to the Gibbs–Marangoni effects, in that the presence of a large quantity of surfactant at the plateau borders allows them to act as a reservoir for surfactant molecules needed to maintain the high surface pressures useful for ensuring foam stability.

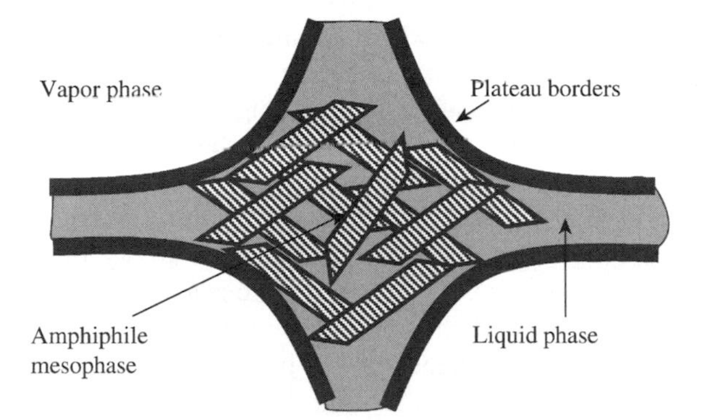

Figure 8.4. A schematic representation of the possible role of amphiphile mesophases in the stabilization of foams.

Obviously, the presence of mesophases or other structures such as, for example, mixed surfactant complexes, can not only increase the stability of the foam from a surface chemical standpoint but can also significantly enhance the physical strength of the system. When thinning reaches the point at which bubble rupture becomes important, the mechanical strength and rigidity of such structures might help the system withstand the thermal and mechanical agitation that would otherwise result in film failure and foam collapse.

8.2.3. Effects of Additives on Surfactant Foaming Properties

As we have seen, the foaming properties of a surfactant can be related to its solution properties through the cmc. It is not surprising, then, that additives in a formulation can affect foaming properties in much the same way that they affect other solution properties. The presence of additives can affect the stability of a foam by influencing any of the mechanisms already discussed for foam stabilization. It may, for example, increase the viscosity of the liquid phase or the interfacial layer, or it may alter the interfacial interactions related to Gibbs–Marangoni effects or electrostatic repulsions. By the proper choice of additive, a normally low-foaming material may produce large amounts of foam in the presence of small amounts of another surface-active material, which itself has few if any useful surfactant properties. Conversely, a high-foaming surfactant can be transformed into one exhibiting little or no foam formation by the judicial (or accidental) addition of the right (or wrong) additive. It is theoretically possible, then, to custom-build a surfactant formulation to achieve the best desirable combination of surfactant actions to suit the individual needs of the system. The use of small amounts of such additives has become the primary way of adjusting the foaming characteristics of a formulation in many practical surfactant applications.

As mentioned earlier, additives that alter the micellization of a surfactant will also affect its foaming properties. Such additives can be divided into three main classes: (1) inorganic electrolytes, which are most effective with ionic surfactants; (2) polar organic additives, which can affect surfactants of all types; and (3) macromolecular materials, which can affect the foaming properties of a system in many ways, some unrelated to the surface properties of the surfactant itself.

It was shown in Chapter 4 that the cmc of ionic surfactants can be very sensitive to the presence of electrolytes in the solution. It should not be surprising to find, then, that their presence also increases the foaming ability of such surfactants, within limits. The addition of excess monovalent ions to an ionic surfactant solution will, as expected, lower the cmc and improve the foaming ability of the amphiphilic material, although that improvement may not be particularly spectacular. If di- and trivalent ions are added in significant amounts to an anionic surfactant system, solubility problems may arise, reducing the foaming characteristics of the system considerably. On the other hand, such ions may enhance the foaming effectiveness of the surfactant in nonaqueous solvent systems. As a general rule of thumb, it can be assumed (or guessed) that any ionic additive that decreases the cmc of an ionic surfactant will increase its effectiveness as a foaming agent.

From a practical standpoint, perhaps the most important class of additives to enhance foaming is that of the polar organic materials, and they have received a corresponding amount of attention both academically and industrially. In the search for foam stabilizers for heavy-duty laundry formulations, investigators found that organic additives that lower the cmc of a surfactant could stabilize foams in the presence of materials that were normally detrimental to foam formation and persistence. Results indicated that increases in foaming and foam persistence induced by the additives could be related directly to the extent to which the material lowered the cmc of the surfactant. Straight-chain hydrocarbon additives with chain lengths approximately the same as that of the surfactant were the most effective at lowering the cmc and increasing foam height. Bulky chains on the additives produced much smaller effects. The effectiveness of polar additives of various types as foam stabilizers was found to be in the following approximate order: *N*-substituted amides > amides > sulfonyl ethers > glyceryl ethers > primary alcohols. This is essentially the same order found for the effects of such materials on the cmc of surfactants. Typical data showing the effects of polar additives on the cmc and foam persistence of sodium dodecylbenzene sulfonate solutions are given in Table 8.3. A quick glance at the ingredients list of almost all shampoos will indicate the presence of long-chain hydrocarbon amides to produce copious foam, perhaps more for aesthetic reasons than for aenhancing the cleaning ability of the product.

Not only does foam stabilization by polar organic additives seem to go hand in hand with the effect of the additive on the cmc of the surfactant; there is also a correlation with the relative amount of additive that is located in the interfacial film. The greater the mole fraction of additive adsorbed at the interface, the more stable is the resulting foam. Many of the most stable foaming systems were found to have surface layers composed of as much as 60–90 mol% additive.

Considering the mechanisms of foam stabilization mentioned above, it is not surprising to find that the addition of polymers to a surfactant solution will often enhance the foaming effectiveness of the system. In such systems, the added

TABLE 8.3. Effects of Organic Additives on the cmc and Foaming Characteristics of Sodium 2-*n*-dodecylbenzene Sulfonate Solutions

Additive	cmc (g/L)	Δ(cmc) (%)	Foam Volume (mL at 2 min)
None	0.59	—	18
Lauryl glycerol ether	0.29	−51	32
Laurylethanolamide	0.31	−48	50
n-Decyl glycerol ether	0.33	−44	34
Laurylsulfolanylamide	0.35	−41	40
n-Octylglycerol ether	0.36	−39	32
n-Decyl alcohol	0.41	−31	26
Caprylamide	0.50	−15	17
Tetradecanol	0.60	0	12

polymer may or may not affect the cmc of the surfactant, but it will undoubtedly affect the rheology of the liquid phase. Since foam production is usually a rapid, high-energy input process, new surface area is produced rapidly and the surfactant molecules must diffuse rapidly to that new interface. A typical water-soluble polymer will diffuse too slowly to be effective in foam formation; the great exceptions are polymers that are themselves surface-active such as proteins and other structures mentioned in Chapter 7. If the polymer additive does not contribute to initial foam formation, the increased solution viscosity in the lamellar film of the foam will naturally retard drainage and therefore enhance foam persistence. If the polymer can, over time, displace adsorbed surfactant from the air–liquid interface, the added interfacial viscosity and monolayer rigidity will normally add to the stability of the foam.

If polyelectrolytes are employed, the addition of di- and trivalent ions such as calcium and aluminum may produce particularly rigid and stable foams. If the foam is desirable, that is all well and good. In systems where foaming is not desirable, however, such effects can significantly complicate life. In wastewater treatment plants, for example, the combination of proteins and metal ions such as Al^{3+} can produce particularly troublesome foams that can almost shut a system down.

8.3. FOAM INHIBITION

Although the presence of some additives can enhance the foaming effectiveness and persistence of a surfactant system, the properly chosen materials can also reduce or eliminate foams. Such materials are termed "foam inhibitors," if they act to prevent the formation of foam, or "foam breakers," if they increase the rate of foam collapse. Foam breakers may include inorganic ions such as calcium, which counteract the effects of electrostatic stabilization or reduce the solubility of many ionic

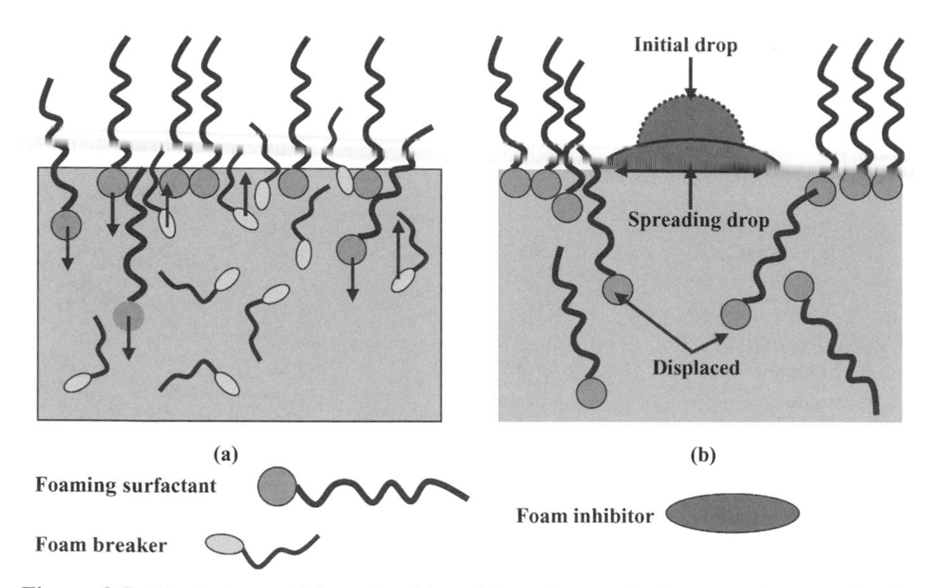

Figure 8.5. Mechanisms of foam breaking: (a) surfactant displacement (monomolecular film); (b) lens formation.

surfactants, or organic or silicone materials that act by spreading on the interface and displacing the stabilizing surfactant species.

A foam breaker that acts by spreading may do so by adsorption as a monolayer, displacing the surfactant molecules that would normally assist in foam formation and stabilization, or as a lens, accumulating in spots along the interface and leaving "weak spots" that can be easily ruptured by mechanical or gravitational forces (Figure 8.5). In addition, such defects in the lamellar walls will not normally be subject to the healing Gibbs–Marangoni effects. In either case, it is assumed that the spreading foam breaker sweeps away the stabilizing layer, leading to rapid bubble collapse. The rate of spreading of the defoamer will, of course, depend on the nature of the adsorbed layer present initially. If the surfactant can be easily displaced from the interface, the defoamer will spread rapidly, resulting in fast foam collapse, or essentially no notable foam formation. If the surfactant is not displaced rapidly, on the other hand, spreading will be retarded, or even halted. Foam collapse will then be a much slower process relying on the thinning or weakening of the lamellae by other drainage mechanisms.

Studies of the relationship between the action of defoamers and the concentration of the surfactant revealed that if the surfactant concentration was below the cmc, the defoamer was most effective if it spread as a lens on the surface rather than as a monolayer film. That is, the defoamer produced a defective lamellar structure that could not withstand the mechanical rigors of the foaming process for sufficient time for the normal stabilizing mechanisms to take hold. In the presence of micelles, the defoamer may be solubilized in the micelles, which can act as a reservoir for extended defoaming action by adsorption as a surface monolayer. When the solubilization limit was reached, additional defoaming effect was obtained by the lens spreading mechanism.

So far the discussion of foams and defoaming has centered on aqueous systems. While most organic liquids will do not form stable foams, the presence of polymers and oil-soluble amphiphiles can result in persistent foams, whether wanted or unwanted.

A classical and commonly encountered example of unwanted foaming in organic liquids is that of used frying oil. While a fresh vegetable oil or fat will not form foam, the frying process brings about several chemical changes that result in foaming systems. One effect is that the components of the material being fried, especially water, will slowly bring about saponification of the oil to produce free fatty acids and monodiglycerides in the system. Both materials are amphiphilic, of course. If neutralization of the fatty acids occurs, soap is formed. A second effect of the frying process is oxidation of the unsaturated fatty acid chains. At frying temperatures, the polyunsaturated acids such as linoleic and linolenic are particularly susceptible to oxidation, introducing peroxide and hydroxyl groups that enhance the amphiphilic character of the materials with the observed result. Such reactions also produce unwanted flavor changes in the fried product, of course. For those reasons, among others, frying oils have limited useful lifetimes and are often stabilized by the addition of antioxidants.

Another basically nonaqueous system that tends to produce unwanted foam is that of lubricating oils. Many such products contain amphiphilic materials for the purposes of reducing friction and corrosion control. However, since the oil is recycled, the presence of foam can foul the recycling mechanism. It has been found that silicone fluids, some of the few materials having the required characteristics of limited solubility and adequate surface tension lowering in organic liquids, act as foaming agents below their solubility limit, but inhibited foam formation when that limit was exceeded.

8.4. CHEMICAL STRUCTURES OF ANTIFOAMING AGENTS

Materials that are effective as antifoaming or defoaming agents can be classified into eight general chemical classifications, with the best choice of material depending on such factors as cost, the nature of the liquid phase, and the nature of the foaming agent present. One of the most common classes of antifoaming agents is polar organic materials such as highly branched aliphatic alcohols. As noted earlier, linear alcohols, in conjunction with surfactants, can increase foam production and stability due to mixed monolayer formation and enhanced film strength. The branched materials, on the other hand, reduce the lateral cohesive strength of the interfacial film, which increases the rate of bubble collapse. The higher alcohols also have limited water solubility and are strongly adsorbed at the air–water interface, displacing surfactant molecules in the process.

Fatty acids and esters with limited water solubility are also often used as foam inhibitors. Their mode of action is similar to that of the analogous alcohols. In addition, their generally low toxicity often makes them attractive for use in food applications. If the acids are neutralized to soaps, however, their antifoaming usefulness disappears. Organic compounds with multiple polar groups are, in general, found to be effective foam inhibitors. The presence of several polar groups generally acts to increase the surface area per molecule of the adsorbed antifoaming material and results in a loss of stabilization.

Metallic soaps of carboxylic acids, especially the water-insoluble polyvalent salts such as calcium, magnesium, and aluminum can be effective as defoamers in both aqueous and nonaqueous systems. In water, they are usually employed as solutions in an organic solvent, or as a fine dispersion in the aqueous phase. Water-insoluble organics containing one or more amide groups are found to be effective antifoaming agents in a number of applications, especially for use in boiler systems. It is generally found that greater effectiveness is obtained with materials containing at least 36 carbon atoms compared to simple fatty acid amides. An example of such a material would be distearoylethylenediamine:

$$C_{17}H_{35}CONHCH_2CH_2NHOCC_{17}H_{35}$$

Alkyl phosphate esters are found to possess good antifoaming characteristics in many systems because of their low water solubility and large spreading coefficient. They also find wide application in nonaqueous systems such as inks and adhesives. Organic silicone compounds are also usually found to be outstanding antifoaming agents in both aqueous and organic systems. Because of their inherently low surface energy and limited solubility in many organics, the silicone materials constitute one of the two types of material that are available to modify the surface properties of most organic liquids.

The final class of materials that have found some application as antifoaming agents are the fluorinated alcohols and acids, which are related to those discussed earlier. As a result of their very low surface energies, they are active in liquids where the hydrocarbon materials have no effect.

8.5. A SUMMARY OF THE FOAMING AND ANTIFOAMING ACTIVITIES OF ADDITIVES

As pointed out above, the mechanisms of action of foaming and antifoaming materials are quite often, understandably, opposite in nature. Foam stabilizers, for example, may increase the surface viscosity, leading to slower liquid drainage and lower gas permeability, or they may lower the critical micelle concentration of the primary foaming agent. They may also enhance the "healing" effects related to surface elasticity and the Gibbs–Marangoni effect. Antifoaming agents, on the other hand, may decrease surface viscosity and elasticity by displacing or disrupting the structure formed by adsorbed surfactant molecules. In addition, they can retard micelle formation and otherwise alter the surfactant properties of the system.

There are, however, similarities between the two types of material in that each usually has limited solubility in the aqueous liquid phase and each can produce a lowering of the surface tension of the system beyond that produced by the primary surfactant. However, the concentration levels at which the two are most effective differ. The antifoaming agents, for example, perform best when present at levels in excess of their solubility limit, being held in reserve by solubilization in micelles. Foam stabilizers, however, are most effective when completely dissolved and may, in fact, become antifoaming agents when present as a separate phase.

As noted earlier, antifoaming agents may act as a result of spreading over the foam surface. In such cases the film elasticity becomes essentially zero (as a lens), or they produce a surface film that provides little opposition to film drainage (as a monolayer). They are rapidly adsorbed at new interfaces so that the surface tension gradients necessary for the "healing" action of the Gibbs and Marangoni effects cannot develop.

It is often found that foam stabilizers can increase the solubility of antifoaming agents by increasing their solubilization in micelles. There is a sort of competition between the foam stabilizer trying to keep the antifoaming agent at bay while the surfactant does its job, and the antifoaming material trying to prevent the surfactant from doing just that.

8.6. THE SPREADING COEFFICIENT

Foaming agents can also prevent or retard the spreading of the antifoaming agents at the interface by lowering the liquid surface tension and producing an unfavorable spreading coefficient, $S^{\sigma}_{2/1}$, given by

$$S^{\sigma}_{2/1} = \sigma_1 - \sigma_2 - \sigma_{12} \tag{8.7}$$

where σ_1 and σ_2 are the surface tensions of the liquid substrate and spreading liquid, respectively, and σ_{12} is the interfacial tension between the two.

The spreading coefficient is a measure of the free-energy change for the spreading of liquid 2 over the surface (solid or liquid) 1 and is called the "spreading coefficient of 2 on 1," $S^{\sigma}_{2/1}$. Essentially, $S^{\sigma}_{2/1}$ is the difference between the thermodynamic work of adhesion of 2 to 1 and the work of cohesion of 2. In this context, the term "adhesion" refers to the interaction of two different materials and "cohesion" that of a material with itself. From Eq. (8.7), it is clear that $S^{\sigma}_{2/1}$ will be positive if there is a decrease in free energy on spreading (i.e., adhesive forces dominate), and the spreading process will be spontaneous. If $S^{\sigma}_{2/1}$ is negative, then cohesive forces will dominate and a drop or lens will result. What Eq. (8.7) also says is that when a liquid of low surface tension such as a hydrocarbon is placed on a liquid or solid of high surface energy such as clean glass or mercury, spontaneous spreading occurs. Conversely, if a liquid of high surface tension such as water is placed on a surface of lower surface energy such as Teflon or paraffin wax, drop or lens formation results.

Unfortunately, complications arise in spreading phenomena because liquids, solids, and gases tend to interact in bulk processes as well as at interfaces, and those bulk-phase interactions may have significant effects on interfacial phenomena. In particular, gases tend to adsorb at solid interfaces and change the free energy of those surfaces, σ_{sv}; they may also become dissolved in liquid phases and thereby alter the liquid surface tension. More importantly, liquids in contact with other liquids tend to become mutually saturated, meaning that the composition of the two phases may not remain "pure" and no longer have the surface characteristics of the original materials. Finally, liquids and solutes, like gases, can adsorb at solid interfaces to alter the surface characteristics of the solid and thereby change the thermodynamics of the spreading process. A classic example of the effects of such complications is that of the benzene–water system.

For a drop of pure benzene ($\sigma_2 = 28.9$ mN/m) placed on a surface of pure water ($\sigma_1 = 72.8$ mN/m) with an interfacial tension, σ_{12} of 35.0 mN/m, Eq. (8.7) predicts a spreading coefficient of

$$S^{\sigma}_{2/1} = 72.8 - 28.9 - 35.0 = 8.9\,\mathrm{mN/m}$$

The positive spreading coefficient indicates that benzene should spread spontaneously on water. When the experiment is carried out, it is found that after an initial rapid spreading, the benzene layer will retract and form a lens on the water. How can this seemingly anomalous result be explained?

In this and many similar cases, it must be remembered that benzene and many other liquids of low water miscibility have, in fact, a small but finite solubility and the water will rapidly become saturated with benzene. Benzene, having a lower surface tension than water, will adsorb at the water–air interface so that the surface will no longer be that of pure water but that of water with a surface excess of benzene. The surface tension of benzene-saturated water can be measured and is found to be 62.2 mN/m, which is now the value that must be used to calculate the spreading coefficient instead of that for pure water, so that

$$S^{\sigma}_{2/1(2)} = 62.2 - 28.9 - 35.0 = -1.7$$

where the subscript 1(2) indicates phase 1 saturated with phase 2. The negative spreading coefficient indicates that lens formation should occur, as is observed. The saturation process occurs, of course, in both phases. However, since water is a material of relatively high surface tension, it will have little tendency to adsorb at the benzene–air interface and will therefore cause little change in the surface tension of the benzene. In this case $\sigma_{2(1)} = 28.8$ mN/m so that

$$S^{\sigma}_{B(A)/A} = 72.8 - 28.8 - 35.0 = 9.0$$

If only the benzene layer were affected by the saturation process, spreading would still occur. Combining the two effects, one obtains

$$S^{\sigma}_{2(1)/1(2)} = 62.2 - 28.8 - 35.0 = -1.6$$

indicating that it is the effect of benzene in water that controls the spreading (or nonspreading) in this system. The interfacial tension of water–benzene is unchanged throughout because it inherently includes the mutual saturation process.

Situations like that for benzene are very general for low-surface-tension liquids on water. There may be initial spreading followed by retraction and lens formation. A similar effect can in principle be achieved if a third component (e.g., a surfactant) that strongly absorbs at the water-air interface, but not the oil–water interface, is added to the system. Conversely, if the material is strongly adsorbed at the oil–water interface, lowering the interfacial tension, spreading may be achieved where it did not occur otherwise. This is, of course, a technologically very important process and will be discussed in more detail in later chapters.

For normal use, it is assumed that the values of σ represent equilibrium saturation values. The antifoaming materials reduce the strength of the surface film by reducing the lateral van der Waals interactions between adsorbed molecules due to branching in the hydrophobic tail. They may also be made to lie flat in the surface by the inclusion of several hydrophilic groups along the chain, by placing the hydrophile in the middle of the chain, and by using the smallest number of methylene groups in the chain consistent with the necessity for limited solubility.

In summary, it can be said that the various aspects of foam formation and persistence are related to the actions of surfactant molecules and additives at the various interfaces in the system, coupled with the rheological characteristics of the system, including the dilational viscosity of the interfacial layers and the bulk rheological properties of the system. Depending on whether foam is wanted, the choice of surfactants and additives for a formulation must address all of those factors in the context of the system being prepared and its end use.

8.7. LIQUID AEROSOLS

Mists and fogs are colloidal dispersions of a liquid in a gas. They may therefore be regarded as being the inverse of foams. The interactions controlling their stability, however, are not generally the same as those involved in foam stabilization, because most mists and fogs do not possess the thin lamellar stabilizing films encountered in foams. In fact, the stability of liquid aerosols is usually more dependent on fluid dynamics than on colloidal factors, as illustrated below.

8.7.1. The Formation of Liquid Aerosols

Liquid aerosols may be formed by one of two processes, depending on whether the dispersed system begins as a liquid or undergoes a phase change from vapor to liquid during the formation process. In the first case, since the dispersed material does not change phases, the aerosol is formed by some process that reduces the particle size of the liquid units. To this class belong spray mists such as those formed at the bottom of a waterfall or by ocean waves (impact), mists produced by vigorous agitation (mechanical breakup), and those formed by some direct spraying or atomization process. Liquid aerosols can also be formed directly by the application of high electrical potentials to the liquid. The second class of mists or fogs is that produced by some process in which the incipient liquid phase is introduced as a vapor and forms droplets as a result of some equilibrium condensation process or the liquid is produced as a result of some chemical reaction. The former mechanism includes, of course, cloud and fog formations, while the latter corresponds to some "chemical" fogs and mists.

8.7.1.1. Spraying and Related Mechanisms of Mist and Fog Formation

Liquid aerosol formation by spraying is a very important industrial process, even though some of the fundamental details of the process are still not very well understood. Major applications include paint application; fuel injection in diesel, gasoline, and jet engines; spray drying of milk, eggs, and other foods; the production of metal and plastic powders (spray cooling); medicinal nose and throat sprays; the application of pesticides to crops; and many more. In all of those applications, it is vitally important that the characteristics of the aerosol produced be optimized to produce the desired particle size and dispersion. Theories related to the formation of drops in spray systems can be very helpful in approximating the conditions necessary to produce an aerosol of defined characteristics. However, because of the nature of the process and the incidence of hard to control external factors, it is usually necessary to arrive at the optimum spraying system by trial-and-error techniques based on previous experience in the field.

Aerosol spays may be formed mechanically by one of four basic processes as illustrated in Figure 8.6. These include

1. Directing a jet of liquid against a solid surface, thereby breaking the liquid up into fine droplets.
2. Ejecting a jet of liquid from an orifice into a stream of air or gas.
3. Ejecting a stream of liquid from a small orifice under high pressure.
4. Dropping liquid onto a solid rotating surface from which small droplets are ejected by centrifugal force.

Other systems exist but are of much less significance in practice. Of the four, the most important industrially are the high-pressure orifice or nozzle sprays and some variation of the rotating disk. For that reason they will be discussed briefly below.

Although the production of aerosols by spray techniques is of great practical importance, the physics of the processes are still not completely understood. Numerous attempts have and are being made to quantify and understand the phenomena involved in order to get a better practical handle on the matter. Most of those treatments are quite complex and beyond the scope of this book. However, it may be instructive to work through two relatively simple approaches in order to see how surface tension forces, and by association surfactants, can come into play.

Spray production by methods involving high-speed ejection of a liquid through an orifice (nozzle atomization) and ejection from a spinning disk by centrifugal

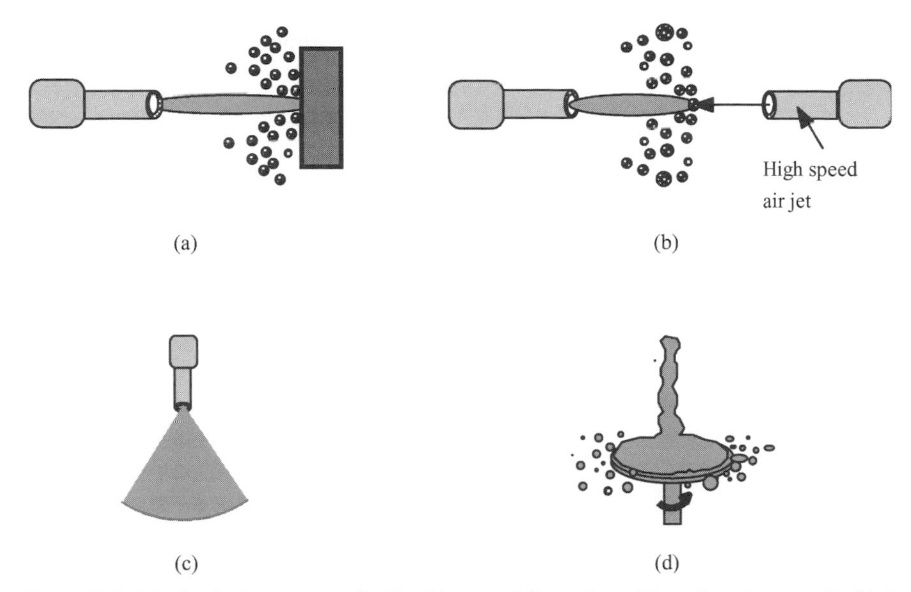

Figure 8.6. Mechanical processes for liquid aerosol formation: (a) surface impact of a high-pressure liquid stream; (b) airjet impact—collision of high-velocity liquid and gas streams; (c) high-pressure spray nozzles; (d) spinning disk centrifugal atomizers.

force (rotary atomization) are the simplest and most important situations because they require knowledge of only one material velocity—that of the liquid. Spray production by the action of an incident air stream on a jet of liquid involves, of course, the velocity of both the liquid and the air.

8.7.1.2. Nozzle Atomization

If a liquid is forced through an orifice (nozzle) under a pressure, the velocity of the liquid in the channel of the orifice becomes so high that turbulent flow is encountered; that is, the liquid will not flow smoothly in lines parallel to the walls of the orifice but will flow in complex patterns with eddies, swirls, and vortices. When the liquid leaves the orifice in this turbulent—or, to use a more fashionable term, chaotic—state, the angular forces in the vortices will act against the surface tension of the liquid to strip off units of liquid to form droplets.

For a simple, classical analysis of the situation, assume that as the liquid leaves the orifice it has not only a linear velocity due to the pressure forcing it through the system but also some angular velocity ω resulting from its chaotic flow pattern. Liquid will therefore rotate within the jet with a period of $2\pi/\omega$. The rotation creates a local centrifugal force F_ω. For a column of exiting liquid of radius r and height dz that force is given by

$$F_\omega = \frac{2}{3}\pi\rho r^3\omega^2\, dz \tag{8.8}$$

where ρ is the density of the liquid. The pressure disrupting the jet will be given by

$$P = \frac{2}{3}\pi\rho r^3\omega^2 dz/2\pi r\, dz = \frac{1}{3}\rho r^2\omega^2 \qquad (8.9)$$

The surface tension forces keeping the jet together will be σ/r. The second radius of curvature for the jet in the direction of travel will be infinitely large. The critical radius at which a continuous jet of liquid becomes unstable and breaks up to form droplets will be

$$r_c = \left(\frac{3\sigma}{r\omega^2}\right)^{1/3} \qquad (8.10)$$

It is difficult, of course, to determine the value of ω in a flowing system, so experimental verification of such an analysis is not a trivial matter. However, if one assumes that ω is proportional to the injection pressure, the product of the pressure and r^3 should be constant. In practice, the agreement is not quite exact. If one were to use an excess pressure—that is, the pressure in excess of that at which chaotic flow begins—the agreement might logically be expected to improve.

Since theories for predicting the drop size of a spray on the basis of the characteristics of the liquid and the apparatus are complex and sometimes unsatisfactory, it is usually necessary to measure sizes for each given situation. In general, however, the following rules hold for most fluid ejection systems:

1. Increasing the surface tension of the liquid will increase the drop size.
2. Increasing the viscosity of the feed liquid will increase the drop size.
3. Increasing jet pressure decreases the average drop radius.
4. Increasing the nozzle diameter increases drop size.

The drop size for a given liquid system can also be controlled somewhat by the use of oval instead of round nozzle orifices, which will induce additional rotational force to the emerging liquid jet.

8.7.1.3. Rotary Atomization

In rotary atomization, a liquid is fed onto the center of a spinning disk or cup and accelerated to high velocity before being ejected into a gaseous atmosphere (Figure 8.7). Under ideal circumstances, the liquid is extended over the entire surface of the spinning element in a thin film. When it reaches the edge, the liquid can suffer one of three fates: (1) droplets may be formed directly at the edge (Figure 8.7a), (2) the liquid may leave the surface in filaments that subsequently break up into droplets (Figure 8.7b), or the liquid may be detached as a sheet which later breaks up to form droplets (Figure 8.7c).

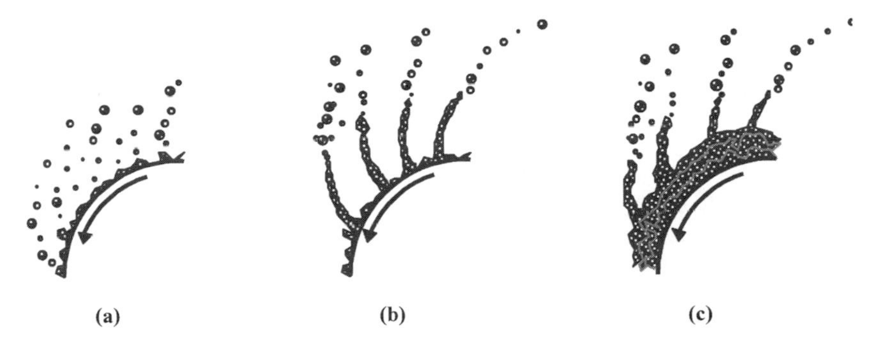

Figure 8.7. Mechanisms and location of drop formation in spinning disk or cup atomizers: (a) direct formation at the disk edge; (b) formation from strings of liquid leaving the disk edge; (c) formation from liquid sheets leaving the disk edge.

The operative mechanism of drop formation will be controlled by various system factors, including

1. The viscosity and surface tension of the liquid.
2. The inertia or kinetic energy of the liquid at the edge.
3. Frictional effects between the liquid and the air it encounters at the edge.
4. Shear stresses present in the liquid as it leaves the spinning edge.

At relatively slow spin speeds and low liquid feed rates, viscosity and surface tension forces predominate. In that case, direct drop formation at the disk edge is usually found. The drops usually consist of a primary drop (relatively large) and several smaller "satellite" drops. Higher spin speeds and feed rates lead to drop formation from strings (mechanisms 2) and sheets (mechanism 3) in which inertial and frictional forces begin to dominate.

If the spinning disk is a smooth, flat surface, the spreading liquid will tend to "slip" over the surface and not attain the maximum theoretical tangential velocity expected on the basis of the mechanical spinning speed. In effect, the liquid film looses traction on the disk surface and is not flung out toward the disk edge with maximum efficiency. The phenomenon is an example of wetting failure found in some linear high-speed coating operations and can lead to a significant amount of film defects. In most spraying operations the spinning disk is not smooth but has a series of vanes that "force" the liquid onto the surface so that more speed is attained before the liquid separates from the edge. That results in a smaller average drop diamcter for the same spin speed. Cup-shaped elements are also employed in situations where very small particles are not required. The effect of changes in various conditions on the average particle size to be expected are given in Table 8.4.

The preceding brief treatment of aerosol drop formation by ejection processes illustrates that theoretical analysis can be used in predicting an approximate result based on a given set of circumstances. However, much more complex analyses are

TABLE 8.4. A Summary of Some Basic Rotary and Nozzle Atomizer and Feed Liquid Characteristics and Their Expected Effects on Average Drop Size of Aerosol Produced

Characteristic	Expected Effect of Increase on Drop Size
	Disk Atomizer
Disk diameter	Decrease
Disk speed	Decrease
Liquid feed rate	Increase
Liquid density	Increase
Liquid viscosity	Increase
Surface tension	Increase
	Nozzle Atomizer
Orifice diameter	Increase
Liquid feed rate	Increase
Liquid feed pressure	Decrease
Liquid viscosity	Increase
Surface tension	Increase (small effect)

necessary to obtain more than a "ballpark" figure, and even then the results may not justify the effort. In liquid aerosol formation, as in many such areas, experience is often the best guide.

8.7.2. Aerosol Formation by Condensation

A "chemical" method for the production of aerosol involves the direct condensation of drops or particles in the air or other gaseous environment. In order for a vapor to condense under conditions far from its critical point, certain conditions must be fulfilled. If the vapor contains no foreign substances that may act as nucleation sites for condensation, the formation of aerosol drops will be controlled by the degree of saturation of the vapor, analogous to the situation for homogeneous crystal formation.

The formation of a new phase by homogeneous nucleation involves first the formation of small clusters of molecules, which then may disperse or grow in size by accretion until some critical size is reached, at which point the cluster becomes recognizable as a liquid drop. The drop may then continue to grow by accretion or by coalescence with other drops to produce the final aerosol. Normally, extensive drop formation is not observed unless the vapor pressure of the incipient liquid is considerably higher than its saturation value, that is, unless the vapor is supersaturated.

The barrier to the condensation of the liquid drop is related to the high surface energy possessed by a small drop relative to its total free energy. Thermodynamically, a simple argument can be given to illustrate the process. If one considers the condensation process as being

$$nA(\text{gas}, P) \leftrightarrow A_n \text{ (liquid drop)}$$

where n denotes the number of molecules of gas A at pressure P involved in the process, then in the absence of surface tension effects, the free-energy change of the process will be given by

$$\Delta G = -nkT \ln\frac{P}{P_0} \tag{8.11}$$

where P is the pressure or activity of A in the vapor phase and P_0 is that in the liquid phase. The ratio P/P_0 is often referred to as the *degree of supersaturation* of the system. A liquid drop of radius r will have a surface energy equal to $4\pi r^2\sigma$, so that the actual free-energy change on drop formation will be

$$\Delta G = -nkT \ln\frac{P}{P_0} + 4\pi r^2\sigma \tag{8.12}$$

Both elements to the right in Eq. (8.12) can be written in terms of the drop radius r. If ρ is the density of the liquid and M its molecular weight, the equation becomes

$$\Delta G = -\frac{4}{3}\pi r^3 \frac{\rho}{M} RT \ln\frac{P}{P_0} + 4\pi r^2\sigma \tag{8.13}$$

where the two terms are of opposite sign and have a different dependence on r. A plot of ΔG versus r exhibits a maximum as illustrated in Figure 8.8 for a hypothetical material with a density of one, molar volume of 20, and pressure or activity ratio of 4 at a given temperature. The radius at which the plot is a maximum may be defined as the critical radius, r_c, which can be determined from Eq. (8.13) by setting

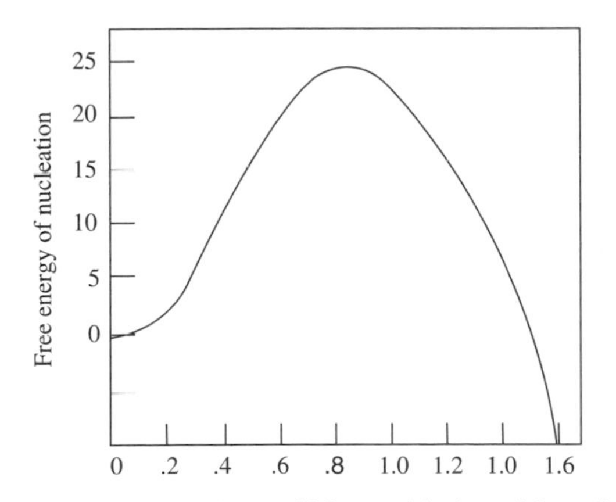

Figure 8.8. In nucleation processes there will be a critical particle radius r_c below which free-energy considerations will cause the incipient aerosol particle to evaporate; above r_c, particle growth will occur.

$(\Delta G)/dr = 0$. That transformation gives the old faithful Kelvin equation, which on rearrangement leads to

$$r_c = \frac{2\sigma V_m}{RT} \ln \frac{P}{P_0} \tag{8.14}$$

where V_m is the molar volume of the liquid.

For water at 25°C and supersaturation (P/P_0) of 6, Eq. (8.14) predicts a critical radius of 0.58 nm, corresponding to a cluster size of about 28 water molecules. It is difficult to say whether a drop of that small size actually has the same properties as the bulk liquid. It is probable, in fact, that the relatively high surface : volume ratio in such an assembly will result in an actual surface tension greater than the "true" bulk value of 72 mN/m. If a larger value for σ is used, the value of r_c decreases. The same occurs as the degree of supersaturation increases. The uncritical quantitative use of Eq. (8.14) can be misleading in that it predicts critical cluster sizes for homogeneous nucleation that are unlikely to occur with much frequency if left to the chance of random fluctuation processes. Qualitatively, however, the equation is useful in explaining the difficulty of forming liquid aerosols by direct condensation in highly purified systems.

If one combines Eqs. (8.13) and (8.14), it is possible to obtain a value for the free energy of formation of a cluster of the critical radius for drop formation, ΔG_{max}:

$$\Delta G_{max} = \frac{4\pi r_c^2 \sigma}{3} = \frac{16\pi\sigma^3 M^2}{3\rho^2 (RT \ln P/P_0)^2} = 16\pi\sigma^3 V_m^{2/3} \left(RT \ln \frac{P}{P_0} \right)^2 \tag{8.15}$$

Conceptually, one can think of the nucleation process in the following terms. If the pressure or activity of the vapor P is small relative to P_0, then ΔG for a given cluster of molecules will increase with each added molecule; that is, the tendency will be for clusters smaller than r_c to return to the vapor phase.

Statistically, one might expect to encounter clusters of all sizes due to random fluctuation processes; however, all except the smallest would be very uncommon. There would therefore be little likelihood of obtaining the critical radius necessary for drop formation to occur. However, as the degree of supersaturation increases, r_c decreases and random fluctuations begin to result in more clusters with that radius. Once that point is reached, the clusters begin to grow spontaneously to form drops. When a specific supersaturation pressure is exceeded, there will develop a steady parade of clusters of the required critical dimensions, resulting in the formation of a visible mist or fog.

8.7.3. Colloidal Properties of Aerosols

While aerosols are "typical" colloids in that they theoretically respond to the same forces as other members of the class—that is, electrostatic, dispersion, and van der

Waals interactions—the special conditions that prevail in terms of the intervening gaseous medium result in a significant qualitative difference from colloids in liquid media. The importance of the intervening medium to the character and interactions of colloidal particles due to the screening effect of the continuous phase on the particle–particle interactions cannot be overemphasized. In aerosols, although the fundamental rules remain the same, the screening effect of the gaseous medium becomes relatively insignificant so that a number of adjustments in thinking must be made in order to reconcile the apparent differences between liquid and solid aerosols and foams, emulsions, and dispersions.

In a first analysis, we can identify at least four basic differences between aerosols and other colloids related to the dispersion medium: (1) buoyancy effects, (2) the effects of movement of the dispersing medium, (3) particle mobility in undisturbed conditions (i.e., free fall), and (4) modification of interactions by the intervening medium. In emulsions, foams, and dispersions buoyancy can be important in determining the stability of a system (i.e., matching the densities of dispersed and continuous phases can retard creaming or sedimentation). In aerosols, where the density of the continuous phase will always be significantly less than that of the dispersed particles, such effects are practically nonexistent—the colloid is essentially left to its own devices: the usual interactions found for all colloids, the "constant" pull of gravity (assuming we are not aboard the International Space Station), and the whims of the winds.

8.7.3.1. The Dynamics of Aerosol Movement

The study of the dynamics of fluid flow is concerned with the forces acting on the bodies in the fluid. In the basic analysis of foams, emulsions, and dispersions, fluid dynamics is largely ignored in favor of "true" colloidal interactions. In aerosols, the nature of the continuous medium makes the subject of fluid dynamics much more important to the understanding of the system, so that the following discussion will introduce a few basic noncolloidal relationships that can be important in their study.

"Winds," in the form of convection currents or other movements of the gaseous medium, are generally more important in gases than in liquids. Small temperature differences or mechanical movements that would be damped out quickly in a more viscous liquid may be translated over large distances in gases and produce a much greater effect in aerosols. (Remember the famous Chinese butterfly that can change the weather in Kansas according to chaos-based theories of weather development?)

In a static system of relatively high viscosity (relative to that of gases), inertial forces due to particle movement are seldom significant; that is, viscous forces dominate. In gases, the forces resulting from particle movement become more important and must be considered in a dynamic analysis of the system. In dynamic fluid flow analysis, the ratio of inertial forces (related to particle mass, velocity, size, etc.) to viscous forces (a characteristic of the medium and not the particles) in a system is a dimensionless number termed the Reynolds number, Re, and is used to define the type of flow occurring in the system (i.e., laminar or turbulent). For spherical

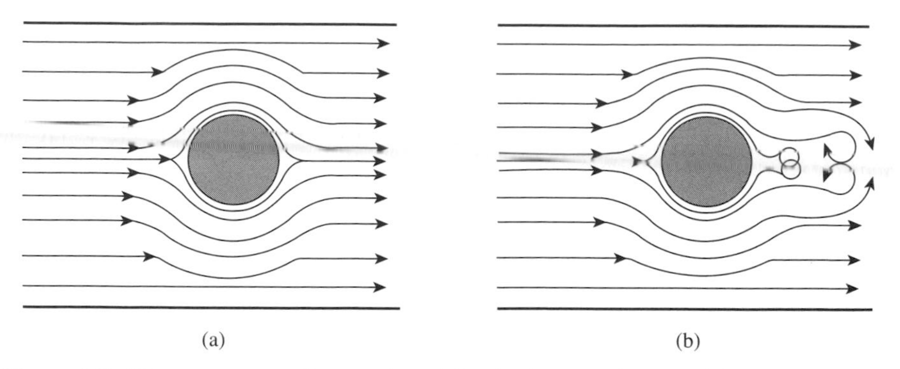

Figure 8.9. The movement of aerosol particles in the gas phase and its relation to the Reynolds number, Re: (a) Re < 1 (laminar flow); (b) Re > 1000 (turbulent flow).

particles of radius R and density ρ moving with a velocity v in a medium of viscosity η, the Reynolds number is given by

$$\mathrm{Re} = \frac{2vR\rho}{\eta} \tag{8.16}$$

When Re < 1, the system is said to be in laminar flow (Figure 8.9a) and the Stokes equation

$$D = \frac{kT}{6\pi\eta a} \tag{8.17}$$

is found to apply. In Eq. (8.9), D is the diffusion coefficient, η is the viscosity of the dispersion medium, a is the average particle radius, and k and T have their usual meanings. When Re $> 10^3$, the system is said to be in fully turbulent flow (Figure 8.9b) and flow resistance is controlled by drag forces due to the medium F_d, given by

$$F_\mathrm{d} \cong 0.2\pi\rho_\mathrm{m}R^2v^2 \tag{8.18}$$

In the region $1 < \mathrm{Re} < 10^3$, a transition occurs from laminar ($F_\mathrm{d} \propto \mathrm{v}$) to turbulent flow ($F_\mathrm{d} \propto \mathrm{v}^2$) and the relationship between F_d and v becomes more complex. Also, since drag forces actually apply only to the relative velocity of the particle to the medium, the effects of drag or viscous resistance to flow for a dispersed particle must be adjusted to account for the flow of the medium.

Even under ideal conditions, the dynamic flow behavior of aerosols versus other colloids can be markedly different. In still air, the average distance that a particle will travel before colliding with another particle, the mean free path, λ, is given by

$$\lambda = [(\sqrt{8})\pi\rho_\mathrm{N}R^2]^{-1} \tag{8.19}$$

where ρ_N is the particle number density. For an aerosol containing 10^8 particles/cm^3 and radius 10^{-4} cm, $\lambda = 0.11$ cm. Thus, a particle in random motion would travel an average of 0.11 cm before colliding with a neighboring particle. Each such collision may result in changes in the characteristics of the system—momentum changes in the case of elastic collisions and possibly size changes for inelastic or "sticky" collisions.

According to the Stokes equation, the velocity of free fall of a particle in an undisturbed gravitational field v_f is given by

$$v_f = \frac{m_a g}{6\pi R\eta} = \frac{2R^2 g\rho}{9\eta} \tag{8.20}$$

For simplicity, it is assumed that the density of the gas phase is small compared to that of the particle. For more accurate results, the density difference between particle and gas ($\rho = \rho_p - \rho_g$) should be employed. At 20°C and atmospheric pressure, the viscosity of air is 1.83×10^{-4} cP (centipoises or g cm^{-1} s^{-1}), so that for an aerosol drop of $R = 10^{-4}$ cm and $\rho = 0.92$ g/cm^3 (e.g., a hydrocarbon) the rate of fall will be approximately 0.011 cm/s. If the particle is emitted by an airplane flying at an altitude of 10,000 m, the hypothetical drop will reach the ground after approximately 2.9 years! If the particle grows to a radius of 10^{-3} cm by coalescing with other drops, its rate of fall increases to 1.1 cm/s, and the same trip will take about 11 days. It is easy to understand why natural and unnatural events that produce high-altitude aerosols can affect not only the color of our sunsets but also other more vital global atmospheric interactions.

8.7.3.2. Colloidal Interactions in Aerosols

Although the rules are the same, particle–particle interactions in aerosols appear to have characteristics significantly different from those of emulsions and dispersions in liquid media. A gaseous medium, because of its very different density, dielectric constant, and other properties, is very ineffective at screening the forces acting between colloidal particles. For that reason, aerosol particles, whether liquid or solid, will tend to have stronger attractive interactions among themselves and with other contacting surfaces than will similar units in a liquid medium. The spontaneous formation of "fuzzballs" and dusty deposits (Figure 8.10) in the cleanest of homes is an all-too-common manifestation of the affinity of dispersed particles in aerosols. The illustration is, of course, for solid aerosols or dusts and smokes, but the concept is the same for liquid aerosol deposits, even though electrostatic effects may be reduced somewhat.

If we use as a measure of the kinetic energy of a aerosol particle the value of kT (Boltzmann's constant times absolute temperature), at ambient temperature that energy will be about 4×10^{-21} J. The colloidal forces in aerosols will be at least an order of magnitude greater, indicating that the attraction between particles will almost always overwhelm the kinetic energy of the particles and inelastic or sticky collisions will commonly occur.

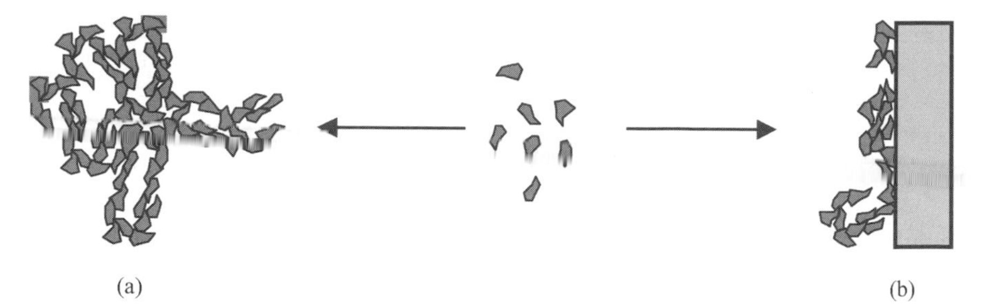

Figure 8.10. The aggregation of aerosol particles to form (a) fuzzballs (homogeneous flocculation) and (b) dust layers or surface dust deposits (heterogeneous Flocculation) due to the overwhelming attractive forces in gaseous media of low dielectric constant.

Such attractive interactions can be particularly important in situations where the presence of even a few extraneous particles on a surface can be highly detrimental, as in the production of microchips for the electronics industry (Figure 8.11). The presence of a single dust particle on the surface of a silicon wafer before coating with the photoresist resin that will be used to engrave the final circuit will, in all probability, result in a defective product in that area. When one considers that modern chips may have circuit line spacings of less than 10^{-4} cm, a particle of that diameter or even smaller will represent a veritable monkeywrench in the works. For that reason, extreme measures must be taken to ensure that aerosol particles are absent (to the extent technologically possible) in production areas.

In a " stable" cloud formation, the water drops lie in a size range in which air currents and other forces allow them to remain dispersed in a more-or-less stable way. Such clouds are usually the characteristics white because the small size of the light scattering particles produces less scattering of the incident sunlight. When droplets coalesce, as in rain formation in clouds, the identity of the small individual drops begins to be lost and larger drops formed, producing the darker, heavier clouds characteristic of rain and thunder showers. The radius of the drop will greatly affect the free-fall rate. If the drop forms under relatively calm conditions (e.g., little vertical convection to retard the drops' fall), small, gentle rainfall will

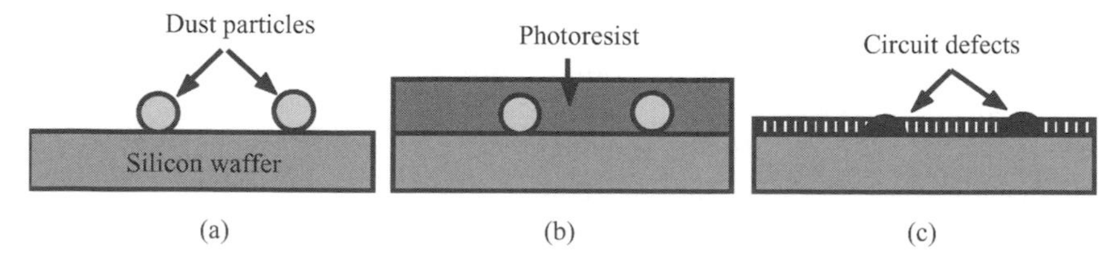

Figure 8.11. The importance of maintaining a clean, aerosol-free atmosphere is vitally important in the electronics industry: (a) dust-confaminated semiconductor surface; (b) photoresist coated over adhering dust particles; (c) developed microcircuit with defects.

result. In cloud formations with high velocity vertical convection currents (as in thunderstorms) the growing drops will be buoyed up by the air currents, allowing more time for drop growth and resulting in larger and more forceful rain. When the cloud formation reaches a high enough altitude and the drops are maintained suspended by strong internal convection currents, the drops may freeze to produce sleet or hail. The theory behind cloud seeding to produce rain is based on the introduction of charged colloidal particles, commonly silver iodide, that possesses a net electrostatic charge (usually positive) that will attract and coalesce small water particles (usually slightly negatively charged) producing drops heavy enough to fall as rain.

As a practical matter, almost all aerosol particles will rapidly acquire an electric charge leading to electrostatic interactions. The mechanisms for acquiring charge include direct ionization or ion exchange, specific-ion adsorption, charge derived from specific crystal structures, and charge acquired as a result of contact or movement such as piezoelectric and impact charging. Because of the absence of a solvent, direct ionization will be of minor importance. Perhaps most important are charge acquisition due to friction (as in walking across a rug on a dry winter day and touching a doorknob), electron gain or loss due to collision with ionizing radiation, and adsorption of ions and charged dust particles from the air.

PROBLEMS

8.1. Many modern washing machines and dishwashers recommend the use of low-foaming detergents for optimum efficiency. Suggest, in general terms, molecular structures and/or characteristics for surfactants that might be expected to combine the requirements for good detergency with low foam formation.

8.2. Two stable soap bubbles were blown from the same soap solution and connected by a tube with a closed stopcock. If the radius of bubble 1 (R_1) is greater that that of bubble 2 (R_2), what changes, if any, will occur in the two bubbles if the stopcock is opened?

8.3. What will be the reversible, thermodynamic work involved in blowing a bubble with a diameter of one meter using a soap/glycerol solution having a surface tension of 28 mN/m?

8.4. Assuming that the surface tension of water versus air follows the equation

$$\sigma = -0.1664\,T + 75.98$$

where T is temperature in degrees Celsius, calculate the critical drop radius for the nucleation of liquid water at 99.9°C, 75°C, 45°C, and 5°C at atmospheric pressure.

8.5. The Gibbs equation relates the amount of surfactant adsorbed at a solution interface to the change in surface tension of the solution with the concentration of the surfactant. Derive an expression for the area occupied by each surfactant molecule at the interface.

8.6. An important practical application of foaming systems is in fighting fires, especially those involving liquid inflammables such as gasoline or jet fuel. Suggest some mechanisms by which an aqueous foam system might assist in controlling and extinguishing such a fire.

8.7. The drainage of a liquid between two stationary soap films may take several hours. Assuming a liquid viscosity $\eta = 10^{-2}$ poise [centigrade; gram, second (c.g.s.)], a film thickness $d = 10^{-4}$ cm, a liquid density $\rho = 1$ g/cm^3, and $g = 980$ cm/s^2, show that such a timeframe is reasonable.

8.8. A vertical soap film can be in mechanical equilibrium only if the force of gravity acting on each film element is balanced by a gradient in the surface tension on both surfaces. Calculate the necessary surface tension gradient for a film with a thickness of 100 nm, given a soap solution density of 1 g/cm^3 ($g = 980$ cm/s^2).

8.9. Calculate the time required for a single gas bubble to disappear completely by shrinkage through diffusion if the original bubble has a radius of 1 mm and is separated from the atmosphere by a film of 1000 nm thickness. Given: $\sigma_{LV} = 30$ mN/m, diffusion coefficient of the gas in the solution $= 10^{-5}$ cm^2/s, gas solubility $= 0.03$ volume gas at STP (standard temperature and pressure) per volume liquid at 1 atm.

8.10. The elasticity of a film element E was defined by Gibbs as

$$E = \frac{2d\sigma}{d\ \ln A_s}$$

where A_s is the surface area of one film surface element. (a) Calculate E for a film element that is covered by an ideal monolayer with a surface pressure $\pi = \Gamma RT$. (b) Show that for a solution containing a single nonionic surfactant, E has the following mathematical form assuming ideal solution behavior

$$E = 4RT\ \frac{\Gamma^2}{c}\ \frac{1}{h + 2d\Gamma/dc}$$

where c and Γ are the volume and surface concentrations of the surfactant in the film and h is the film thickness.

8.11. The concentration of a surface-active impurity in an aqueous solution is 10^{-5} M; the surface concentration at the liquid–air interface is 10^{-10} mol/cm^2. To remove the impurity, air is blown into the solution, causing the formation of a foam that contains bubbles with volumes of about 0.1 cm^3 and thickness of about 10^{-4} cm. (a) Estimate the amount of impurity contained in 1 cm^2 of foam film and the approximate ratio of the amount adsorbed to that remaining in the bulk solution. (b) How many square meters of film should be produced and removed to reduce the impurity content of 1 L of solution by 50%? (c) What will be the volume of foam produced to attain that goal?

8.12. Explain why the addition of a small amount of *n*-octanol to a soap foam breaks it, but does not prevent its formation if added to the solution before foam formation.

9 Emulsions

It has been repeatedly pointed out that one of the most significant results of the physical phenomenon of "surface activity" is the preferential adsorption of amphiphilic molecules at interfaces, resulting in potentially dramatic changes in the characteristics of those interfaces. The ability to reproducibly control such adsorption and interfacial modifications is of immeasurable technological importance, not to mention the fact that our very existence as living organisms would be impossible had such a phenomenon not been a direct consequence of natural laws as we understand them. This chapter is concerned with one of the most important overall areas of impact of surfactants on our technological existence: emulsion formation and stabilization.

The preparation, stabilization, and use of systems of one fluid dispersed in a second, immiscible phase impacts almost every aspect of our lives, from the food we eat to the pharmaceutical formulations that make our lives longer and more comfortable. As is always the case in any discussion of surfactants and their applications, definitions and nomenclature can play a significant role in the way the material is presented. Although by some definitions "foams," dispersions of a vapor in a liquid or solid, could logically be considered a subclass of the general class of "emulsions," the nature of such systems and the requirements for their preparation and stabilization make them sufficiently different from liquid–liquid systems to warrant their discussion in a separate chapter. For our present purposes, we follow the definition that an emulsion is a heterogeneous system, consisting of at least one immiscible liquid dispersed in another in the form of small droplets, usually with a diameter of < 0.1 mm. Such systems possess a minimal stability, which may be enhanced by the addition of amphiphilic materials, polymers, or finely divided solids. Obviously, such a definition excludes foams and solid particulate dispersions from classification as emulsions, although it is possible that systems prepared as emulsions may, at some subsequent time, become dispersions of solid particles or foams.

When discussing emulsions, it is always necessary to specify the physical condition of each immiscible phase of the system. In almost all cases, since at least one liquid will be water or an aqueous solution, it is common practice to describe an

Surfactant Science and Technology, Third Edition by Drew Myers

emulsion as being either oil-in-water (O/W) or water-in-oil (W/O), where the first phase mentioned represents the dispersed phase and the second the continuous phase. Although oil-in-oil (O/O) emulsions are not impossible, they are seldom found. The generally high miscibility of most organic liquids is, of course, one limitation to the availability of such systems. More important, however, is the fact that few materials are sufficiently surface-active at such O/O interfaces to impart the minimal stability necessary for their preparation and maintenance.

Oil-in-oil emulsions do constitute an intermediate step in the preparation of nonaqueous emulsion polymers, although their existence as such is transient. Block polymers and comb polymers probably represent the most fertile area for research into emulsifiers and stabilizers for the preparation of O/O emulsions. They have so far found more application as stabilizers for dispersions of solids in organic solvents.

Before discussing the role of surfactants and their chemical structure in the preparation and stabilization of emulsions, it is important to have in mind a clear concept of the physical and theoretical aspects of liquid–liquid interfaces and the stabilization mechanisms available. Detailed discussions of these topics can be found in the works cited in the Bibliography. What follows is only a brief summary of some current thinking on the subject; no attempt is made to provide a comprehensive overview.

9.1. THE LIQUID–LIQUID INTERFACE

As discussed in Chapter 3, the presence of an interface induces an imbalance of forces that alters the energetic situation of molecules at or near that interface, usually giving molecules in that region a higher net energy than those in the bulk. The drive to lower the energy of the system resulting from the presence of the interface is one factor that results in the preferential adsorption of materials such as surfactants at such interfaces. The action of the adsorbed materials in lowering the free energy of the two-phase system reduces the work required to generate new interfacial area and therefore facilitates the preparation of emulsified systems.

The preparation of an emulsion requires the formation of a very large amount of interfacial area between two immiscible phases. If, for example, a sample of 10 mL of an oil is emulsified in water to give a droplet diameter of 0.2 mm, the resulting O/W interfacial area will have been increased by a factor of approximately 10^6. As shown in Chapter 8 [Eq. (8.1)] the work required to generate one square centimeter of new interface is given by

$$W = \sigma_i \, \Delta A \tag{9.1}$$

where σ_i is the interfacial tension between the two liquid phases and ΔA is the change in interfacial area. If the interfacial tension between the oil and water is assumed to be 52 mN/m (as for a hydrocarbon liquid), the reversible work required to carry out the dispersion process will be on the order of 20 J. Since that amount of work remains in the systems as potential energy, the system will obviously be thermodynamically unstable and will rapidly undergo whatever transformations are possible to attain a minimum in potential energy (i.e., minimum interfacial area), including coalescence, creaming, and sedimentation. If some material can be added to the system that reduces the value of σ_i to approximately 1 mN/m, the magnitude of W will be reduced to 0.3 J—a substantial reduction in W, but still a thermodynamically unfavorable situation. Only if the interfacial tension (and therefore W) is zero can a truly stable system be obtained. Luckily, although thermodynamics will be the factor controlling the long-term stability of such an emulsion, kinetics can play a dominant role over the short term, and it is through kinetic pathways that most useful emulsions achieve their needed stability. It is clear, then, that while lowering the interfacial tension between phases is an important factor in the formation and stabilization of emulsions, it may not always represent the most important role of surfactants and emulsifiers in such systems.

The relationship between the adsorption of a molecule at an oil–water interface and the resulting interfacial tension is an important one and is briefly restated here. The Gibbs equation for a system composed of one phase containing a nonionic solute adsorbing at the interface with a second phase is written as follows

$$\Gamma_i = -\left(\frac{1}{RT}\right)\left(\frac{\delta\sigma_i}{\delta \ln a}\right)_T \tag{9.2}$$

where the terms are as defined in Chapter 8. This equation shows that at a liquid–liquid interface, as in the liquid–vapor case, the amount of surfactant adsorbed can be determined from the slope of the curve of σ_i versus ln a. In dilute surfactant systems, the concentration C (mol/L) can be substituted for activity without serious loss of accuracy. As seen in Chapter 8, the simple relationship of the Gibbs equation can have significant practical application in the preparation of emulsions, especially in defining the relationship between emulsion droplet size and total surfactant concentration.

9.2. GENERAL CONSIDERATIONS OF EMULSION STABILITY

Even though emulsions as defined above have been in use for thousands of years (even longer if natural emulsions are considered), no comprehensive theory of emulsion formation and stabilization has yet been developed that quantitatively describes, and predicts, the characteristics of many of the complex emulsions and formulations that may be encountered while working in this field. Except in very limited and specialized areas, the accurate prediction of such aspects of emulsion technology as droplet size and distribution and stability remains more in the realm of art than science.

To serve as a complete theoretical description of emulsions, a theory must be able to explain and predict all aspects of emulsion formation, stability, and type (O/W or W/O), the influence of environmental factors such as temperature and pressure, the role of emulsifiers and stabilizers and their chemical structures, the role of the chemical structures of the immiscible phases, and the effects of additives in each phase. That represents a very tall order, as illustrated by the fact that even though vast amounts of experimental data relating to each of those questions are available, no generally applicable theory has yet appeared.

When discussing the stability of an emulsified system, it is important to have a clear idea of the physical condition of the components and the terminology employed. Four terms commonly encountered in emulsion science and technology related to stability are "breaking," "coalescence," "creaming," and "flocculation." Although they are sometimes used almost interchangeably, those terms are in fact quite distinct in meaning as far as the condition of an emulsion is concerned. "Coalescence," for example, refers to the joining of two (or more) drops to form a single drop of greater volume, but smaller interfacial area (Figure 9.1a). Such a process is obviously energetically favorable in almost all cases. Although coalescence will result in significant microscopic changes in the condition of the dispersed phase, such as changes in average particle size and distribution, it may not immediately result in a macroscopically apparent alteration of the system. The "breaking of an emulsion" (Figure 9.1b) refers to a process in which a gross separation of the two phases occurs. The process is a macroscopically apparent consequence of the microscopic process of drop coalescence. In such an event, the identity of individual drops is lost, along with the physical and chemical properties of the emulsion. Such a process obviously represents a true loss in the stability of the emulsion.

"Flocculation" refers to the mutual attachment of individual emulsion drops to form flocs or loose assemblies of particles in which the identity of each is maintained (Figure 9.1c), a condition that clearly differentiates it from the action of coalescence. Flocculation can be, in many cases, a reversible process, overcome by the input of much less energy than was required in the original emulsification process. Finally, "creaming" is related to flocculation in that it occurs without the

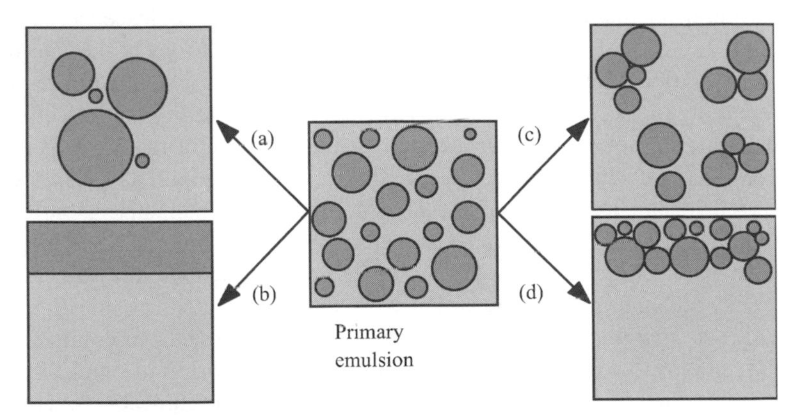

Figure 9.1. The ultimate fates of emulsions related to colloidal stability: (a) coalescence; (b) breaking; (c) flocculation; (d) creaming.

loss of individual drop identities (Figure 9.1d). It will occur over time with almost all emulsion systems in which there is a difference in the density of the two phases. The rate of creaming will depend on the physical characteristics of the system, especially the viscosity of the continuous phase and the density difference between the two phases. It does not necessarily represent a change in the dispersed state of the system, however, and it can often be reversed with minimal energy input. Obviously, both flocculation and creaming represent conditions in which drops "touch" but do not combine to form a single unit. The key to understanding the true stability of emulsions, then, lies on the line separating the processes of flocculation and coalescence.

Even in the infancy of emulsion technology several thousand years ago, it was recognized that to obtain a useful emulsion with any long-term stability it was necessary to include a third component, at least, that served some "magical" purpose and imparted the required degree of stability. Such additives included simple inorganic electrolytes; natural resins and other macromolecular compounds; finely divided, insoluble solid particles located at the interface between the two phases; and amphiphilic or surface-active materials that were soluble in one or both phases and significantly altered the interfacial characteristics of the system. Although the focus of this chapter is the role of surface-active, monomeric materials in emulsion preparation and stabilization, it may be of interest to briefly discuss some aspects of the influence of the three other stabilizing classes since, in practice, combinations are most often employed. The four mechanisms are illustrated schematically in Figure 9.2.

The least effective additives for the enhancement of emulsion stability are the inorganic electrolytes. Materials such as potassium thiocyanate (KCNS), when included in an emulsion formulation at the proper levels, may facilitate the preparation of dilute O/W emulsions. The stability of such emulsions, while greater than that of the system in the absence of the electrolyte, will be very limited. It is generally assumed that such action by simple electrolytes stems from the limited adsorption of the anionic species at the oil–water interface, imparting a weak electrical double layer that retards the close approach, and thus coalescence, of individual drops in the emulsion. Because the adsorption of such ions at the interface is small, the electrical effect is slight and provides only limited, short-term resistance to the breaking of the emulsion. Because of their limited utility and effectiveness, such materials are not included in the general classification of materials as emulsifying agents or emulsifiers. The remaining three classes of materials constitute what are referred to as "true emulsifiers and stabilizers. By their strong preferential adsorption at the O/W interface, these materials provide a much greater interfacial effect than do the simple electrolytes and make possible the extremely diverse technological applications of emulsions that we see today.

In nature as well as in human-made technology, macromolecular emulsifiers and stabilizers play major roles in the preparation and stabilization of emulsions. Natural materials such as proteins, starches, gums, and their modifications, as well as totally synthetic compounds such as polyvinyl alcohol, polyacrylic acid and other polyelectrolytes, and polyvinylpyrrolidone, have several characteristics that make

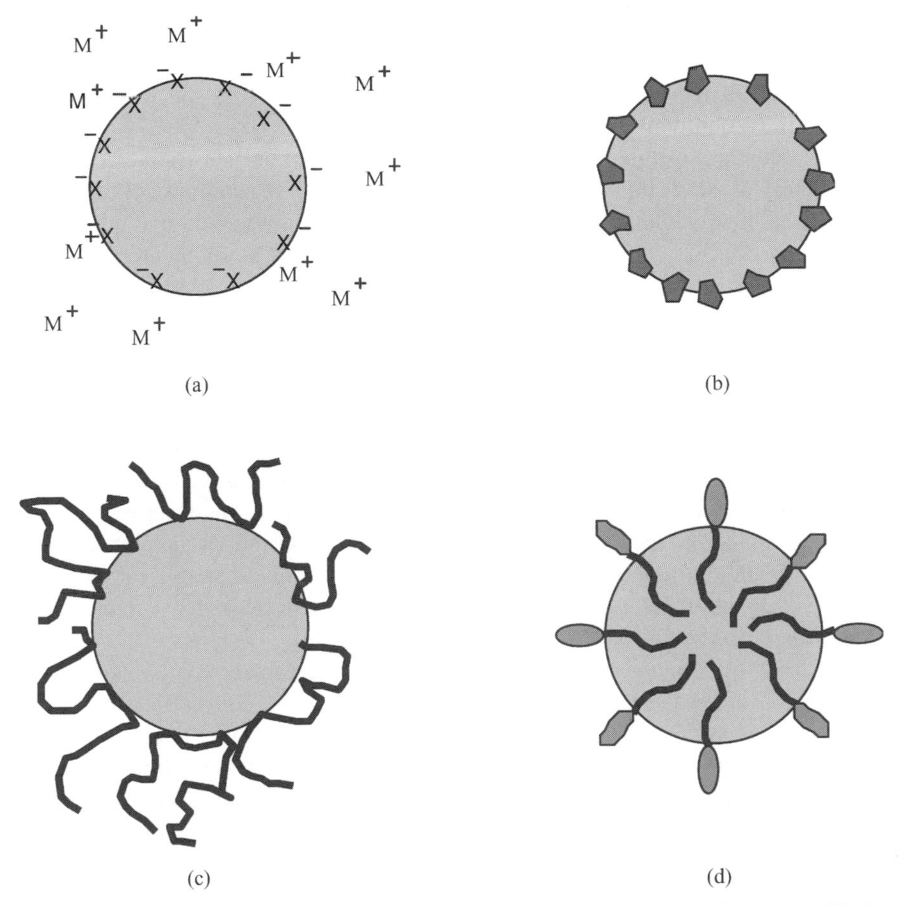

Figure 9.2. Mechanisms for the stabilization of emulsions: (a) adsorbed ions—specific-ion adsorption; (b) colloidal sols—adsorption of polymer chains; (c) polymeric stabilizers—solid particles; (d) surfactants—amphiphile adsorption.

them extremely useful in emulsion technology. By the proper choice of chemical composition, such materials can be made to adsorb strongly at the interface between the continuous and dispersed phases. By their presence, such materials can reduce the energetic driving force to coalescence by lowering the interfacial tension and/or forming a mechanical barrier between drops.

The effectiveness of polymeric materials at lowering interfacial tensions is often limited by their relatively slow rate of diffusion to the interface, but that is not always the case. More important to their function is the fact that polymers can form a substantial mechanical and thermodynamic barrier at the interface that retards the approach and coalescence of individual emulsion droplets. The polymeric nature of the materials means that each molecule can be strongly adsorbed at many sites on the interface. As a result, the probability of desorption or depletion is very small, and the interfacial layer attains a degree of strength and rigidity not easily found in systems of monomeric materials. In addition, the presence of polymeric materials in the system can retard processes such as creaming by increasing the viscosity of the continuous phase in addition to reducing the rate of droplet encounters, which could lead to flocculation or coalescence.

A second class of effective emulsifying agents commonly encountered consists of the finely divided solid particles. It has been known for some time that particles of colloidal dimensions (e.g., <1 mm in diameter) that are wetted by both aqueous and organic liquids can form stabilizing films and produce both O/W and W/O emulsions with significant stability. Emulsion stabilization by solid particles relies on the specific location of the particles at the interface to produce a strong, rigid barrier that prevents or inhibits the coalescence of drops. If the solid has a native electrical charge, it may also impart a degree of electrostatic repulsion that enhances the overall stabilizing power of the system.

There are three keys to the use of particulate solids as emulsion stabilizers: particle size, the state of stabilizer particle dispersion, and the relative wettability of the particles by each liquid component of the emulsion system. The stabilizer particles must be small compared to the emulsion droplet and in a state of incipient flocculation; that is, the particle dispersion must be near the limit of stability so that their location at the interface will result in some attractive particle–particle interaction to give strength to the system.

For the third condition, the solid must exhibit a significant contact angle at the three-phase (oil–water–solid) contact line, as measured conventionally through the aqueous phase. For maximum efficiency, it is usually found that the stabilizer should be preferentially wetted by the continuous phase. If the solid particles are too strongly wetted by either of the two liquid phases, the optimum stabilizing action will not be attained. It is usually necessary, therefore, to closely control such factors by controlling the system pH or by adding materials that adsorb onto the particles and impart the required surface characteristics. For example, if a mineral particle is used, but it is wetted too well by the aqueous phase, it can be treated with amphiphilic materials that will partially or completely coat the particle surface and thereby modify its interactions with water and/or the oil phase. A similar effect may, in some cases, be achieved by modifying the surface charge of the particles through specific-ion adsorption, pH changes, adsorption, and other variables. More details on those processes are given in Chapter 10.

The last major class of emulsifiers and stabilizers is that of the monomeric surfactants that adsorb at interfaces and produce electrical, mechanical, and steric barriers to drop coalescence, in addition to their role in lowering the interfacial free energy between the dispersed and continuous phases. Since these materials are the central concern of this work, they are addressed in detail below.

9.2.1. Lifetimes of Typical Emulsions

Any discussion of the stability of emulsions must be concerned not only with the mechanism of stabilization but also with the timeframe of the stability requirements and the conditions of preparation. The rates of degradation of emulsions vary immensely, and it is not possible to define a single number that can be used as a measure of acceptable or unacceptable stability—that must be defined by end-use requirements. In any emulsion, especially one that has no specific stabilization mechanism incorporated or one in which the stabilization is minimal, the degradation or breaking process will involve the coalescence of droplets brought together by Brownian motion, thermal convection currents, and random mechanical disturbances. Their stability can usually be measured on the order of seconds or minutes. In the presence of gentle agitation the process may be accelerated, while more vigorous stirring may result in the occurrence of competitive processes of coalescence and new droplet formation. Therefore, moderate agitation may result in the development of a steady-state or equilibrium particle size distribution that will be highly dependent on the rate of agitation, the concentration of the dispersed phase, and the conditions of disturbance.

Emulsions that contain more effective stabilizing additives such as one of those described above may be stable for hours, days, or months. In such systems the action of random or induced motion and droplet collision will continue, but the interfacial layers will possess sufficient strength and rigidity to prevent coalescence in most cases. When emulsion creaming (or sedimentation) occurs, additional pressures are applied to the interfacial area. At extreme pressures, as in the process of centrifugation, the drops may be deformed into the shape of polyhedra. In such a case, the interfacial area per unit volume will increase and the stabilizing layer will be stretched, reducing its strength and possibly leading to rupture and breaking (Figure 9.3).

In addition to the mechanical actions and interfacial potential-energy considerations that will act to reduce the degree of dispersion of an emulsion, other considerations act to limit their stability. One such factor affecting long-term stability is the phenomenon, commonly termed "Ostwald ripening" in crystalline systems, in which large drops (or crystals) are found to grow at the expense of smaller ones. Such action, whether in a crystalline or emulsion system, results from differences in

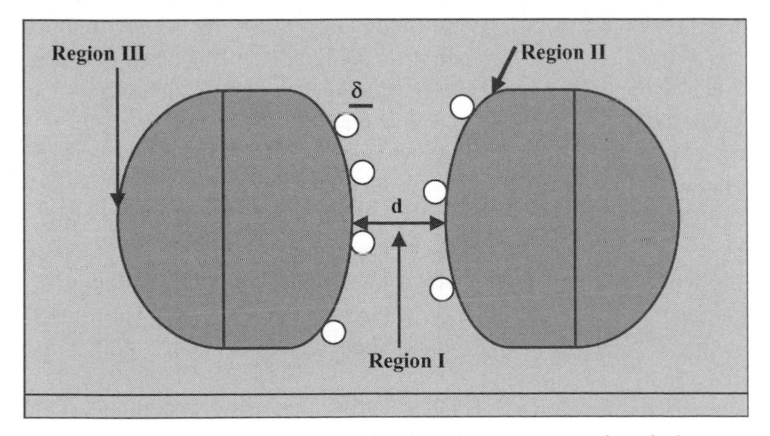

Figure 9.3. The deformation of emulsion droplets due to compaction during creaming or sedimentation stretches the stabilizing interfacial membrane and potentially reduces stability.

the chemical potential (and therefore solubility) of molecules in small particles relative to those in larger ones. Such differences arise from the fact that the Laplace pressure Δp inside a drop is inversely proportional to the drop radius r:

$$\Delta p = \frac{2\sigma_i}{r} \tag{9.3}$$

In terms of the Kelvin equation and solubility, the effect of radius can be given as

$$\ln\left(\frac{S_1}{S_2}\right) = \frac{\sigma_i V}{RT}\left(\frac{1}{r_1} - \frac{1}{r_2}\right) \tag{9.4}$$

where S_1 and S_2 (not to be confused with the spreading coefficient, $S^{\sigma}_{2/1}$, from Chapter 8) are the solubilities of the particles of radii r_1 and r_2 and V is the molar volume of the phase inside the droplets or crystals. The effect of the Kelvin relationship is often readily apparent in foam systems, where the solubility of gases in the liquid phase can be substantial. In emulsion systems, on the other hand, the solubility of the dispersed phase may be so low that diffusion from small to large droplets will be exceedingly slow. The process will occur even in such circumstances, but at such a rate that it will not be apparent for long periods. In that context, it is often possible to greatly reduce the rate of droplet growth due to Laplace pressures by employing emulsifiers that form a barrier to the passage of dispersed-phase molecules into the continuous phase. The presence of a large excess of surfactant in the form of micelles capable of solubilizing the dispersed phase could obviously be detrimental to long-term stability due to the enhancement of the Ostwald ripening effect.

Other external factors affecting the stability of emulsions include the actions of bacteria and other microorganisms that can degrade components and significantly alter the characteristics of the system, and freezing, especially in O/W emulsions. During the process of freezing, the formation of ice crystals in the continuous phase forces the emulsion droplets together under significant pressures, often resulting in the rupture of the interfacial film and drop coalescence. It is obvious, then, that stability to such action will require an interfacial film of considerable strength. Even though the protection of emulsions from breaking due to freezing action is of considerable economic importance, there has been relatively little fundamental research published in the area.

Bacterial action can be of importance in food, pharmaceutical, and cosmetic emulsions. Such systems are obviously of great economic importance, and a great deal of research has been devoted to the problem. When biological stability is important, some advantages can be gained by the proper choice of surfactant in the stabilizing formulation, since many such materials show significant bacteriostatic activity.

From the discussion above, it is clear that, from a surface chemical point of view, one aspect of emulsion stability is of primary importance—the protection of the emulsion droplets from coalescence. As a result, that area has received the greatest overall amount of attention from both academic and industrial research efforts. Flocculation without coalescence, however, is an important consideration and has begun to draw closer and more frequent attention, especially concerning the role of polymeric species that may act as bridging agents between droplets. The physical characteristics of the interfacial film—its strength and rheological properties—are beginning to come under closer examination as new techniques for such studies become available. Several excellent reviews and monographs published more recently cover in detail the questions raised above. Some of those are listed in the Bibliography.

9.2.2. Theories of Emulsion Stability

Early "practical" theories of emulsion stability recognized the importance of additives such as surfactants, polymers, and particulates to the processes of emulsion preparation, the type of emulsion produced, and the overall stability of the final system. However, a reasonably sound theoretical picture began to evolve only once an understanding of the concepts and principles of interfaces and monolayers began to become clear. Studies of oriented amphiphilic monolayers at interfaces led to the conclusion that such structures, in which each portion of the adsorbed molecules showed a strong preference for association with one of the two liquid phases, offered the best explanation for observed experimental results. As a result, it became possible to schematically represent the emulsion droplet as shown in Figure 9.2d.

The concept of an adsorbed monolayer film acting as an emulsion droplet stabilizer found early experimental support in work showing that the equilibrium area per molecule of surfactant at an oil–water interface, as determined from final droplet diameters, approached a constant value, regardless of initial surfactant concentrations; that is, the area occupied by adsorbed amphiphilic molecules, initially carboxylate soaps, had a lower limit related to the nature of the molecule. It was found that emulsions in which the level of surfactant at the oil–water interface corresponded to an "expanded" monomolecular film were much less stable than those in which sufficient stabilizer was present for the formation of a "condensed" film. For purposes of the present discussion, an expanded monomolecular film may be roughly defined as one that has a relatively high compressibility compared to the bulk liquid but exists as a continuous interfacial monolayer (Figure 9.4a); that is, it has the basic characteristics of a compressible gas, rather than being a collection of isolated "islands" of amphiphilic molecules (Figure 9.4b). The "condensed" film, on the other hand, will have relatively low compressibility, existing as a close–packed monomolecular array (Figure 9.4c).

Similar studies explored the role of film tenacity in emulsion stabilization. Such studies compared the stability of emulsions prepared with mixtures of surfactants to the ability of monomolecular films of the same compositions at the air–water interface to resist high surface pressures without film breakdown. The results indicated that the more resistant the film was to high surface pressures, the greater was the stability of the related emulsion. It was clear to investigators some time ago,

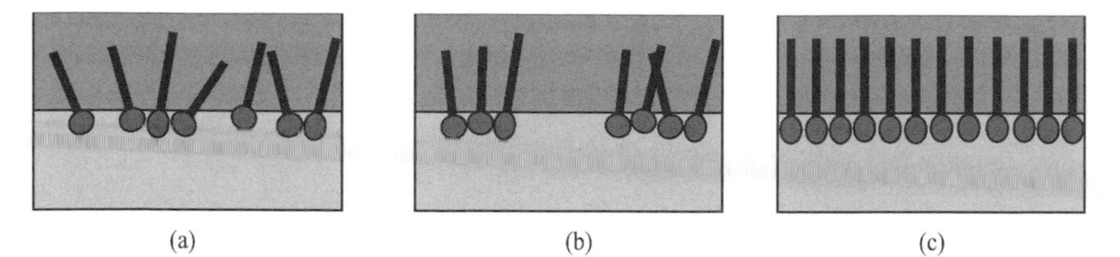

Figure 9.4. A schematic representation of adsorbed surfactant monolayers: (a) a compressible monolayer film; (b) hypothetical "islands" of amphiphile; (c) a condensed monolayer.

therefore, that the nature of the interfacial film stabilizing an emulsion played an important role in determining the ultimate stability of the dispersed system. To this day, however, there is still some question as to some of the exact details of the stabilization process.

The effectiveness of any adsorbed film of amphiphilic materials in retarding the inevitable movement of emulsified systems toward a minimum in total energy may be considered in at least three contexts. The adsorbed molecules can (1) reduce the potential energy of the dispersed system by lowering the interfacial tension; (2) erect a rigid or highly viscous barrier at the interface capable of preventing or retarding the coalescence of droplets that collide as a result of random Brownian motion, thermal convection, or mechanical agitation; and (3) in cases where the adsorbed molecules carry an electric charge, impart that charge to the surface of droplets, resulting in the formation of an electrical double layer that lessens the frequency and effectiveness of close droplet approach and contact leading to droplet growth.

Although it may be tempting to attribute emulsion stability to the existence of a low interfacial tension, Eq. (9.1) shows that even a low value of σ_i will still result in a relatively high value for the work required for the formation of fresh interface. It is generally felt today that interfacial tension effects are less important to overall long-term emulsion stability than are the effects of the nature of the interfacial film. The ability of the interfacial film to withstand the pressures of droplet contacts (its tenacity), its properties as a barrier to the passage of dispersed phase into the continuous phase (to limit Ostwald ripening), and its ability to erect a steric or electrical barrier to droplet approach and contact appear to be the major characteristics determining the ultimate stability of an emulsion. With those concepts in mind, we now turn more specifically to the role of surfactants in the preparation and stabilization of emulsions.

9.3. EMULSION TYPE AND NATURE OF THE SURFACTANT

The concept that surfactant molecules preferentially orient at the oil–water interface not only clarified the picture of monomolecular film stabilization but also shed light on the problem of explaining the emulsion type obtained as a function of the chemical structure of the adsorbed species. It was recognized early that the

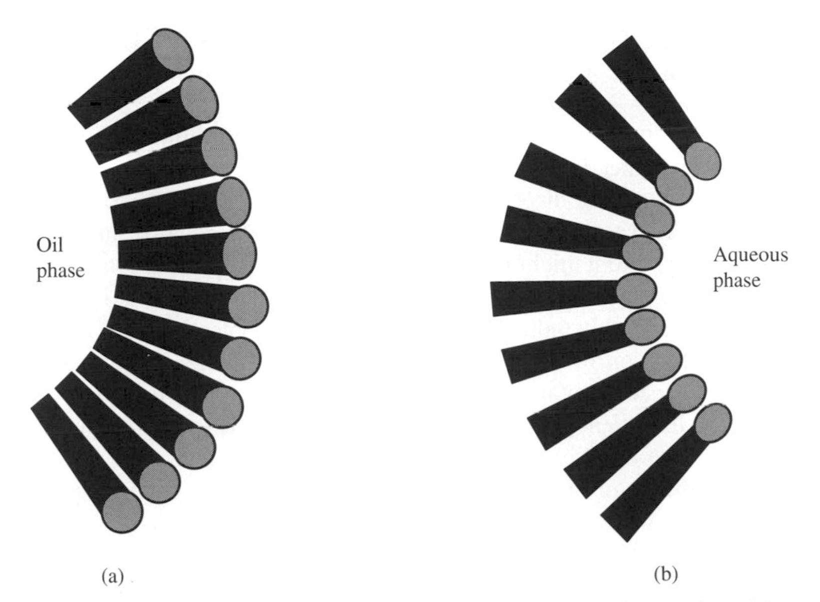

Figure 9.5. A schematic representation of the "oriented wedge" picture of emulsion stabilization by adsorbed surfactant monolayers: wedges favoring (a) O/W emulsions and (b) W/O systems.

nature of the surfactant employed in the preparation of an emulsion could control the type of emulsion formed. It was found, for example, that while the alkali metal salts of fatty acid soaps produced O/W emulsions, the use of di- and trivalent soaps resulted in the formation of W/O systems. The adsorbed monolayer mechanism for the stabilization of emulsion droplets requires the formation of a relatively close-packed surfactant film at the interface. It is clear that the geometry of the adsorbed molecules must play an important role in the effect obtained. For efficiency of packing, it can be seen from Figure 9.5 that the formation of W/O systems with polyvalent soaps is almost inevitable, since they can be seen as double-tailed surfactants, while the monovalent analogs are single-tailed. Such a steric requirement in terms of the orientation of surfactant molecules at the interface has classically been referred to as the "oriented wedge" theory. Such a concept has now been given theoretical validity by the molecular geometry and critical packing parameter concepts discussed in previous chapters.

Although such a simple view of the role of the monolayer in determining the nature of the emulsion can be quite useful, exceptions are known. Such exceptions probably reflect the conflicting role of solubility in stabilization, since some monovalent salts with relatively low water solubility produce W/O emulsions. A rule of thumb for predicting the type of emulsion formed on the basis of the relative solubility of the surfactant employed, often referred to as the "Bancroft rule," states that the liquid in which the surfactant was most soluble would form the external or continuous phase. The rule was extended by the assertion that the presence of an absorbed interfacial film required the existence of two interfacial tensions:

one at the oil–monolayer interface and a second at the water–monolayer interface. Since the two tensions would not, except in very unusual circumstances, be equal, the interfacial layer would be curved, with the direction of curvature determined by the relative magnitudes of the two tensions. Logically, the film will curve in the direction of the higher interfacial tension so that the phase associated with that interface will become the dispersed phase in the system (Figure 9.5).

Aside from the nature of the emulsifier employed, the relative amounts of the two phases in the system might be expected to affect the type of emulsion obtained. If one assumes that an emulsion is composed of more or less rigid, spherical droplets of equal size (highly unlikely in reality), simple geometry shows that the maximum volume fraction of dispersed phase that can be obtained is approximately 74%. It was suggested that any emulsified system in which that level was exceeded would produce less stable, deformed droplets and would lead to phase inversion to an emulsion of the opposite type. Practice has shown, however, that it is possible to prepare emulsions of dispersed-phase volume fractions far exceeding that theoretical limit. Seen with the (sometimes) 20/20 vision of hindsight, there are several possible ways to explain the failure of such a simple geometric approach.

In the first place, emulsion droplets are not and can likely never be perfectly monodisperse; as a result, it is possible for smaller droplets to insert themselves in the void spaces between close-packed, larger droplets (Figure 9.6), increasing the total potential packing density of the system. In addition, emulsion droplets are not rigid but highly deformable, spheres; thus they can be easily deformed from spherical to various oval or polyhedral shapes to fit the demands of the system. Large excursions from a spherical shape are, of course, generally unfavorable, since they entail the formation of additional interfacial area for a given dispersed volume fraction. As mentioned above, such an increase in interfacial area could strain the ability of the adsorbed emulsifier film to the point of droplet coalescence.

In the past there have been some suggestions that the mechanical process of emulsification could also play a role in determining the type of emulsion produced. A number of studies have verified that, in some cases at least, such a mechanical effect on emulsion characteristics does seem to exist for some specific formulations.

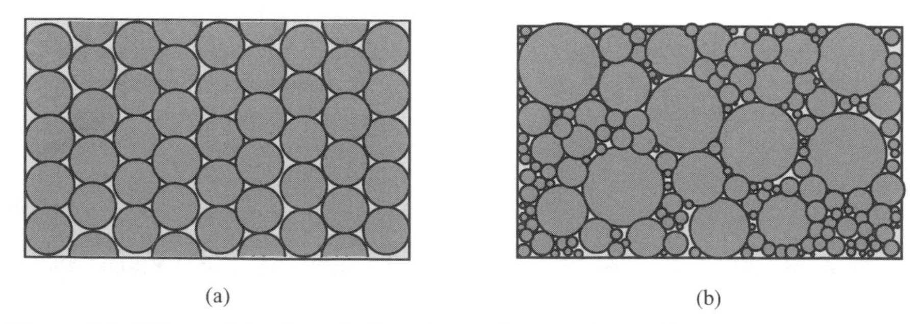

Figure 9.6. Effects of droplet polydispersity on the potential packing density of emulsions; interstitial spaces between larger drops may be filled by smaller units: (a) ideal hexagonal close packing; (b) high-density polydisperse packing.

A fully satisfying theoretical explanation for such an effect, however, has not been worked out, but the order and manner of mixing the components of the two phases have been found to significantly affect the quality and stability of the resulting emulsion. Some cosmetic formulations, for example, give better results if the aqueous phase is added to the organic phase rather than the reverse, even though the final emulsion is of the O/W type. Possible mechanical and procedural questions aside, most theories of emulsion formation place the most emphasis on the natures of the continuous and dispersed phases and the nature of the emulsifying agent employed.

9.4. SURFACE ACTIVITY AND EMULSION STABILITY

To be a generally applicable description of emulsions, a theory must not only explain and predict the consequences of the system composition and conditions of preparation on emulsion type but also be able to accurately relate the long-term stability of the system to all those factors. Even in the light of the vast amounts of experimental data on emulsions published to date, no generally applicable theory has been developed that can handle all modern emulsion formulations. That should not be taken as a mark against the science, however, since a glance at the list of ingredients for most modern emulsions may be as long as your arm! With so many actors on the stage, the theory may become as complex as the precise calculation of all of the gravitational interactions of all of the planets and satellites in our solar system.

Modern attempts to formulate a quantitative theory of emulsions and emulsion stability have looked most closely at the nature of the interfacial region separating the two immiscible phases, especially the chemical and physical nature of the adsorbed film, the role of mixed films and complex formation, interfacial rheology, and steric and electronic factors at the interface. The theoretical foundations for current ideas concerning emulsion formation and stability are presented in several of the references cited in the Bibliography. A few of the most basic ideas, however, are presented below.

In order to picture what is happening when two emulsion droplets undergo close approach, one can think of two water-filled balloons being pushed together. As they begin to interact, the approaching surfaces begin to deform or flatten. Air between the balloons is also forced out of the interstitial region. The membranes separating the contents of the balloons will be stretched, but their strength is (usually) sufficient so that rupture does not occur. As indicated in Figure 9.3, a similar deformation occurs as two emulsion droplets undergo close approach, except that the interfacial membranes involved do not have the same physical strength as the balloons. So other mechanisms must act to prevent a complete coalescence of the drops.

As the emulsion droplets approach during the process of flocculation or coalescence, a thin lamellar or interstitial film of the continuous phase will form between them (Figure 9.7a). When the film begins to reach a critical thickness, solvent and

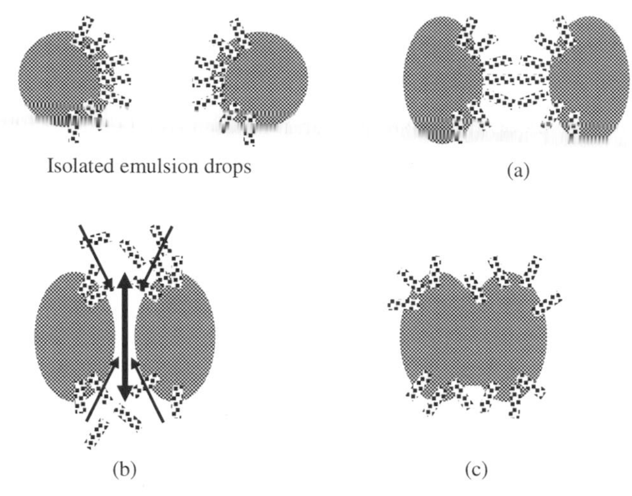

Figure 9.7. A schematic representation of the mechanisms leading to the weakening of adsorbed monolayers and the possible coalescence of the drops: (a) as drops undergo close packing, deformation begins; (b) surfactantant forced out (heavy arrow) as osmotic forces attempt to reestablish the stabilizing layer (light arrows); (c) membrane rupture and drop coalescence.

counter ions associated with the adsorbed surfactant molecules begin to be squeezed out (Figure 9.7b). If the surfactant in the monolayer is not strongly adsorbed, it may desorb and its interfacial concentration may become depleted, significantly reducing its stabilizing effectiveness to such an extent that drop coalescence occurs. If the adsorption if sufficiently strong, osmotic forces will try to bring the counterions and associated solvent back into the interstitial region, forcing the drops apart and maintaining the stability of the emulsion. In the case of flocculation, there may be an optimum separation distance at which the forces of attraction between drops and the osmotic forces and electrostatic repulsion are balanced and the drops remain closely associated without coalescence.

If rupture and coalescence take place (Figure 9.7c), they occur in an area of the adsorbed monolayer thinned out by the mechanical action of the approaching drops in stretching the monolayer as a result of deformation and increases in the interfacial area and/or of depletion of emulsifier due to desorption. Understanding the behavior of such lamellar films and the role of surfactants in their action requires an understanding of the forces involved in interactions across the film and the kinetic aspects of film fluctuations.

The coalescence of liquid droplets, therefore, is intimately related to the nature of the thin lamellar film formed between them as they are brought into close encounters as a result of thermal convection, Brownian motion, or mechanical agitation. It is important to understand the nature of the forces acting across the film in order to obtain information about the thermodynamic stability, metastability, or instability of the film, and the kinetic processes that will control the rate of film breakdown. Comprehensive reviews of those aspects of emulsion stability can be

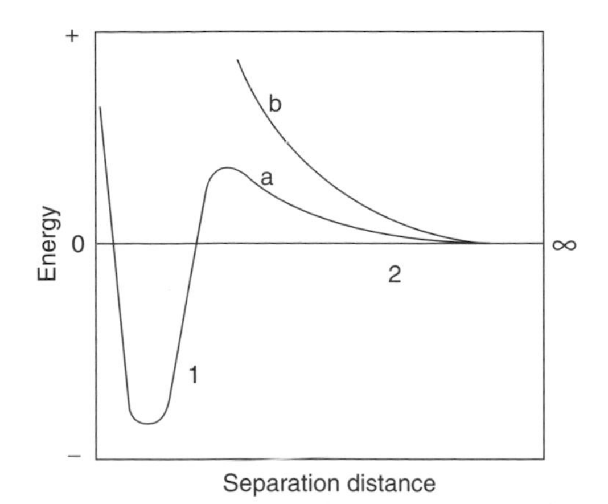

Figure 9.8. Representative potential-energy diagrams: (a) electrostatic stabilization with primary (1) and sometimes secondary (2) minima; (b) steric stabilization.

found in the comprehensive works edited by Becher (2001) and references cited therein.

When two liquid droplets possessing adsorbed monomolecular stabilizing films come into close proximity on a colloidal scale (Figure 9.8), the thin-film region will develop in the area of closest approach, forming two essentially flat, parallel monolayers separated by a distance d (region 1), a transition region at the ends of the parallel layers where the surface curvature is large relative to that of unaffected droplets (region 2), and regions well away from the lamellar region where curvature is that of the undeformed interface (region 3).

When discussing the forces acting across the lamellar film, it is necessary to consider two ranges for the value of the separation distance d: (1) that in which d is greater than twice the thickness of the monomolecular films d ($d > 2d$) and (2) that in which d is less than $2d$. In the first case, the significant forces are those considered to be long-range in the colloidal sense: van der Waals and electrical double−layer interactions. Such interactions can be characterized by conventional potential-energy diagrams (Figure 9.8), which relate the potential energy of interaction of the two drops to their distance of separation. The shape of the diagram will be a function of the nature of the attractive and repulsive forces acting between the approaching drops. In Figure 9.8a, the combination of attractive van der Waals and repulsive electrostatic forces will normally show the presence of a maximum and primary and secondary minima as a function of the distance of separation. In general, the secondary minimum, when present, will represent a state of reversible flocculation. The primary minimum, on the other hand, represents a state of irreversible coalescence.

If the lamellar film between approaching emulsion droplets thins beyond $d > 2\delta$ (Figure 9.3), with no drop coalescence taking place, interactions generally referred to as "steric repulsions" come into play (Figure 9.8a). Such repulsions normally result in a steep maximum in the potential-energy curve. Interfacial film rupture, then, requires that those repulsive forces be overcome by deformation of the interfacial film of adsorbed surfactant or emulsifier. The so-called steric or enthalpic stabilization mechanism will generally produce a monotonic curve of increasing potential energy (Figure 9.8b). Conceptually, that process may be visualized as arising because the space between the stabilizing layers is reduced to such an extent that the two layers become compacted and further compression becomes more difficult. If the components of those layers were rigidly fixed and unable to move, there should develop an essentially infinite barrier to drop coalescence. In some colloidal systems, particularly solid particles having adsorbed polymeric stabilizing layers, such a mechanism can often provide much greater stability than electrostatic forces. In emulsion systems employing normal surfactants, the effectiveness of steric stabilization process is limited by the fluidity and deformability of the dispersed phase, as well as by the fact that monomeric surfactant layers are seldom rigid. Adsorbed molecules may move around relatively freely, especially under the influence of the forces developed during drop approach. As will be seen in later sections, situations do arise in which emulsion stability can be enhanced by the presence of less fluid interfacial structures such as molecular complexes and liquid crystals. In the present context, we are interested only in the specific roles that surfactants can play in affecting film drainage and rupture.

The presence of almost any surface-active species will result in an increase in the stability of most oil–water systems subjected to agitation, as is often painfully obvious to any organic chemist attempting to purify a reaction product by solvent extraction. The action of many organic materials, even those not normally considered to be surfactants, in lowering the interfacial tension between the water and oil phases affects both the ease with which the interfacial area can be increased and the rate at which individual droplets will coalesce and lead to complete phase separation.

The exact role of interfacial forces in emulsion stabilization is the subject of some question. In the view of many researchers in the field of emulsion stabilization, the rate of failure of the adsorbed lamellar film at the droplet interfaces as they approach and touch is a primary factor in the stabilization of an emulsion system. Such phenomena are related to the elasticity of the film and have been addressed in the theories of Gibbs and Marangoni. As pointed out previously, the Gibbs adsorption isotherm relates the lowering of the interfacial tension between two phases, $d\sigma_i$, resulting from the presence of a surface-active solute to the concentration C_i of the solute in the system

$$-d\sigma_i = \Gamma RT\, d\,(\ln C_i) \tag{9.5}$$

where the symbols are as defined previously. The assumption that the concentration of the surfactant can be equated to its activity, while common and useful for simplifying the mathematics, is not always valid at surfactant concentrations far above the critical micelle concentration. Although the Gibbs equation has been employed to determine the concentration of surfactants at the oil–water interface, with reasonably good agreement found between experiment and theory, its use is not without difficulties. This is because the attainment of the final equilibrium interfacial concentration of surfactant may be slow, and most common experimental methods such as the du Nouy ring, the Wilhelmy plate, and the drop and bubble procedures are dynamic and rely on an increasing interfacial area during the measurement. Such methods, then, will usually produce results larger than the true equilibrium value.

At equilibrium, the interfacial tension in a system will be uniform. However, in the dynamic environment of an emulsion system, nonuniformities will arise as a result of particle deformations in which new surface area will be produced by deviation of the droplet from a perfectly spherical shape. Since the diffusion of new surfactant molecules to the interface to lower the interfacial tension will require a finite amount of time, interfacial tension gradients will develop, leading to the presence of surface elastic response. If sufficient differences in local interfacial tensions develop, a rapid spreading of surfactant molecules into regions of higher tension will occur. Concurrent with the movement of surfactant into regions of high σ_i, underlying layers of liquid associated with the surfactant may be dragged along.

Surface elasticity in the sense under consideration cannot exist in a system of pure liquid phases. In a system containing surfactant molecules, gradients in interfacial tension can arise as a result of the formation of new area, as mentioned above, or because of the loss of interfacial area. In the former case, the time lag between the formation of new interface and the diffusion of surfactant to that interface will produce an interfacial tension that is higher than equilibrium. The local value of the surface excess Γ_i will fall and the value of σ_i will approach that of the pure system. The net effect will be a tendency for the interface to contract, providing a "healing" effect to reduce the chance of droplet coalescence. In the case of loss of interfacial area, there will be a time lag from the point of compression of the interfacial film until the excess surfactant molecules can desorb and diffuse away from the interface.

In addition to the Marangoni effect, surface elasticity is affected by the Gibbs effect, which is concerned with changes in the physical condition of the liquid lamella as two drops approach and begin to touch in the process of flocculation and coalescence. Not only do interfacial tension gradients occur in the film as a result of the finite time required for the adsorption of surfactant molecules at newly formed interface, but the film will have a limit to which it can be stretched before the lamellar interfacial tension increases to the point where the stabilizing effect of the film is lost. The coefficient of elasticity E for an interfacial film under such conditions was given by Gibbs as

$$E = 2A\left(\frac{\delta\sigma_i}{\delta A}\right) = 4RT(\sigma_i^2/C)\frac{1 + \delta\ln\sigma_i/\delta\ln C}{h + 2\delta\sigma_i/dC} \tag{9.6}$$

where A is the interfacial area occupied by a given quantity of surfactant of concentration C and h is the thickness of the adsorbed film. Calculations using Eq. (9.6) indicate that in a 0.1 M solution of surfactant with a lamellar thickness of 100 nm, the Gibbs coefficient of elasticity will be on the order of 100 mN/m. An extension of the film of 1%, therefore, will result in an increase in the interfacial tension of each side of the film of 1 mN/m. As in the case of the Marangoni effect, the Gibbs elasticity will be significantly affected by the surface activity of the adsorbing species, as indicated by the $(\delta\sigma_i/d \ln a)$ term.

The experimental determination of the elasticity of lamellar films has, until relatively recently, been difficult, as have studies of the rates of diffusion of surfactant molecules to newly formed interface. It is difficult, therefore, to determine the relative importance of each mechanism in the stabilization of O/W emulsions. New techniques using photon correlation spectroscopy, which can measure the duration and amplitude of surface and interfacial waves, promise to provide a great deal of useful information about the physical properties of such regions as surfaces, interfaces, and lamellar films.

9.5. MIXED SURFACTANT SYSTEMS AND INTERFACIAL COMPLEXES

We have seen that the presence of small amounts of surface-active materials can have a dramatic effect on the surface tension of solutions, even those containing relatively large amounts of other surface-active materials. The classical example of such effects is the minimum in the surface tension–concentration curves found for many anionic sulfate surfactants that contain small amounts of the starting alcohol as an impurity. The action of the impurity may be seen as being twofold: (1) because of its less hydrophilic head group (the hydroxyl), the alcohol will be more efficiently adsorbed at the surface; and (2) because of the smaller size of the head group, the impurity can be packed into the adsorbed layer between adjacent molecules of the primary surfactant, resulting in a greater surface excess and a lower surface tension. Once the primary surfactant concentration has reached its cmc, the less soluble impurity can be solubilized into micelles so that the surface tension will be determined more directly by the primary surfactant species.

In the discussion of foams and foam stability in Chapter 8 it was shown that the presence of small amount of a surface-active impurity can contribute greatly to foam stability as a result of its effect on lamellar film elasticity. Such a dramatic effect has not been found in the case of emulsion stability against droplet coalescence. Although a limited amount of data are available on the interfacial tension of ternary mixtures, it is generally assumed that surface-active impurities of this type are probably too soluble in the oil phase to remain at the interface. They will be extracted into the oil phase, where they are not particularly surface-active and are not strongly adsorbed back into the interface.

In contrast to the insignificant effect of surface-active impurities on emulsion stability, it has been found that the presence of two primary surfactant species, one soluble in water and the other in oil, can greatly enhance the stability of an emulsion system. The effect has been related to the production of very low interfacial tensions and the formation of cooperative surfactant "complexes" that impart greater strength and coalescence resistance to the O/W interface. A broad definition of the term "complex" should be inferred in this context. It is not used, necessarily,

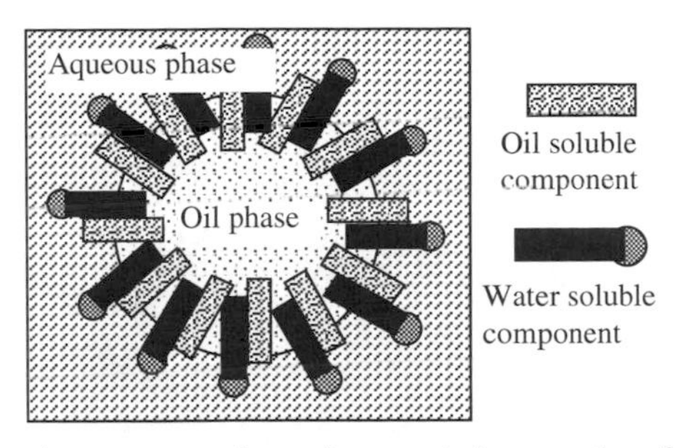

Figure 9.9. A schematic representation of synergistic complex formation at emulsion interfaces.

to imply some formally identifiable, fixed composition combination of two amphiphiles, but rather a general synergistic effect produced by the combination.

Investigations of the effects of oil-soluble surfactants on the emulsification of paraffins in aqueous surfactant solutions led to the proposal that the formation of interfacial complexes at the oil–water interface could increase the ease with which emulsions could be formed and, possibly, explain the enhanced stability often found in such systems (Figure 9.9). By definition, an interfacial complex is an association of two or more amphiphilic molecules at an interface in a relationship that will not exist in either of the bulk phases. Each bulk phase must contain at least one component of the complex, although the presence of both in any one phase is not ruled out. The complex can be distinguished from such species as mixed micelles by the fact that micelles (and therefore mixed micelles) are not adsorbed at interfaces. According to the Le Chatelier principle, the formation of an interfacial complex will increase the Gibbs interfacial excess Γ_i [Eq. (9.2)] for each individual solute involved, and consequently, the interfacial tension of the system will decrease more rapidly with increasing concentration of either component.

The existence of the interfacial complex is distinct from the situation of simple coadsorption of oil-soluble and water-soluble surfactants. In the case of coadsorption, each component will be competing for available space in the interfacial region and will contribute a weighted effect to the overall energy of the system. One could even think of coadsorption as producing individual and independent "islands" of the two different adsorbed species at the interface with their resulting local influences (Figure 9.10). The interfacial complex, on the other hand, implies a more uniform and ordered adsorption pattern for each amphiphilic component resulting in a general interface wide impact, with the net synergistic interfacial effect exceeding that produced by either component or by a simple combination of the two.

Another possible beneficial effect of interfacial complex formation, in addition to the improved surface energetic just mentioned, is that such structures may possess a greater mechanical strength than will a simple mixed interfacial layer. Closer molecular packing densities and a greater extent of lateral interaction between hydrophobic chains may result in significant decreases in the mobility of molecules

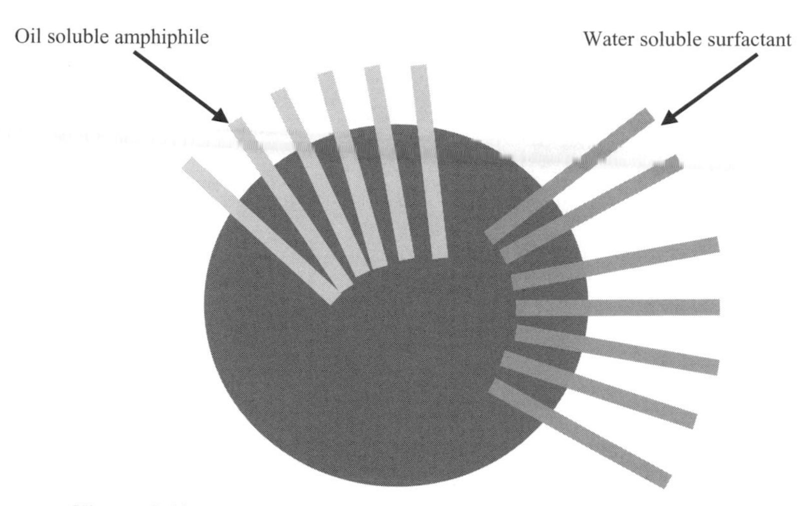

Figure 9.10. "Islands" of coadsorbed amphiphiles in the interface.

at the interface and a decrease in the rate of drop coalescence. Such an effect has often been mentioned in terms of increased interfacial viscosity or elasticity.

When one discusses the phenomenon of surface elasticity, it is easy to think in terms of the effects of bulk rheological phenomena such as viscosity on the characteristics of a system. In fact, it has been assumed by many that the presence of a high surface viscosity alone can contribute significantly to the stability of an emulsion system. Much of the evidence used to argue in favor of such a role for surface viscosity is related to the rate of film drainage between approaching droplets. In essence it is argued that if a higher interfacial viscosity slows the loss of intervening material between drops, it must therefore enhance the stability of the system to coalescence. In some instances, that stabilizing mechanism may be active; in others, not so much so.

In an emulsion system stabilized only by the presence of adsorbed monomeric surfactant species, there have been several arguments against the role of surface viscosity in emulsion stability: (1) high interfacial viscosities have very rarely been found in O/W systems not containing a polymeric emulsifier component and (2) because emulsion droplets are very small, tangential shear stresses will produce localized interfacial tension gradients, which will immediately be counteracted by the Marangoni effect discussed previously. The actual flow of the interfacial film in the thinning process, which could be affected by viscosity, will not occur. If an isolated emulsion droplet is exposed to shear in a flowing field, the front and rear portions of the drop will deform much like the deformation of a water-filled balloon (Figure 9.11), but again, substantial flow in the interfacial region will not occur. If drops are closely spaced, however, the additional interactions induced by the hydrodynamic flow may result in partial destabilization of the interacting drops leading to flocculation or coalescence. In that case, a given emulsion formulation may, and probably will, have an optimum energy

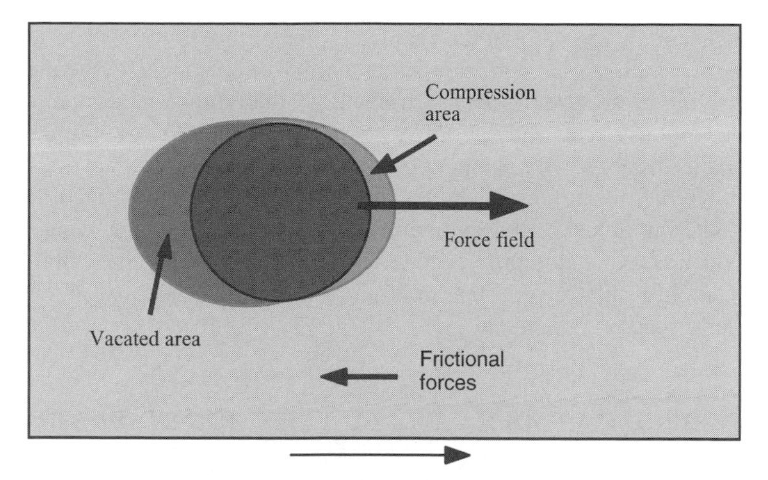

(a)

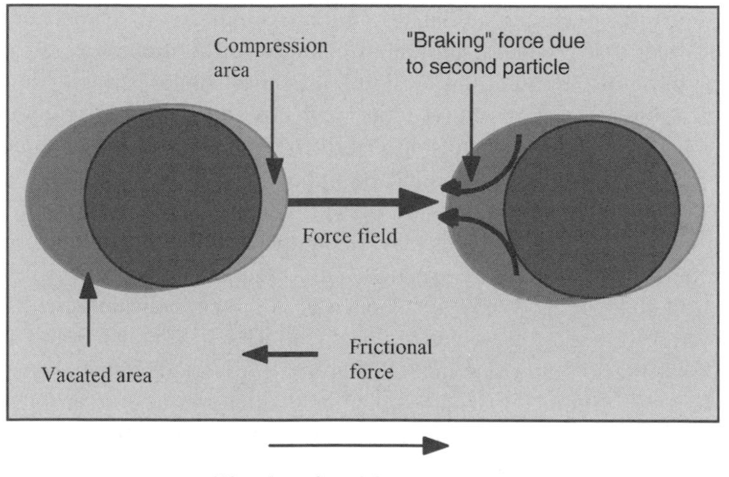

(b)

Figure 9.11. The deformation of an emulsion drop in a flowing field: (a) isolated particle; (b) interacting particles.

input for formation of the best emulsion beyond which the excess energy results in drop coalescence rather than disruption.

In the presence of polymeric stabilizers, the situation is much more complex and is poorly understood. It can reasonably be assumed, however, that if the stabilizing activity of the adsorbed polymer is based on steric interactions, both the viscosity of the interfacial layer and osmotic forces in the interstitial region between drops will contribute to the overall stabilizing effect.

In conclusion, it appears that the characteristics of the oil–water interfacial film that enhance the kinetic properties or metastability of emulsions in the presence of monomeric surfactant species can be related to their function in the context of Gibbs–Marangoni effects and the formation of specific interfacial complexes leading to enhanced stability. By damping out local variations in interfacial tension and distortions in the intervening interstitial film, surfactants help maintain a uniform lamellar thinning process between approaching droplets, "healing" local areas of weakness that could lead to droplet coalescence, and increasing the stability of the system. Current results suggest that the role of interfacial viscosity, in the absence of polymeric species, is minimal.

9.6. AMPHIPHILE MESOPHASES AND EMULSION STABILITY

The mechanical strength, elasticity, and rheological properties of the interfacial film stabilizing an emulsion obviously have a significant impact on the overall stability of the system. Chapters 4 and 5 introduced the concept of liquid crystal and mesophase formation in surfactant systems, usually in the context of increases in the concentration of surface-active material in solution. As indicated, such phases possess a degree of order that produces substantial changes in the properties of the system relative to those of the molecular or simple micellar solutions, including a higher degree of rigidity, larger structural units, and less fluctuation in composition. In view of the mechanisms of emulsion stabilization discussed above, such phases, if present at the O/W interface, might be expected to impart an added degree of stability to systems in which they are present (Figure 9.12). In a practical sense, liquid crystals or other mesophases may be compared to the mixed interfacial complexes discussed in the preceding section, only producing a more complex interfacial situation on a larger scale. The presence of liquid crystals in the region of high drop curvature (the plateau border region mentioned in Chapter 8 and shown in, Figure 8.4) will also help overcome the Laplace pressure differential.

The presence of liquid crystals or other such structures at or near the oil–water interface has been shown to produce improvements in the stability of numerous emulsions, although the exact mechanism of their action is still subject to some question. Even in the absence of complete understanding (as is often the case for surfactant-related topics), the usefulness of such structures at O/W interfaces has been demonstrated in practical applications.

By analogy with monomolecular films at liquid–air interfaces, surfactants at the liquid–liquid interface will normally form monolayers with various molecular packing densities ranging from relatively loosely packed arrangements normally associated with fluid phases (gases and liquids) to the close-packed solid phase. Classically, amphiphilic adsorption at interfaces has been roughly classified in terms of film types related to normal states of matter (Figure 9.13):

1. *Liquid-expanded* (L1), in which the adsorbed molecules may assume a variety of orientations with regard to the dividing surface ranging from perpendicular

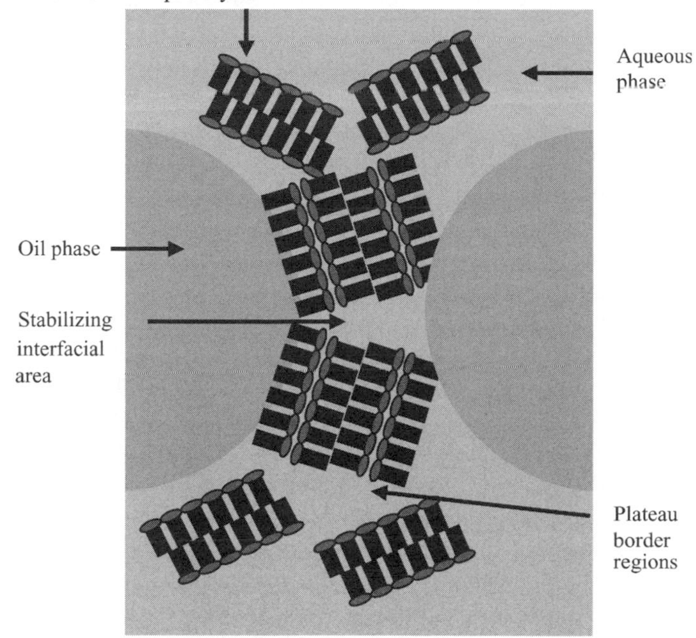

Figure 9.12. Some potential roles of amphiphilic mesophases at or near emulsion interfaces.

to almost parallel (Figure 9.13a), but with average molecular distances much greater than in bulk liquids and significant rotational disorder in the vertical chains.

2. *Liquid-condensed* (L2) (Figure 9.13b), in which the adsorbed molecules are relatively close-packed with a tilted orientation and reduced chain mobility.
3. *Condensed solid* (CS) (Figure 9.13c), in which the adsorbed molecules are close-packed with essentially vertical orientation to the interface and minimum mobility—essentially incompressible unless layer collapse occurs.

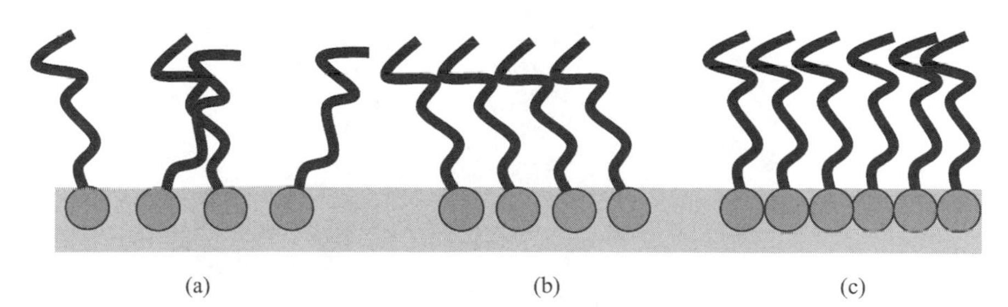

Figure 9.13. Typical forms of monolayer films at L/L and L/V interfaces: (a) moderately close-packed with significant chain mobility (liquid expanded); (b) close-packed with tilted orientation and reduced chain mobility (liquid condensed); (c) close-packed with essentially vertical orientation and very limited chain mobility (condensed solid).

Although the experimental basis for such a classification stems from monolayer work at the L/V (liquid–vapor) interface, enough correspondence has been found with L/L (liquid–liquid) systems that a high degree of confidence can be placed in the translation of the same concepts to emulsion systems, at least as a first approximation to reality. The presence of liquid phases on each side of the monolayer, the potential for amphiphile movement across the interface, and the complications of solubility in the two phases, all of which will no doubt affect the properties of the monolayer structures, requires some flexibility in the interpretation of experimental results.

In L/V monolayers, multilayer formation can be found once the available surface area has been saturated with surfactant molecules or when surface pressures exceed the limit of the system and monolayer collapse occurs. The normal growth process will result in the addition of consecutive layers with alternating surface characteristics; thus, the first monolayer will expose hydrophobic groups, the second hydrophilic, and so on (Figure 9.14, left side). In a L/L situation, growth may be more directly by bilayer units so that the relationship between the two liquid phases and the exposed surfactant groups is maintained (Figure 9.14, right side). The presence of such multilayer film formation between emulsion droplets will have obvious consequences for the stability of a system.

While there is still some controversy in the literature concerning the exact role of liquid crystals and interfacial complexes in emulsion stabilization. Whether specific association complexes between mixed surfactant or surfactant–additive systems occur at the interface, or whether the results so interpreted are actually produced

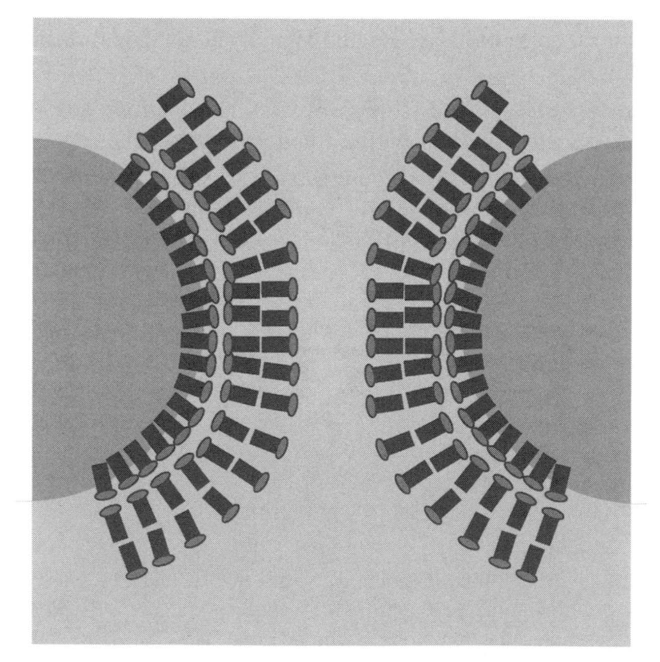

Figure 9.14. The growth of surfactant monolayers in L/V and L/L environments.

by some other less specific phenomenon such as enhanced film cohesiveness due to more favorable van der Waals interactions, the net result is that enhanced emulsion stability can be obtained by the proper choice of the materials employed in the formulation.

9.7. SURFACTANT STRUCTURE AND EMULSION STABILITY

Ideally, the world of emulsion formulation would be such that a simple correlation could be obtained between the chemical structure of a surfactant and its performance in practice. Unfortunately, the complicated nature of typical emulsion formulations—such as the nature of the oil phase, additives in the aqueous phase, specific surfactant interactions, and end-use requirements—makes correlations between surfactant structure and properties in emulsification processes very empirical.

In the absence of a truly quantitative and absolute method for choosing a surfactant for a given application, it is possible to outline a few rules of thumb that have historically prove useful for narrowing down the possibilities and limiting the amount of experimentation that will be required for the final selection of surfactant(s) for a given application. For example, the surfactant(s) must exhibit sufficient surface activity to ensure effective and efficient adsorption at the oil–water interface. That activity must also be related to the actual conditions of use, including the nature of the oil phase, possible additives in each phase, and conditions of temperature and pressure, and seldom can be inferred directly from activity in water alone. As pointed out previously and again below, the presence of materials such as electrolytes and polymers can greatly alter the functioning of surfactants in stabilizing an emulsion as well as in controlling the type of emulsion formed.

The basic role of the interfacial film between the two phases has already been pointed out. It is clear, then, that the surfactant(s) employed should produce as strong an interfacial film as possible, one with high viscosity and tenacity, consistent with their ability to produce the required droplet size under the conditions of emulsification. It is useful, therefore, to choose a surfactant system with maximum lateral interaction among the surfactant molecules concurrent with efficient and effective lowering of the interfacial tension.

On a molecular level, the choice of surfactant for a given application must also take into consideration the type of emulsion desired and the nature of the oil phase. As a general rule, water-soluble surfactants yield O/W systems while oil-soluble materials preferentially produce W/O emulsions. Because of the role of the interfacial layer in emulsion stabilization, a mixture of surfactants with widely differing solubility properties will often be found to produce emulsions with better stability than will ones with equivalent concentrations of either material alone. Looking for synergism in mixed surfactant systems can be a very valuable rule of thumb in the absence of more specific guidance. Finally, it is usually safe to say that the more polar the oil phase, the more polar will be the surfactant required to provide optimum emulsification and stability. Such rules of thumb, while having great practical utility, are less than satisfying on a scientific level. One would really like to have a neat, quantitative formula for the design of complete emulsion systems. A number of attempts have been made over the years to develop just such a quantitative approach to surfactant selection, and a brief discussion of some such approaches follows.

9.7.1. Hydrophile–Lipophile Balance (HLB)

It has been a long-term goal of surfactant chemists and formulators to devise a quantitative way of correlating the chemical structure of surfactant molecules with their surface activity through some quantitative relationship that would facilitate the choice of material for use in a given formulation. Perhaps the greatest success along these lines has been achieved in the field of emulsions; therefore, it is appropriate to discuss the subject in some detail in that context.

The first reasonably successful attempt to quantitatively correlate surfactant structures with their effectiveness as emulsifiers was the hydrophile–lipophile balance (HLB) system, in which the objective is to calculate a number that "measures" the emulsifying potential, in terms of emulsion quality and stability or the so-called HLB, of a surfactant from its chemical structure, and to match that number with the corresponding HLB of the oil phase to be dispersed. The system employs certain empirical formulas to calculate the HLB number, normally giving answers within a range of 0–20 on some arbitrary scale. At the high end of the scale lie hydrophilic surfactants, which possess high water solubility and generally act as good solubilizing agents, detergents, and stabilizers for O/W emulsions; at the low end are surfactants with low water solubility, which act as solubilizers of water in oils and good W/O emulsion stabilizers. The effectiveness of a given surfactant in stabilizing a particular emulsion system would then depend on the balance between the HLBs of the surfactant and the oil phase involved.

For nonionic surfactants with polyoxyethylene solubilizing groups, the HLB was calculated from the formula

$$\mathrm{HLB} = \frac{\text{mol\% hydrophilic group}}{5} \tag{9.7}$$

In such a scheme, an unsubstituted polyoxyethylene glycol would have an HLB of 20. HLB values for some typical nonionic surfactants are given in Table 9.1.

Surfactants based on polyhydric alcohol fatty acid esters such as glycerol monostearate can be handled by the relationship

$$\mathrm{HLB} = 20\left(1 - \frac{S}{A}\right) \tag{9.8}$$

where S is the saponification number of the ester and A is the acid number of the acid. A typical surfactant of this type, commercially known as Tween 20 [polyoxyethylene (20) sorbitan monolaurate], with $S = 45.5$ and $A = 276$, would have an

TABLE 9.1. Some Calculated HLB Values for Typical Nonionic Surfactant Structures

Surfactant[a]	HLB
Sorbitan trioleate	1.8
Sorbitan tristearate	2.1
Propylene glycol monostearate	3.4
Glycerol monostearate	3.8
Sorbitan monooleate	4.3
Sorbitan monostearate	4.7
Polyoxyethylene(2) cetyl ether	5.3
Diethylene glycol monolaurate	6.1
Sorbitan monolaurate	8.6
Polyoxyethylene(10) cetyl ether	12.9
Polyoxyethylene(20) cetyl ether	15.7
Polyoxyethylene(6) tridecyl ether	11.4
Polyoxyethylene(12) tridecyl ether	14.5
Polyoxyethylene(15) tridecyl ether	15.4

[a]Numbers in parentheses indicate the average number of OE units in the hydrophilic chain.

HLB of 16.7. For materials that cannot be saponified, an empirical formula of the form

$$\text{HLB} = \frac{E + P}{5} \tag{9.9}$$

can be employed. In the equation, E is the weight percent of oxyethylene chains and P is the weight percent of polyhydric alcohol (glycerol, sorbitan, etc.) in the molecule.

Although the HLB system has proved to be very useful from a formulation chemist's point of view, its empirical nature did not satisfy the desire of many for a sounder theoretical basis for surfactant characterization. It was subsequently suggested that HLB numbers could be calculated on the basis of group contributions according to the formula

$$\text{HLB} = 7 + \sum(\text{hydrophilic group numbers}) - \sum(\text{hydrophobic group numbers}) \tag{9.10}$$

Some typical group numbers are listed in Table 9.2.

The use of the HLB system for choosing the best emulsifier system for a given application originally required the performance of a number of experiments in which surfactants or surfactant mixtures with a range of HLB numbers are employed to prepare emulsions of the oil in question, and the stability of the resulting emulsions is evaluated by measuring the amount of creaming that occurred with time. The use of surfactant mixtures can become complicated by the fact that such

TABLE 9.2. Typical Group Numbers for Calculation of HLB Numbers

Group	HLB Number	Group	HLB Number
Hydrophilic		Hydrophobic	
$-SO_4Na$	38.7	–CH–	−0.475
–COOK	21.1	$-CH_2-$	−0.475
–COONa	19.1	$-CH_3$	−0.475
–N (tertiary amine)	9.4	=CH–	−0.475
Ester (sorbitan)	6.8	$-CF_2-$	−0.87
Ester (free)	2.4	$-CF_3$	−0.87
–COOH	2.1	Miscellaneous	
–OH (free)	1.9	$-(CH_2CH_2O)-$	0.33
–O–	1.3	$-(CH_2CH_2CH_2O)-$	−0.15
–OH (sorbitan)	0.5		

mixtures often produce more stable emulsions than would a single surfactant with the same nominal HLB number. The HLB of a mixture is assumed to be an algebraic mean of the HLBs of the components

$$\mathrm{HLB_{mix}} = f_A \times \mathrm{HLB_A} + (1 - f_A) \times \mathrm{HLB_B} \tag{9.11}$$

where f_A is the weight fraction of surfactant A in the mixture. It has been found experimentally, however, that such a linear relationship for nonionic mixtures is observed only when each component of the mixture is able to act independently of the other, with no specific molecular interactions occurring. A number of studies have indicated significant deviations from linearity, both positive and negative, based on emulsion stability tests, interfacial tension measurements, and correlations using gas–liquid chromatography, cloud point determinations, and phase inversion temperature (PIT) data (see text below).

Although going a long way toward simplifying the choice of surfactants for the stabilization of a given oil in water, the HLB system does not always provide a clear-cut answer for a given system. It does not, for example, account for the effects of a surfactant on the physical properties of the continuous phase, especially its rheological characteristics. As noted previously, the viscosity of the continuous phase will significantly affect the rate of creaming, as will alterations in the relative densities of the two phases. As a result, it is possible to prepare very stable emulsions with surfactants whose HLB numbers lie well away from the "optimum" that would be predicted by the performance of a series of HLB-related experiments. Its obvious faults notwithstanding, the HLB system as originally derived and subsequently expanded has found extensive practical use. Especially when applied to nonionic surfactants and their mixtures, the concept appears to possess some fundamental rationality that has yet to be revealed completely at the molecular level.

One approach to the establishment of a sounder theoretical basis for the HLB number concept (in aqueous systems, at least) has been the relationship between HLB and the degree of hydration of the surfactant molecule, as expressed by the

Hildebrand solubility parameter or cohesive energy density (δ). The concept is that the solubility and surface activity of a given surfactant structure can be quantitatively related to the "critical" HLB number for the emulsification of a given oil phase in aqueous solution by its solubility parameter or cohesive energy density in water. The efficiency of a surfactant at emulsifying and stabilizing an oil phase will be a function of the relative degrees of interaction of the various portions of the surfactant molecule with the oil and aqueous phases. The original concepts of HLB were, as seen above, based on a simple ratio of hydrophilic to hydrophobic groups in the molecule. With the introduction of Hildebrand solubility parameters, usually denoted as δ, as a quantitative way to calculate the solubility of materials, especially polymers, in various solvents, an attempt was made to relate the HLBs of a number of surfactants to their calculated δ. Their results fitted the relationship

$$\delta = \frac{243}{54 - \text{HLB}} + 12.3 \tag{9.12}$$

where δ, in SI units, is in (megapascals)$^{1/2}$ (MPa$^{1/2}$). More complicated treatments used three-dimensional Hansen parameters to relate the emulsifying tendencies of surfactants with oils in terms of the ratios of the cohesive energies of the oil and the hydrophobic portion of the surfactant molecule and that of the water with the hydrophilic portion. In that concept, the best result can be expected when the various components of the cohesive energy density of the three phases—oil, water, and surfactant—are matched. The dispersion (d), polar (p), and hydrogen bonding (h) portions of the three-dimensional solubility parameters are, to a first approximation, related by

$$\delta^2 = \delta_d^2 + \delta_p^2 + \delta_h^2 \tag{9.13}$$

so that division of Eq. (9.13) by δ^2 will lead to

$$1 = f_d + f_p + f_h \tag{9.14}$$

where f is that fraction of the total solubility parameter attributable to each type of interaction. Using that relationship, it is possible to construct a triangular diagram that can relate the δ values of each component and serve as a useful predictive tool for emulsion formulation. Unfortunately, the utility of such an approach is somewhat limited by the scarcity of experimental data for most surfactant structures and the complications introduced by the complex and sometimes varied compositions of many industrial surfactants.

When all of the component solubility parameters of the oil and the surfactant are matched, a relationship between HLB and chemical composition of the form

$$\text{HLB} = \frac{20M_h}{M_l + M_h} \tag{9.15}$$

results, where M_h is the molecular weight of the hydrophilic portion of the molecule (including the carbon atom of a carboxyl group) and M_l is that of the hydrophobe or lipophile.

Because the "effective" HLB of a given surfactant will depend on the nature of the solvent, HLB numbers cannot be considered to be absolute, realistic measures of the emulsifying ability of a material under all conditions. The actual HLB of a surfactant in a system will depend on the nature of the solvent, the temperature, and the presence of additives such as cosolvents, electrolytes, and polymers. Although the relationship will not always be linear, the HLB may be expected to vary in a manner analogous to that found for the critical micelle concentration of the surfactant under the same conditions.

Using the HLB group calculation approach, the HLB and cmc of a surfactant can be related through a relationship of the form

$$\ln\,(\mathrm{cmc}) = C_1 + C_2(\mathrm{HLB}) \tag{9.16}$$

where the constants C_1 and C_2 are characteristics related to a homologous series of surfactants. Using that formulation, it is possible to define a relationship between the free energy of micellization for a surfactant and its HLB in water. Combining the relevant equations yields

$$\mathrm{HLB} = C_1 + C_2 \frac{\Delta G_m}{RT} \tag{9.17}$$

Further separation of ΔG_m into its hydrophilic and hydrophobic components allows one to write

$$\mathrm{HLB} = C_1 + C_2 \frac{\Delta G_\mathrm{mh}}{RT} + C_2 \frac{\Delta G_\mathrm{ml}}{RT} \tag{9.18}$$

It can be seen that Eq. (9.18) is formally equivalent to Eq. (4.21). In this way, the concept of the HLB can be associated with the tendency of a surfactant to aggregate or adsorb at interfaces. It has been found that a reasonably linear relationship between HLB and ΔG_m can be obtained, with slopes varying according to the nature of the hydrophilic and hydrophobic groups, as would be expected from the relationships discussed in Chapter 4.

As already pointed out in Chapters 4 and 5, geometric constraints imposed by the particular molecular characteristics of a surfactant molecule control the formation of aggregates (size, shape, curvature, etc.). The previously defined geometric packing parameter $P_\mathrm{c} = v/a_\mathrm{o}l_\mathrm{c}$, where v is the volume of the hydrophobic group, a_o is the optimum head group area, and l_c is the critical length of the hydrophobe, can be viewed as a type of HLB number, based on volume fraction instead of weight fractions of hydrophobe and the geometry of the hydrophobic chain. In that case, $\mathrm{P_c}$ could be considered an "inverse" HLB ($\mathrm{HLB_i}$), where

$$\mathrm{HLB_i} = 20 - \mathrm{HLB} \tag{9.19}$$

By use of geometric considerations, it can be seen that the value of P_c determined from molecular geometry should predict the type of emulsion formed by a particular surfactant. For instance, for $P_c < 1$, the curvature of the oil–water interface should be concave toward the oil phase, leading to an O/W emulsion. For $P_c > 1$, the reverse would be expected. At $P_c = 1$, a critical condition would be expected where phase inversion would occur or multiple emulsion formation would be favored (see text below).

From the discussion above, it should be clear that, theoretical "desires" notwithstanding, the goal of a quantitative magic formula for calculating the surfactant characteristics needed for a given emulsion remains elusive. As has been seen for other surfactant applications, simple answers apply only in simple systems, and simple systems are not, unfortunately, the rule in practice. While HLB numbers, solubility parameters, and geometric factors provide extremely useful approaches for making an educated guess about a surfactant or surfactants to use in a formulation, the final answer will still require old-fashioned, hands-on experiments to find the best of all worlds. And that best answer will no doubt be significantly influenced by the presence of other actors on the stage.

9.7.2. Phase Inversion Temperature (PIT)

Several references were made above to the term "phase inversion temperature." With the exceptions of Eqs. (9.17) and (9.18), however, no specific reference was made to the effect of temperature on the HLB of a surfactant. From the discussions in Chapter 4, it is clear that temperature can play a role in determining the surface activity of a surfactant, especially nonionic amphiphiles in which hydration is the principal mechanism of solubilization. The importance of temperature effects on surfactant solution properties, especially the solubility or cloud point of nonionic surfactants, led to the evolution of the concept of using that property as a tool for predicting the activity of such materials in emulsions. Since the cloud point is defined as the temperature, or temperature range, at which a given amphiphile loses sufficient solubility in water to produce a "normal" surfactant solution, it was assumed that such a temperature-driven transition would also be reflected in the role of the surfactant in emulsion formation and stabilization.

In this case, it was felt that the phenomenon would result in an "inversion" of the role of the material in terms of the type of emulsion favored by its presence. For example, at low temperatures a given material would be expected to be an O/W emulsifier, while at temperatures above the cloud point it would become a W/O emulsifier. In the context of emulsion technology, therefore, the cloud point phenomenon became known as the phase inversion temperature (PIT) of the surfactant and was proposed as a quantitative approach to the evaluation of surfactants in emulsion systems. In effect, the PIT is not a characteristic of a surfactant, but rather a characteristic of the complete emulsion system.

The general procedure developed for the evaluation system was as follows. Emulsions of oil, aqueous phase, and approximately 5% surfactant were prepared by shaking at various temperatures. The temperature at which the emulsion was found to be inverted from O/W to W/O was then defined as the PIT of the system. Since the effect of temperature on the solubility of nonionic surfactants is reasonably well understood, the physical principles underlying the PIT phenomenon followed directly.

It is generally found that the same circumstances that affect the solution characteristics of nonionic surfactants (cmc, micelle size, cloud point, etc.) will also affect the PIT of emulsions prepared with the same materials. For typical POE surfactants, increasing the length of the POE chain will result in a higher PIT, as will a broadening of the POE chain length distribution. The use of phase inversion temperatures, therefore, represents a potentially useful tool for the comparative evaluation of emulsion stability. Many reports of its use in the emulsion technology field are available.

Because the PIT approach to surfactant evaluation is newer than the HLB method, the effects of variables on the relationship between PITs, surfactant structures, and emulsion stability have not been as clearly defined in a quantitative way. It has been found, however, that there is an almost linear correlation between the HLB of a surfactant under a given set of conditions and its PIT under the same circumstances. In essence, the higher the HLB of the surfactant system, the higher will be its PIT.

The sensitivity of emulsions to temperature led to the suggestion that the PIT phenomenon could also be used as a convenient method for emulsion preparation. In such a procedure, an emulsion is prepared very near the PIT of the system (routinely $\pm 4^{\circ}$), where minimum droplet sizes can normally be obtained. The emulsion is then cooled to its storage or use temperature, where enhanced stability usually results. In nonionic systems particularly, such an approach to emulsion preparation can provide advantages in terms of the energy required for emulsification in addition to providing enhanced stability.

Because the PIT appears to be directly related to the HLB of the surfactant, the effects of such factors as surfactant concentration, oil-phase polarity, additives, and phase ratios would be expected to parallel what has been observed for HLB determinations. Certainly, it can be expected that much more information will be published in the coming years, which will allow for a better understanding of the relationships between these two concepts of surfactant evaluation.

9.7.3. Application of HLB and PIT in Emulsion Formulation

The choice of a particular emulsifier system for an application will depend on several factors, some of which will be chemically related (optimum HLB, PIT, etc.), while others will be driven by economic, environmental, and aesthetic factors. The relative value of the latter will depend mostly on price and value-added considerations for each individual system. Here we are concerned only with the chemical aspects of emulsion formation and stabilization. In most general applications, the HLB system has been found most useful in guiding the formulator to a choice of surfactant most suited to individual needs. Table 9.3 lists the ranges of HLB numbers that have proved to be most useful for various applications.

TABLE 9.3. HLB Ranges and Their General Areas of Application

HLB Range	General applications
2–6	W/O emulsions
7–9	Wetting and spreading
8–18	O/W emulsions
3–15	Detergency
15–18	Solubilization

Obviously, the ranges in which surfactants of various HLBs can be employed are quite broad. Specific requirements for many systems have been tabulated in the works cited in the Bibliography list for this chapter. While such tabulations can be very useful to the formulations chemist, it must be kept in mind that there is nothing particularly magic about a given HLB number. Many surfactants or surfactant mixtures may possess the same HLB, yet subtle differences in their chemical structures, solution chemistry, and specific interactions with other system components may result in significant differences in performance. Particularly important may be the formation of interfacial complexes, as noted above. Even though the additive nature of surfactant mixture HLBs [(Eq. (9.11)] has not been found to be linear over a wide range of compositions, over the short range of one or two HLB units usually encountered in formulation work, linearity can usually be assumed with little risk. It is therefore possible, in most cases, to fine-tune a surfactant mixture with a minimum of experimental effort.

As alluded to above, one approach to the application of surfactant HLB to formulation is to match that of the surfactant to the oil phase being employed. The HLB of the oil can be determined empirically or calculated using the procedures discussed previously. It is usually found that the additivity principle will hold for mixtures of oils in a way similar to that for surfactants, possibly even to the extent of nonlinearity in cases where oil structures differ significantly. Therefore, in formulating an emulsion, it is possible to determine the HLB of the oil phase and to vary the surfactant or mixture HLB to achieve the optimum performance. HLB numbers of some commonly used oil phases are given in Table 9.4.

It should be noted that HLB numbers are most often used in connection with nonionic surfactants. While ionic surfactants are included in the HLB system, the more complex nature of the solution properties of the ionic materials makes them less amenable for the normal approaches to HLB classification. In cases where an electrical charge is desirable for stability or for some other functional reason, it has been suggested that surfactants having limited water solubility and a bulky hydrophobic structure that inhibits efficient packing into micelles should be most effective as emulsifiers. Surfactants such as the sodium trialkylnaphthalene sulfonates and dialkylsulfosuccinates, which do not readily form large micelles in aqueous solution, have found use in that context, usually providing advantages in droplet size and stability over simpler materials such as sodium dodecylsulfate.

TABLE 9.4. HLB Numbers for Typically Encountered Oil Phases

Oil Phase	Nominal HLB
Lauric acid	16
Oleic acid	17
Cetyl alcohol	15
Decyl alcohol	14
Benzene	15
Castor oil	14
Kerosene	14
Soybean oil	13
Lanolin	12
Carnauba wax	12
Paraffin wax	10
Beeswax	9

Clearly, the process of selecting the best surfactant(s) for the preparation of an emulsion has been greatly simplified by the development of the generally empirical, semiquantitative approaches exemplified by the HLB and PIT methods described above. Unfortunately, each method has its limitations and cannot eliminate the need for some amount of trial-and-error experimentation. As our fundamental understanding of the complex phenomena occurring at oil–water interfaces improves, and the effects of additives and environmental factors on those phenomena become more clear, it may become possible for a single, comprehensive theory of emulsion formation and stabilization to lead to a single, quantitative scheme for the selection of the proper surfactant system.

9.7.4. Effects of Additives on the "Effective" HLB of Surfactants

The exact mechanisms by which various additives affect the effective HLBs of surfactants are not fully understood. For nonionic POE surfactants, in which hydration of the POE chain is the primary solubilizing mechanism in aqueous solution, the extent of chain hydration has seldom been found to be increased by the addition of materials that "salt in" the surfactants. That conclusion is based on the observation that the viscosity of the solutions is not significantly affected, indicating that the hydrodynamic radius of the molecules is not increased by increased hydration. In fact, the actions of such additives are in all probability related to their effects on the structure of the solvent, altering the thermodynamics of solvent–solute interactions.

A proposed relationship between the HLB and the heat of hydration Q_h of a surfactant is given by

$$\text{HLB} = 0.42\, Q_h + 7.5 \tag{9.20}$$

where Q_h is given in calories per gram. From this equation, it seems that the addition of materials that increase the heat of hydration, such as sodium thiocyanate, should produce an increase in the effective HLB of the surfactant. Such an effect is found for nonionic surfactants in aqueous thiocyanate solution compared to those in the presence of "salting–out" additives such as sodium chloride.

In the case of nonionic surfactants, the steric or enthalpic barrier to droplet coalescence produced by the hydrated hydrophilic chain is a major factor in emulsion stability. The addition of additives that increase the heat of hydration should result in an increase in the enthalpic contribution to stability. Experimentally, unambiguous evidence to support such a hypothesis has been difficult to obtain.

The difficulty of obtaining good experimental evidence for the correlation of increased cloud points for nonionic surfactants and their effectiveness as emulsion stabilizers may be related to the fact that surfactant interactions at interfaces, as well as their own self-aggregation characteristics, are affected in very complex ways by the presence of additives. As discussed in Chapter 4, the solubility and cmc of a given surfactant depend strongly on the nature of the solvent and additives, as does the adsorption of molecules at various interfaces. If there is an optimum HLB for a given oil that produces maximum stability, any additive that alters the solution properties of the surfactant should also alter the effective HLB of the surfactant and, therefore, shift it away from the optimum value.

Some studies have tried to follow the factors controlling the effective HLB of specific systems, varying the surfactant, the oil, and the continuous phase in a regular way. By monitoring such factors as emulsion droplet size, emulsion rheology, creaming, and inversion temperatures, they determined that additives that acted as "salting-in" agents produced a decrease in the critical HLB needed for maximum stability. "Salting-out" additives had the opposite effect.

9.8. MULTIPLE EMULSIONS

While a great deal of information has been published over the years on the theoretical and practical aspects of emulsion formation and stabilization, until relatively recently little has been said about more complex systems generally referred to as "multiple emulsions". Multiple emulsions, as the name implies, are composed of droplets of one liquid dispersed in larger droplets of a second liquid, which are then dispersed in a final continuous phase (Figure 9.15). Typically, the internal droplet phase will be miscible with or identical to the final continuous phase. Such systems may consist of a W/O/W dispersion, where the internal and external phases are aqueous, or O/W/O multiple emulsions, which have the reverse structure. Although known for almost a century, such systems have only relatively recently become of practical interest for possible use in controlled drug delivery, emergency drug overdose treatment, wastewater treatment, and separations technology. Other useful applications will no doubt become evident as our understanding of the physical chemistry of such systems improves.

Because they involve a variety of phases and interfaces, multiple emulsions are inherently more unstable than are simple emulsions. Their surfactant requirements are such that two stabilizing systems must be employed—one for each oil–water

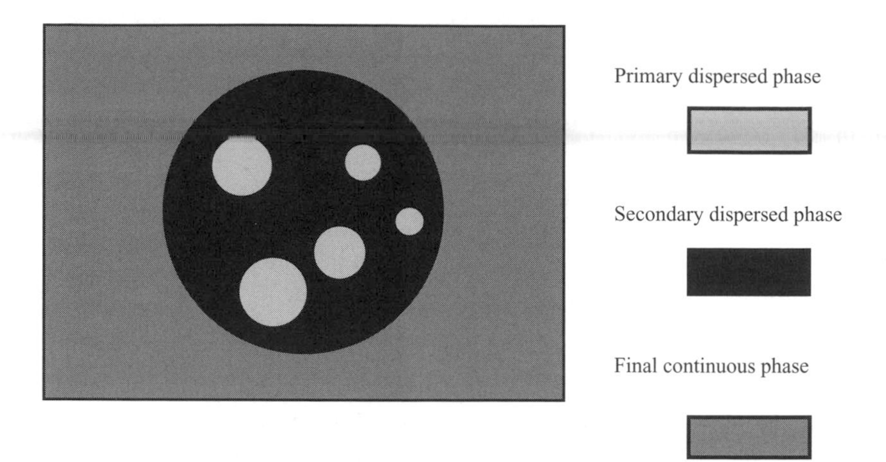

Figure 9.15. A schematic representation of a multiple emulsion.

interface. Each surfactant or mixture must be optimized for the type of emulsion being prepared, but must not interfere with the companion system designed for the opposite interface. Long-term stability, therefore, requires careful consideration of the characteristics of the various phases and surfactant solubilities.

9.8.1. Nomenclature for Multiple Emulsions

For systems as potentially complex as multiple emulsions, it is very important that a clear and consistent system of nomenclature be employed. For a W/O/W system, for example, in which the final continuous phase is aqueous, the primary emulsion will be a W/O emulsion, which is then emulsified into the final aqueous phase. The surfactant or emulsifier system used to prepare the primary emulsion is referred to as the *primary surfactant*, and the volume fraction of the primary dispersed phase is the internal aqueous phase of the final multiple emulsion. Subscripts are used to further avoid ambiguities as to components or their locations in the system. For example, in a W/O/W system the aqueous phase of the primary emulsion would be denoted as W_1 and the primary emulsion as W_1/O. After the primary emulsion has been further dispersed in the second aqueous phase W_2, the complete system is denoted $W_1/O/W_2$. In the case of a O/W/O multiple emulsion, the notation is $O_1/W/O_2$. Additional refinements to fit even more complex systems, including the "order" of multiple emulsions, have been suggested.

9.8.2. Preparation and Stability of Multiple Emulsions

In principle, multiple emulsions can be prepared by any of the numerous methods for the preparation of conventional emulsion systems, including sonication, agitation, and phase inversion. Great care must be exercised in the preparation of the final system, however, because vigorous treatments normally employed for the preparation of primary emulsions will often break the primary emulsion, resulting in loss of phase identity.

Multiple emulsions have reportedly been prepared conveniently by the phase inversion technique mentioned earlier; however, such systems have generally been found to have limited stability. It generally requires a very judicious choice of surfactant or surfactant combinations to produce a multiple emulsion system that has useful characteristics of formation and stability. A general procedure for the preparation of a W/O/W multiple emulsion may involve the formation of a primary emulsion of water in oil using a surfactant suitable for the stabilization of such W/O systems. Generally, that will involve the use of an oil-soluble surfactant with a low HLB (2–8). The primary emulsion will then be emulsified in a second aqueous solution containing a second surfactant system appropriate for the stabilization of the secondary O/W emulsion (HLB 6–16). As noted above, because of the possible instability of the primary emulsion, great care must be taken in the choice of the secondary dispersion method. Excessive mechanical agitation such as in high-speed mixers and sonication could result in gross coalescence of the primary emulsion and the production of essentially "empty" oil droplets. The evaluation of the yield of filled secondary emulsion drops, therefore, is very important in assessing the value of different preparation methods and surfactant combinations.

The nature of the droplets in a multiple emulsion will depend on the size and stability of the primary emulsion. A system of classification has been proposed dividing W/O/W multiple emulsions into three classes according to the nature of the oil-phase droplets (Figure 9.16). Type A systems are characterized as having one large internal drop essentially encapsulated by the oil phase. Type B systems contain several small, well-separated internal drops, and systems of type C contain many small internal drops in close proximity. It is understood that any given system will in all probability contain all three classes of drops, but one will be found to dominate, depending primarily on the surfactant system employed.

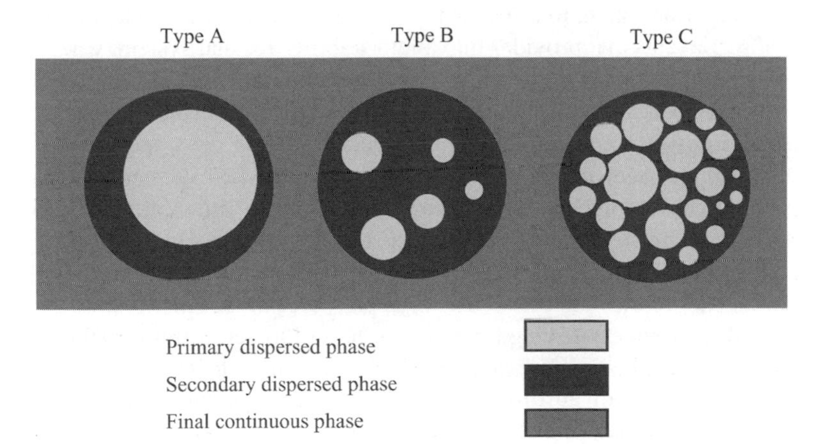

Figure 9.16. Multiple emulsion classification based on droplet characteristics in the primary emulsion.

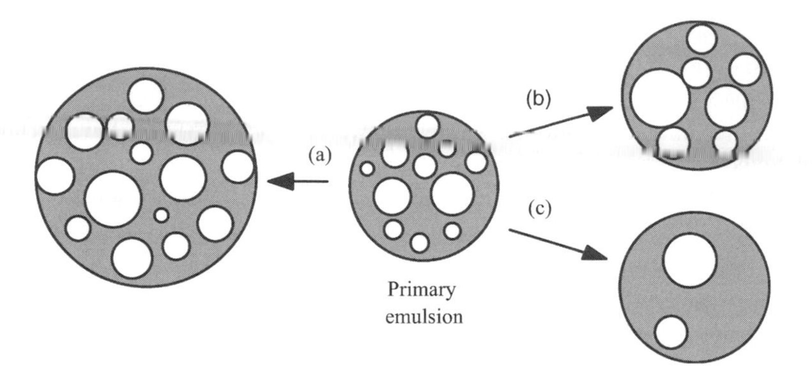

Figure 9.17. Some possible pathways for the breakdown of multiple emulsions: (a) coalescence of secondary emulsion drops; (b) coalescence of primary emulsion drops; (c) loss of primary emulsion dispersed phase to external phase.

9.8.3. Pathways for Primary Emulsion Breakdown

There are several possible pathways for the breakdown of multiple emulsions. A few are shown schematically in Figure 9.17. Although all possible mechanisms for droplet coalescence cannot be conveniently illustrated in a single figure, a consideration of just a few possibilities can help clarify the reasons for instability in a given system. Even though there may be a number of factors involved, one of the primary driving forces will be, as always, a reduction in the free energy of the system through a decrease in the total interfacial area. As has been noted previously, a major role of surfactants at any interface is to reduce the interfacial energy through adsorption. In a typical multiple-emulsion system, the primary mechanism for short-term instability will usually be droplet coalescence in the primary emulsion. It will be important, then, to select as the primary emulsifier a surfactant or combination of surfactants that provides maximum stability for that system, whether W/O or O/W.

A second important pathway for the loss of "filled" emulsion droplets is the loss of internal drops by the rupture of the oil layer separating the small drops from the continuous phase. Such an expulsion mechanism would be expected to account for the loss of larger internal droplets. Unless the two phases are totally immiscible (in fact, a rare situation), there will always exist the possibility that osmotic pressure differences between the internal and continuous portions of the system will cause material transfer to the bulk phase. The high pressures in the smaller droplets would be expected to provide a driving force for the loss of material from smaller drops in favor of larger neighbors (Ostwald ripening), as well as to the continuous phase. Finally, the presence of an oil-soluble surfactant always suggests the possibility of nonaqueous reversed micelle formation and the subsequent solubilization of internal aqueous phase in the oil. Such a solubilization process also represents a convenient mechanism for the transport of material between the two similar phases.

In the context of a critical application such as controlled drug delivery, in which the mechanism of delivery is diffusion-controlled, such breakdown mechanisms would be very detrimental to the action of the system, since they could result in a rapid release of active solute with possibly dangerous effects.

The proposed mechanisms of emulsion breakdown, as well as others, must be addressed in order to understand and control a particular multiple-emulsion system. In all cases, the final stability of the system will depend on the nature of the oil phase of interest, the characteristics of the primary and secondary emulsifier systems, and the relationship between the internal and continuous phases.

9.8.4. Surfactants and Phase Components

Choice of surfactants for the preparation of multiple emulsions can, in principle, be made from any of the four classes of surfactants discussed in Chapter 2. The choice will be determined by the characteristics of the final emulsion type desired: the natures of the various phases, additives, solubilities, and so on. In many applications (e.g., foods, drugs, cosmetics), the choice may be further influenced by such questions as toxicity, interaction with other addenda, and biological degradation. For that reason, well-studied nonionic surfactants have received a great deal of attention for such applications. In a given system, different types of surfactant may produce different types of multiple emulsion (A, B, or C types), so that such questions must also be considered.

As stated earlier, two surfactants or combinations of surfactants must be employed for the formation of multiple emulsions: one for the preparation of the primary emulsion (e.g., an oil-soluble system for a final W/O/W emulsion), and a second of a significantly different nature for the final emulsification step. Most attempts to design a system for surfactant choice employ the HLB system to determine the optimum surfactants for both the primary and secondary emulsions. For the production of a W/O/W emulsions they employ a series of surfactants with HLB numbers in the range of 2–8, typical for the formation of W/O emulsions, as the primary emulsifier, and those in the range of 6–16 for the secondary stage. The evaluation of emulsion stability based on creaming leads to the optimum choice of surfactant for each stage of the preparation. It has been noted, however, that, although useful, the HLB concept as a basis for surfactant choice is limited by the fact that other factors such as viscosity and concentration effects cannot be easily evaluated.

As stated previously, the nature of the oil phase and its specific interactions with the surfactant can significantly affect the emulsion characteristics of the system. The polarity of the oil phase (e.g., fatty esters vs. hydrocarbon oils) will determine the proper surfactant solubility requirements and HLB for the primary emulsion, the mutual solubility of the various components, and the transport of materials from the internal to the external phases. The choice of the optimum HLB for the secondary emulsifier can be strongly affected by the concentration of the primary emulsifier employed. For a single-component secondary emulsifier in a $W_1/O/W_2$ multiple emulsion, the HLB required to provide maximum multiple emulsion stability is found to increase as the concentration of primary emulsifier in the system increases. Since the HLB of the oil phase of the primary emulsion is not changed, the results are explained in terms of the migration of excess or "free" primary emulsifier to the secondary O/W_2 interface. The result is the production of an "effective" HLB at the secondary emulsion interface, which can be related to the final mixed emulsifier composition by Eq. (9.11).

The composition of the primary emulsion dispersed phase may have a significant effect on the overall stability of a system, especially when interactions between the components and surfactant are possible, or when the components themselves may be somewhat surface-active. In most instances of multiple-emulsion formulation, the internal primary and external secondary phases will be similar in that each will be aqueous or an oil, but the nature of addenda included in each will differ. In particular, there may be significant differences in the level and nature of organic additives and electrolytes present that could alter the stability of the total system.

Electrolytes in particular can exhibit significant effects on the stability of emulsions prepared with one or more ionic surfactants. There are multiple potential effects, including (1) changes in the role of the surfactant at the various interfaces as a result of changes in their electrical properties, (2) changes in the nature of the interfacial films due to the presence of specific ionic interactions between surfactant and electrolyte, and (3) alterations in the transport properties of the intervening phase due to differences in the osmotic pressure between the two phases.

The role of electrolyte in affecting the interfacial characteristics of surfactants has already been mentioned in several contexts. In the second instance, the presence of excess ionic species in the internal phase of a $W_1/O/W_2$ emulsion could lead to a closer packing of surfactant molecules at the interface and the erection of a more rigid barrier to the transport of molecules into the oil phase, a potentially useful device for the control of additive release from W_1.

A major imbalance in the osmotic pressure between internal and external aqueous phases as a result of a high electrolyte content internally would result in a driving force for the movement of water into the primary emulsion. The net result would then be a swelling of the internal droplets and, ultimately, rupture to release the components to the continuous phase. When such catastrophic release of the internal components is unacceptable, the introduction of a neutral electrolyte into the outer phase can reduce the osmotic pressure differences and retard the overall process. Other additives such as synthetic polymers and proteins can perform a similar role in either W/O/W or O/W/O emulsions.

An additional aspect of the nature of the surfactant and its role in stabilizing a multiple emulsion is the formation of liquid crystal phases at the interfaces. It has been found, for example, that the presence of liquid crystal phases greatly enhances the stability of certain O/W/O and W/O/W emulsions. Such systems will generally impart improved stability to coalescence as well as a barrier to the transport of material from one phase to another. Typically, an emulsion may be prepared at a temperature where the surfactant has favorable emulsifying properties followed by cooling to a temperature where liquid crystal formation occurs.

Clearly, multiple emulsions represent a fertile field of research in both applied and academic surface science. Although there are an ever-increasing number of publications appearing on the subject, the area remains somewhat empirical in that each system is highly specific. As yet there are few general rules to guide the interested formulator in the selection of the optimum surfactants for a given application. Cubic bicontinuous phases discussed in Chapter 5 appear to offer some of the same potential advantages sought in multiple emulsion systems. A great deal remains to be done in understanding the colloidal stability of such complex systems and the effects of the various components in each phase on overall multiple-emulsion preparation and stability. A sound understanding of the role of surfactants in simple emulsions and an intuitive feel for the effect of the multiple interfaces present on system operation seem to provide the best guidance at the present time.

PROBLEMS

9.1. Tetradecane was emulsified at 25°C in two 0.5% (w/w) surfactant solutions: (a) $C_{12}H_{25}$–$(OCH_2CH_2)_5OH$ and (b) $C_{12}H_{25}OSO_3Na$. What class of emulsion would you expect in each case? What would you expect to be the natures of the two emulsions when heated to 50°C?

9.2. Two common surfactants, SDS and sodium dioctylsulfosuccinate, have been found to occupy areas of 0.50 and 1.11 nm^2, respectively, when adsorbed at the oil–water interface. Calculate the amount of each surfactant required to completely saturate the interface formed by the emulsification of 500 g of tetradecane in one liter of aqueous surfactant solution. Assume that the emulsion is monodisperse with a droplet diameter of 1000 nm.

9.3. One process by which an emulsion, like a foam, destroys itself is Ostwald ripening—the diffusion of liquid from small to large droplets. Estimate the time required for a benzene droplet to disappear when it is positioned near much larger droplets at a distance comparable to its radius. Assume droplet radii of 100 and 1000 nm. The solubility of benzene in water may be taken as 0.2% (v/v); the diffusion constant of benzene in water $D = 10^{-5}$ cm^2/ s, the interfacial tension of water–benzene $\sigma = 25$ mN/m, and the molar volume of benzene $V_m = 100$ cm^3.

9.4. Calculate (to the nearest whole number) the maximum possible value for the dispersed phase fraction ϕ in an emulsion consisting of uniform spherical particles.

9.5. A simple geometric theory for the stabilization of emulsions is that of the "oriented wedge," in which the adsorbed surfactant molecules are assumed to form a uniform structure of "wedges" around the emulsion droplet. If an emulsion of 1000-nm-diameter droplets is stabilized by a surfactant whose head group occupies a surface area of 0.45 nm^2, what must be the cross-sectional area of the hydrophobic tail for maximum effectiveness?

9.6. A mixture of 70% sorbitan monostearate and 30% sorbitan tristearate was found to give optimum stability to a particular emulsion system. What composition of a mixture of sodium dodecyl sulfate and cetyl alcohol could be theoretically expected to produce the same result?

9.7. A surfactant mixture with an HLB of 8 is expected to produce a stable emulsion with lanolin. A new chemist in the firm is given the job of formulating a suitable emulsion, with the requirement that the mixture contain 10% cetyl alcohol. Suggest at least two alternative surfactant compositions that meet the stated requirements.

9.8. An emulsion of white petroleum oil is prepared in 0.001 M KCl using sodium dodecylsulfate as emulsifier. If the surfactant adsorbs at the oil–water interface occupying 0.5 nm^2 at equilibrium concentrations above 0.001 M, how much surfactant is needed per liter of emulsion if the average drop size is to be 50 nm in diameter?

9.9. Using the information provided in Table 9.2, calculate the theoretical molecular composition of a totally fluorinated sodium carboxylate surfactant that has the same HLB in as sodium decanoate.

9.10. Suggest possible surfactant combinations for preparing the following emulsions: (a) paraffin wax in water, (b) water in kerosene, (c) decyl alcohol in water, (d) soybean oil in water, and (e) water in benzene.

9.11. Suggest possible surfactant combinations for the preparation of W/O/W multiple emulsions in which the oil phases are paraffin wax and benzene.

10 Solid Surfaces and Dispersions

The technological, environmental, and biological importance of adsorption from solution onto a solid surface can hardly be overestimated. The impact of such phenomena on our everyday lives is evident in such areas as foods and food science, agriculture, cosmetics, pharmaceuticals, mineral ore froth flotation, cleaning and detergency, the extraction of petroleum resources, lubrication, surface protection, and the use of paints and inks. Each of these applications, and many more, would be difficult if not impossible in the absence of the effects of adsorbed surfactants and stabilizers at the solid–liquid interface.

The presence of amphiphilic materials adsorbed at a colloidal interface may reduce interfacial energies to promote the formation of small particle sizes, enhance the wetting of the solid by a liquid phase, and provide a stabilizing surface layer to prevent or retard particle coagulation or flocculation. In addition, the action of surfactants in conjunction with dispersed drugs can greatly enhance the efficacy of drug delivery.

At the macroscopic level, the modification of a solid surface by adsorption can greatly affect its subsequent rewetting characteristics (waterproofing), its electrical properties (antistatic agents), its physical "feel" against human skin (fabric softeners), its interaction with contacting surfaces (lubrication), or its ability to adsorb other solutes (dyeing modifiers), just to name only a few commonly encountered examples. Comprehensive discussions of some of the theoretical and practical aspects of surfactant adsorption at solid interfaces can be found in the references cited in the Bibliography. In practically all the above mentioned examples of surface modification by adsorption, and many more, the primary physical phenomena involved are pretty much the same. The following discussion introduces the basic concepts involved in terms of the natures of the adsorbing species and the surface being affected.

10.1. THE NATURE OF SOLID SURFACES

While the fluid interfaces discussed so far are relatively easy to treat from a thermodynamic standpoint because of the assumptions of molecular smoothness and

Surfactant Science and Technology, Third Edition by Drew Myers

homogeneity, assumptions that are not really accurate at the molecular level, solid surfaces present a number of formidable obstacles to the achievement of a clear, reasonably simple description of their surface energies and interactions with adjacent bulk phases and adsorbing species. As a result, many of the phenomenological aspects of adsorption at solid–liquid (S/L) and solid–vapor (S/V) interfaces are significantly more complicated and difficult to interpret than are those at liquid–liquid (L/L) and liquid–vapor (L/V) interfaces. In order to provide a clearer picture of the mechanics of surfactant adsorption onto solids, then, it will be useful to briefly describe some of the special aspects of solid surfaces that differentiate them from fluid interfaces.

To better understand the properties of a solid surface, it is necessary to consider the possible structures of solids and their exposed surfaces. Solids and solid surfaces may be roughly divided into two main categories, based on the nature of the arrangement of the constituent units. A solid may be crystalline, having reproducible intrinsic bulk properties such as a sharp melting point and a uniform pattern of packing of its constituent units (atoms, ions, or molecules) into a lattice structure, or it may be noncrystalline or amorphous, in which case the properties may be highly dependent on the previous history of the sample, and so tend to be variable and indefinite. A third class of solids is a heterogeneous mix of crystalline and amorphous structures scattered throughout the solid. Sodium chloride, for example, has a sharp melting point at 804°C at atmospheric pressure, while a typical soda glass (sodium silicate) begins to soften at $\sim$500°C and becomes less rigid with increasing temperature. At 1500°C the glass becomes completely molten and flows pretty much like any other liquid. There is no one temperature, however, that can be accurately identified as the true melting point, although one is generally defined in glass science on the basis of a somewhat arbitrarily chosen bulk viscosity value. Sodium chloride has a very well-defined crystal lattice structure, while the structure of glass is essentially random and without a defined repeat unit or unit cell. As examples of the third possibility, one can find certain polymers, such as polyethylenepterphthalate (PET), which can be prepared in such a way that crystalline and amorphous regions are present within the same bulk solid.

Another simple way of classifying solids for purposes of the present discussion is to define four types of solid structures according to the chemical nature of the structure:

1. *Ionic solids*, composed of a lattice structure of discrete ionic species. In an ionic solid (Figure 10.1a), the constituent units are ions and each ion of a given charge is located approximately equidistant from a small number of ions of the opposite charge symmetrically arranged around it. The structure is thus held together primarily by the electrostatic attraction between unlike charges, as reflected in the crystal lattice energy. Sodium chloride and most minerals fall into this general classification.
2. *Homopolar solids*, in which the various units making up the solid are essentially identical, but do not involve ionic species. In a homopolar solid (Figure 10.1b), the basic units are neutral atoms bonded to a number of

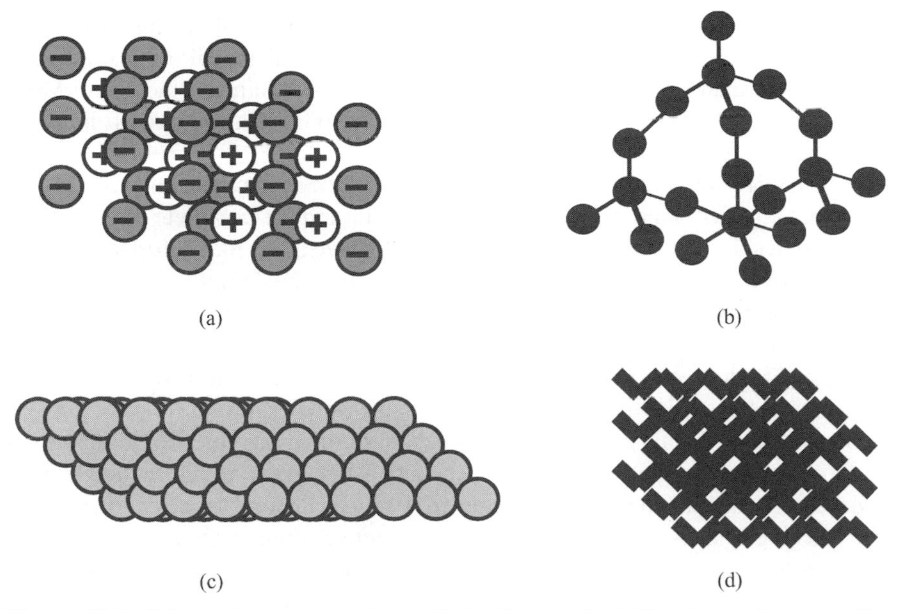

Figure 10.1. Schematic representations of the four main lattice types for crystalline materials: (a) ionic crystal lattice; (b) homopolar diamond lattice; (c) metallic lattic; (d) molecular lattice.

neighbors by covalent chemical bonds, as for example, in diamond, in which each carbon atom is bound to its four nearest neighbors to form a repeating tetrahedral unit, and graphite, in which carbons atoms are arranged in layered sheets of a hexagonally bonded flat structure.

3. *Metallic solids*, in which the constituent units, the metal atoms, are repeated, but in which the presence of mobile electrons imparts special characteristics in terms of interfacial interactions. In a metallic lattice (Figure 10.1c), all of the atomic nuclei may be packed into a uniform pattern resembling a homopolar solid, but the valence electrons of each atom are hybridized into a range of energies allowing for the free movement of electrons throughout the crystal, thus producing the observed electrical conductivity of metals and other conductive materials.
4. *Molecular solids*, in which the basic unit is a nonionic molecule that may present various surface characteristics depending on the atomic makeup and exact unit packing arrangement. A molecular solid (Figure 10.1d) is composed of an arrangement of molecules held together by attractive forces weaker in magnitude than the ionic and covalent forces active in the structures described above. The "weaker" forces involved in molecular lattices, including van der Waals attractions, dipolar interactions, and hydrogen bonding or acid–base interactions, may also be present in the other lattice types but are generally of much less importance than the primary covalent and

ionic forces. Because the interactions in molecular lattices are weaker, the distance of closest approach of adjacent molecules will be considerably greater than the bond distances between atoms within the molecule. Molecular solids may be either crystalline, amorphous, or a mixture of the two.

The distinctions among ionic, homopolar, and metallic structures are not, of course, absolute. A perfect ionic bond, for example, would require complete rigidity of the ionic species. In fact, the ions in such crystals behave more like deformed spheres, with the degree of deformation, or polarization, following definite patterns. Anions, for instance, are generally more highly polarized than are cations, and polarization susceptibility increases with the ionic radius. Because of the wide variability in the strength of ionic interactions in such materials, ionic solids exhibit a wide range of intensive and extensive properties, including surface energies.

Stearic acid crystals are a typical molecular solid within the current definition. In those materials, the shortest distance between carbon atoms in neighboring molecules is about 0.35 nm, compared to a covalent bond length of 0.154 nm. The packing density of molecules in such a lattice, then, would be expected to be lower than that found in the ionic, homopolar, or metallic cases. As was pointed out in Chapter 3, the surface energy of a material is directly related to the difference in the magnitude of the forces acting on an atom or molecule at the surface and that in the bulk. It is not surprising to find that ionic and metallic crystals, materials in which the magnitude of the lattice energies is high, possess high surface energies, ranging from several hundred to several thousand millijoules per square meter (mJ/m^2). Table 10.1 lists the surface energies of several types of solid. Because the lattice forces in most molecular solids are much lower than those in the other systems, such materials exhibit much lower surface energies.

The study of the surface chemistry of solids is concerned with specific properties of the atomic or molecular layers of material within a few molecular diameters of the interface with a vapor (or vacuum), a liquid, or another solid. To visualize the unique character of solid surfaces, it is helpful to compare the similarities and differences between solid and liquid surfaces.

TABLE 10.1. Experimentally Determined Surface Energies of Representative Solids

Surface	Surface Energy (mJ/m^2)
Polyhexafluoropropylene	18
Polytetrafluoroethylene	19.5
Paraffin wax	25.5
Polyethylene	35.5
Polyethylene terephthalate	43
Quartz (SiO_2)	325
Tin oxide (SnO_2)	440
Platinum	1840

10.2. LIQUID VERSUS SOLID SURFACES

We know from Chapter 3 that the surface tension of a liquid may be seen as arising from an imbalance of forces acting on molecules or atoms of the liquid at the surface relative to those in the bulk. A drop of liquid in equilibrium with its vapor, in the absence of any external forces, will assume a spherical shape that corresponds to a minimum surface : volume ratio. Work must be done on the drop to increase its surface area, implying that the surface molecules are in a higher energy state than are those in the bulk. Such a state can be viewed as resulting from the fact that surface atoms or molecules have fewer nearest neighbors and, therefore, fewer intermolecular interactions. There is a free-energy change associated with the reversible, isothermal formation of new liquid surface termed the "excess surface free energy." The excess surface free energy is not the total energy of the surface, but the excess over that of the bulk material resulting from the units being located at the surface. The concept is illustrated in another way using the spring model shown in Figure 10.2.

Because of the thermodynamic imperative to attain a state of minimum free energy for the system as a whole, surface units are subjected to a net inward attraction normal to the surface. Geometrically, that can be equivalent to saying that the surface is in a state of net lateral tension defined as a force acting tangent to the surface at each point on it. It is this apparent tangential force that leads to the concept of a surface tension. The units of surface tension and of the excess surface free energy are dimensionally equivalent and, for pure liquids in equilibrium with their

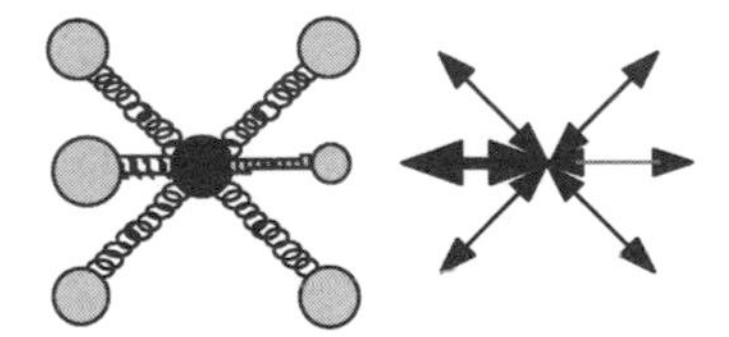

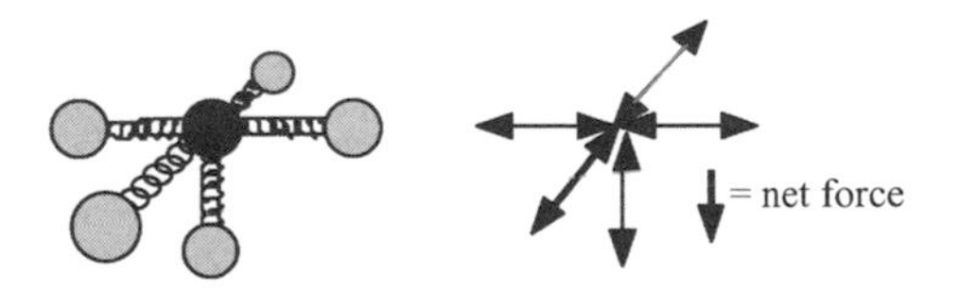

Bulk atom: net force = 0.

Surface atom: net attraction of surface atom into the bulk phase.

(a)

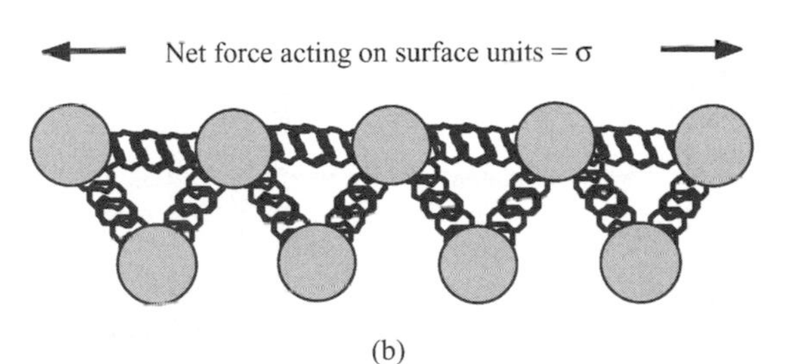

(b)

Figure 10.2. Schematic representation of a "spring" model of the origins of excess interfacial energies due to unbalanced, nonhomogeneous atomic (or molecular) forces acting on atoms (molecules) in the interfacial region.

vapor, they are numerically equal when used to denote specific (i.e., per unit area) values.

In principle, all the concepts related to the surface energy of liquids can also be applied to that of solid surfaces. In practice, however, the situation is much more complex. When a fresh solid surface is formed by cleaving in a vacuum or its own vapor, the atoms, ions, or molecules in the new surface will seldom be able to immediately assume their equilibrium location or configuration because of the greatly reduced mobility in the solid state. Only mica is generally found to produce an approximately homogeneous, molecularly smooth surface under normal cleavage conditions. The surface of a solid, therefore, will usually have a nonequilibrium, nonuniform structure that may contain one or all of a number of defects, the presence of which will produce variations in energy across the surface. Some examples of such defects are shown in Figure 10.3. The surface energy of any given solid will therefore depend somewhat on the history of the solid and cannot be equated directly to the intensive thermodynamic properties of surface tension or specific excess surface free energy. Such variations in the surface energy of solids will also be reflected in the interaction of other phases with that surface, including the adsorption of surfactants. The problems of describing heterogeneous solid surfaces and related adsorption phenomena is most pronounced in inorganic surfaces, especially where electrostatic effects (charged species) may be present. As we shall see, the presence of charge sites on the solid surface will greatly affect (and often dominate) the adsorption of surfactants. In organic solids, surface heterogeneities

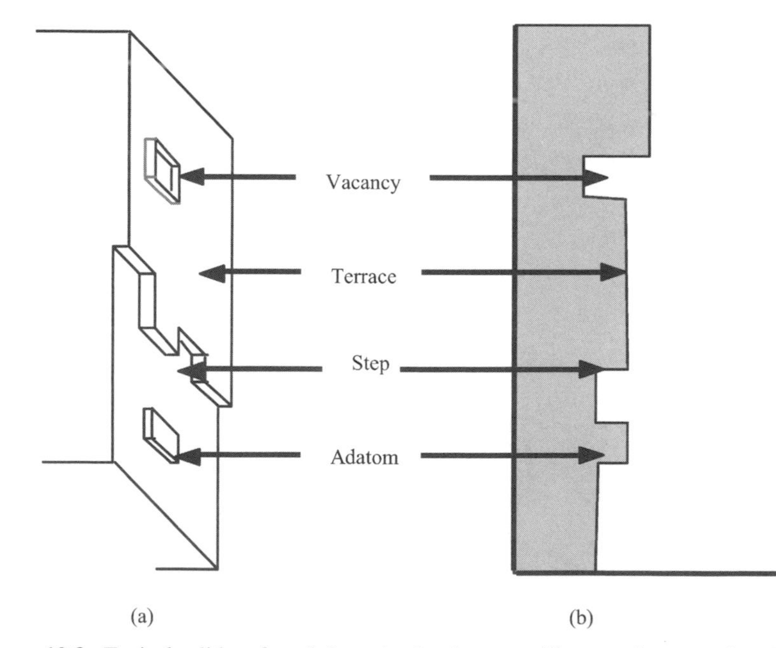

Figure 10.3. Typical solid surface defects that lead to nonuniform surface energies: (a) angled surface view; (b) cross-sectional view.

are somewhat less important, although they cannot be ignored. As in human society, history can play an important role in the characteristics of all solid surfaces.

10.3. ADSORPTION AT THE SOLID–LIQUID INTERFACE

Because of their inherently high surface energies, most newly formed solid surfaces, except those of very low energy polymers or waxes, will exhibit a strong tendency to adsorb almost any available material, including gases normally considered to be "inert" such as nitrogen, helium, and argon. The adsorption of such materials has been extensively studied since around 1905 and has given rise to a broad quantitative understanding of the nature of solid surfaces and their adsorption characteristics. Such adsorption, however, differs significantly from that of surfactants and other amphiphiles, and the details are left for the adventurous reader to pursue.

The adsorption of surface-active materials onto a solid surface from solution is an important process in many situations, including those in which we may want to remove unwanted materials from a system (detergency), change the wetting characteristics of a surface (waterproofing), control the triboelectric properties of a surface (static control), or stabilize a finely divided solid system in a liquid where stability may otherwise be absent (dispersion stabilization). In these and many other related applications of surfactants or amphiphilic materials, the ability of the surface-active molecule to situate itself at the solid–liquid interface and produce the desired effect is controlled by the chemical natures of the components of the system: the solid, the surfactant, and the solvent. The following discussions summarize some of the factors related to chemical structures that significantly affect the mechanisms of surfactant adsorption and the orientation with which adsorption occurs.

10.3.1. Adsorption Isotherms

The experimental evaluation of the adsorption from solution of amphiphilic materials at the solid–fluid interface usually involves the measurement of changes in the concentration of the adsorbed material in the contacting solution, for S/L interfaces, or the amount actually adsorbed onto the solid for S/V systems. The usual method for evaluating the adsorption mechanism is through the adsorption isotherm. The important factors to be considered are (1) the nature of the interaction between the adsorbate (the amphiphile) and the adsorbent (the solid); (2) the rate of adsorption; (3) the shape of the adsorption isotherm and the significance of plateaus in it; (4) the extent of adsorption (i.e., monolayer or multilayer formation); (5) the interaction of solvent, if present, with the solid surface; (6) the geometric orientation of the adsorbed molecules at the interface; and (7) the effect of environmental factors such as temperature, solvent composition, and pH on the adsorption process and equilibrium.

Early classifications of adsorption isotherms at solid–vapor interfaces were found to fit one of five basic shapes (Figure 10.4) at temperatures below the critical temperature (T_c) of the adsorbate. Type I isotherms were originally interpreted to

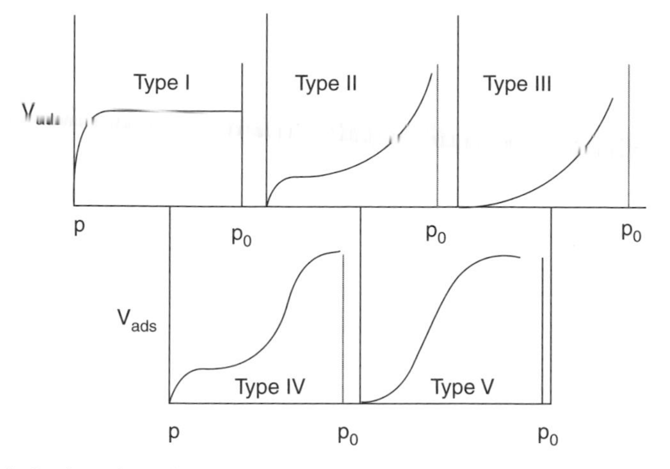

Figure 10.4. Basic adsorption isotherm shapes for vapors on solids below critical temperature T_c of the adsorbate.

represent monolayer adsorption of gas onto the solid surface, although it is was subsequently realized that microporous adsorbents will produce similar isotherms. The remaining isotherms (II–V) represent various processes of multilayer adsorption.

Although most adsorption isotherms fall into two main categories, many subtle and not so subtle shape variations have been reported, leading to a more complex classification system as illustrated in Figure 10.5. That classification system gives

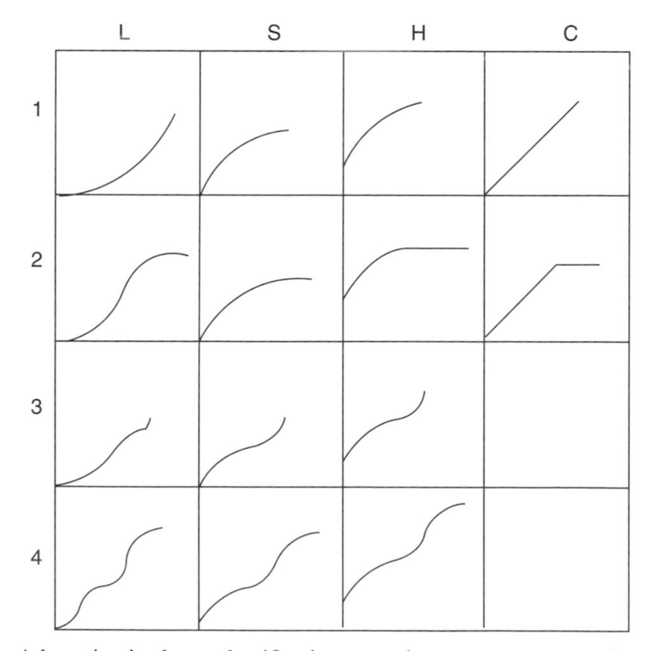

Figure 10.5. Adsorption isotherm classifications covering greater concentrations of adsorbate.

four basic isotherm shapes based on the form of the isotherm at low concentrations; the subgroups are then determined by their behavior at higher concentrations.

The Langmuir (L class) isotherm is the most common found and is characterized by concavity of its initial region (L1) to the concentration or horizontal axis. As the concentration of adsorbate increases, the isotherm may reach a plateau (L2), followed by a section convex to the concentration axis (L3). If the L3 region attains a second plateau, the region is designated L4. Under some circumstances, such as surfactants and some dyes that aggregate in solution and contain highly-surface active contaminants, a maximum may be obtained (L5: not shown in Figure 10.5). Since it is thermodynamically impossible for a pure system to exhibit such a maximum, its existence can represent a tell–tale sign of the presence of impurities.

In the S class of isotherms, the initial slope is convex to the concentration axis (S1) and is often broken by a point of inflection (S2), leading to the characteristic S shape (S3). Further concentration increases may then parallel those of the L class, including the presence of a maximum. The high affinity or H class of adsorption isotherm occurs as a result of very strong adsorption at low adsorbate concentrations. The result is that the isotherm appears to have a positive intercept on the ordinate. Higher concentrations lead to similar changes to those found in the L and S classes.

The final type of isotherm is the C class. Such systems exhibit an initial linear portion of the isotherm, indicating a constant partitioning of the adsorbate between the solution and the solid. Such isotherms are not found for homogeneous solid surfaces but are found in systems in which the solid is microporous. The classification system illustrated in Figure 10.5 has proved very useful in providing information about the mechanism of adsorption in many systems. An excellent discussion of many of the details is given by Parfitt and Rochester listed in the Bibliography.

10.3.2. Mechanisms of Surfactant Adsorption

The adsorption of surface-active agents at the interface of a solid and a liquid phase is a fundamentally important phenomenon, both scientifically and technologically. The facility and strength of that adsorption are very largely controlled by three factors, which are related to the materials in question: (1) the chemical nature of the species being adsorbed, including the nature of the head group (anionic, cationic, nonionic, etc.) and that of the hydrophobe (length and nature of the chain, degree of branching, etc.), (2) the nature of the solid surface onto which the surfactant is being adsorbed (highly charged, nonpolar, etc.), and (3) the nature of the liquid environment (in water, the pH, electrolyte content, temperature, additives, etc.). A slight change in one of these or other factors can result in a significant change in the adsorption characteristics of the system.

The adsorption of a surfactant molecule onto a solid surface can be significantly affected by relatively small changes in the characteristics of the system. Such sensitivity results from the wide range and strengths of adsorption mechanisms that may be operative. Except in some cases of surfactant–polymer interaction discussed in Chapter 7, it is usually found that adsorption occurs on a molecular level. The mechanisms of adsorption available generally parallel those operative in all intermolecular interactions, such as electrostatic and dipolar attractions, and van der Waals forces. The strengths of such interactions range from the relatively weak but long distance dispersion forces resulting from the nature of electron motion in molecular (and atomic) orbitals, to the stronger, more specific, but shorter-range interactions between surfactants and solid surfaces of opposite electrical charge.

10.3.2.1. Dispersion Forces

Since dispersion force interactions are essentially statistical-mechanical, they have been extensively analyzed from a theoretical standpoint and are reasonably well understood, conceptually at least. Although relatively weak ($\approx$8 kJ/mol), they are generally considered to be long-range forces, since their influence can sometimes be detected over several hundreds of nanometers. The effect of hydrocarbon chain length within a homologous series of ionic surfactants observed on adsorption onto a given solid surface is generally considered to indicate that such systems follow Traube's rule; that is, many of the characteristics of a material, including surface tension and adsorption, will vary regularly with the molecular weight of the species, other things being equal.

The adherence of surfactants to Traube's rule has been the subject of many investigations. It has generally been found that adsorption increases with an increase in the molecular weight of the hydrophobic portion of the surfactant molecule. The longer the hydrocarbon chain in a given series, the larger the amount of surfactant adsorbed at saturation and the lower the total surfactant concentration at which saturation occurs. As seen in Chapter 4, the critical micelle concentrations of the same series will decline in a similar way. Since saturation adsorption often occurs coincident with the initiation of micelle formation, it is inferred that the two phenomena are related. Some authors have pointed out that if adsorption isotherms for a homologous series of surfactants are plotted as a function of the concentration at saturation, C_a, divided by the cmc, the results will be almost super imposable. Such a relationship is not always clear-cut, however, since some systems show a maximum in adsorption, with the total amount of adsorbed material decreasing as the cmc is exceeded. Such systems usually reflect the presence of a surface-active impurity or a mixture of homologs in which the more surface-active component is adsorbed initially, but is desorbed and solubilized in micelles once present.

Regardless of the specifics of the adsorption mechanism due to dispersion forces, the result is important scientifically and technologically because it is always present, whether as an independent effect or acting in conjunction with the other mechanisms discussed below. The importance of the effect in conjunction with other forces is illustrated by the propensity for surfactants with longer hydrophobic tails to displace similarly charged lower-molecular-weight materials and inorganic ions from solid surfaces.

Dispersion forces are usually discussed in terms of being attractive and commonly are so for identical or similar materials, although many situations arise in which they may actually have a net repulsive effect, especially in three–component

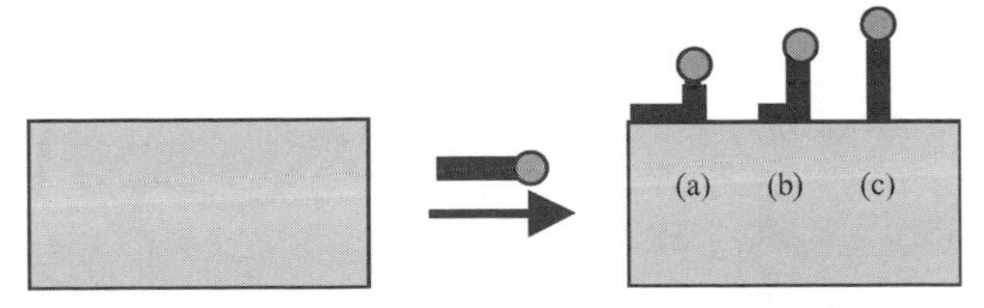

Figure 10.6. Modes of surfactant adsorption through nonpolar dispersion forces: (a) trains; (b) "L"s; (c) perpendicular (or almost so).

ponent systems. The theoretical aspects of such repulsive dispersion interactions is not covered here. However, in relation to the situation of attractive dispersion forces between surfactant molecules and a solid surface leading to adsorption, hydrophobic bonding may be considered to be "pushing" in character, in that it might be interpreted conceptually as pushing the surfactant molecule out of the aqueous environment onto the adsorbent surface or into micelles. As mentioned previously and discussed in more depth in Chapter 4, one of the main driving forces for the formation of micelles in aqueous solution, and the adsorption of surfactants at interfaces, is the tendency of hydrophobic groups to want to "escape" from the aqueous environment (or for the aqueous environment to push the hydrophobic groups out). The thermodynamically unfavorable situation of the structured water molecules necessary for the solvation of the hydrophobic tail of the surfactant and the mutual attraction of the tails due to dispersion attractions leads to micelle formation. Similarly, in the presence of a solid surface the adsorption of surfactant molecules may be enhanced by the "push" to remove themselves from the water structure. The aggregation of surfactant molecules on the solid surface would then be a manifestation of both these mechanisms (Figure 10.6).

10.3.2.2. Polarization and Dipolar Interactions

Somewhere between the strictly nonpolar dispersion and hydrophobic adsorption mechanisms and the dipolar and electrostatic effects discussed below lie the molecular interactions resulting from induced dipolar or polarization effects. When one component of an adsorbing systems—usually (not always) the surfactant—contains an electron-rich group (e.g., an aromatic nucleus) and the other, a strongly polarizing site such as a positive charge, there exists the possibility for the induction of a dipole in one species leading to a significant improvement in the ease of adsorption. The strength of such interactions usually lies in the range of 4–8 kJ/mol, in the same range as dispersion force interactions. Such induced polarization usually is encountered in conjunction with adsorption resulting from other forces as well.

Less well-defined than most atomic and molecular interactions, but potentially more important as a mechanism of adsorption in aqueous systems, is that resulting from acid–base interactions and hydrogen bond formation between the adsorbent and water or surfactant (Figure 10.7). The presence (or absence) of such interactions can greatly alter the extent and mode of surfactant adsorption, and therefore greatly alter the macroscopic affects of the process.

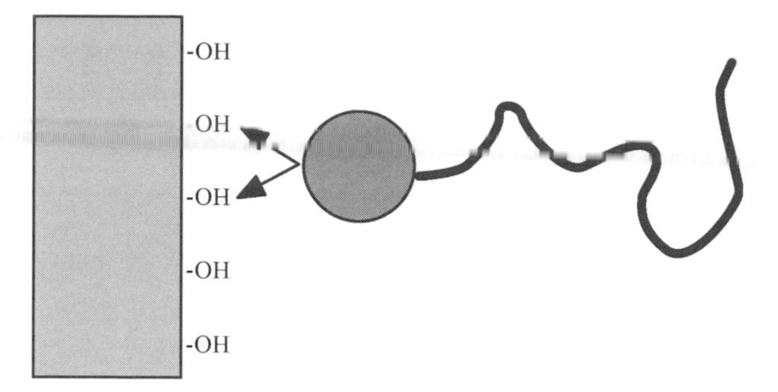

Figure 10.7. Adsorption via polar, hydrogen bonding, or acid–base interactions.

An interesting example of the potential importance of such interactions can be seen in the example of a pure, clean quartz surface. If the surface has been heated to a sufficiently high temperature (>200°C), a drop of distilled water placed on the surface will form a sessile drop with a non-zero-contact angle, usually in the range of 30°. The heat treatment of the quartz produces a dehydrated surface composed of Si–O–Si bonds that, while polar, are not strongly interacting in terms of hydrogen bonding. If the same surface is allowed to hydrate so that the surface layer is now basically composed of –SiOH groups, a drop of water will spread completely on the surface. The effect is illustrated schematically in Figure 10.8.

10.3.2.3. Electrostatic Interactions

At the end of the spectrum of adsorption mechanisms opposite to the universal dispersion forces lie those interactions resulting from the presence of discrete electrical charges on the surfactant molecules and the solid adsorbent. Those interactions may be described as being either ion pairing or ion exchange (Figure 10.9). The distinction between the two mechanisms lies in the fact that in ion exchange,

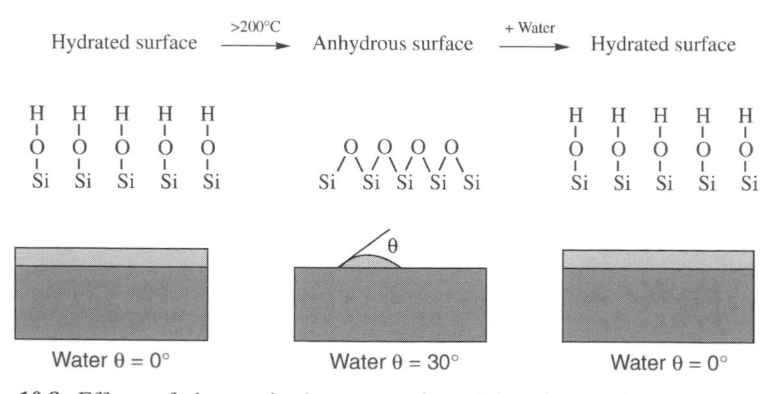

Figure 10.8. Effects of changes in the nature of a solid surface on interactions involving hydrogen bonding or acid–base chemistry, as illustrated for the contact angle (θ) of water on a silica surface.

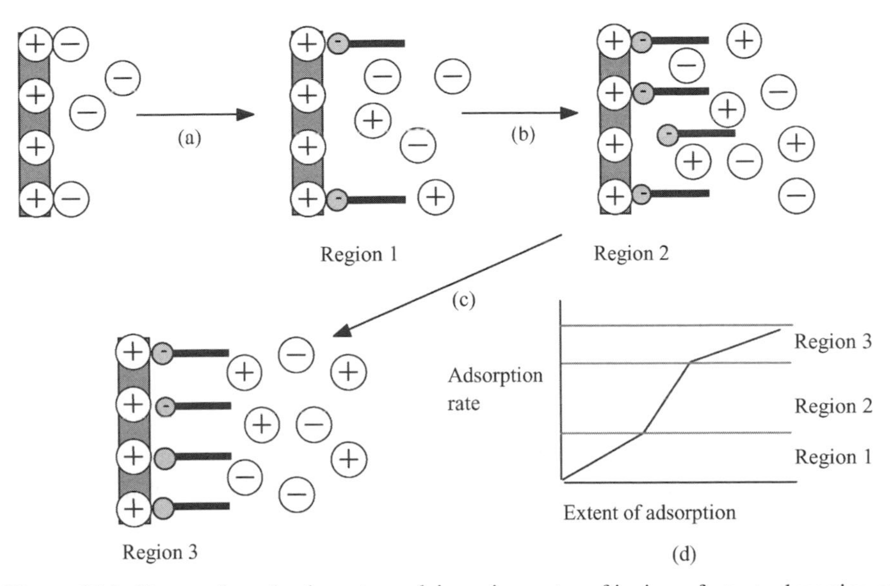

Figure 10.9. Proposed mechanisms to explain various rates of ionic surfactant adsorption as a function of surface coverage and mode of adsorption: (a) native surface; (b) ion exchange; (c) ion pairing; (d) charge neutralization; (e) representative isotherm.

the adsorbing surfactant molecule displaces a similarly charged counterion adsorbed on the substrate surface, while in ion pairing, the surfactant adsorbs onto an oppositely charged site, which was not previously occupied by a counterion. This distinction may seem rather fine at first glance, but it can be significant in determining and understanding the resultant effects of such adsorption.

When one considers adsorption as a result of dispersion, hydrophobic, and, to some extent, polarization mechanisms, it is generally assumed that the solid substrate presents a homogeneous surface on which adsorption is essentially random. As seen above, the assumption of homogeneity for solid surfaces is not necessarily valid, although it may be difficult to isolate the individual effects of surface defects. In the event that discrete electronic charges are present, the situation is further complicated by the presence of an electrical double layer, which can significantly alter the adsorption process.

10.3.3. The Electrical Double Layer

The existence of electrical charges at any interface will give rise to electrical effects, which will, in many cases, determine the major characteristics of that interface. Those characteristics will affect many of the properties of a multicomponent system, including emulsion and foam formation and stability, solid dispersions, and aerosols. The theoretical and practical aspects of electrical double layers are the subject of a vast amount of literature and for that reason have not been addressed in any detail so far. Such details can be found in bibliographic references cited for this chapter.

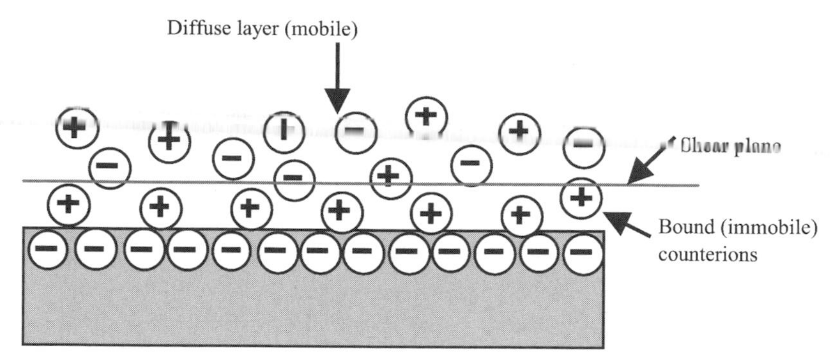

Figure 10.10. A schematic representation of the origin of the diffuse double layer.

Any interface that has an unbalanced electrical charge distribution will result in the formation of a net electrical charge of one sign on one side of the interface and a charge of opposite sign on the other side. Such a situation gives rise to the so-called electrical double layer. Since electrical neutrality must be conserved, the net charge on both sides of the interface will be zero. However, differences in the mobility of the different charges may give rise to wide differences in the distribution of the charges, especially those in a liquid phase associated with a charged solid surface (Figure 10.10).

A major field of investigation in modern colloid and interface science has been the search for a means to predict and determine the exact distribution of electrical charges at or near a solid–solution interface. Building on the original theories of Helmholtz, a model has developed that allows for a reasonable description of how the electrical potential behaves in relation to the distance from the solid charged surface. The model includes the disposition of counterions tightly bound to the surface (the Stern layer) and those more weakly held in the diffuse portion of the double-layer region.

The effective thickness of the diffuse layer can be calculated to determine the distance from the surface at which the electrical properties of the layer are essentially those of the bulk solution. Often termed the *Debye length*, κ, the effective thickness is given by

$$\frac{1}{\kappa} = \left(\frac{\epsilon_r \epsilon_0 \, RT}{4\pi F^2 \sum C_i Z_i^2} \right)^{1/2} \tag{10.1}$$

where $\epsilon_r = \epsilon/\epsilon_o$ is the relative static dielectric constant of the solution (ϵ is the static permittivity of the solution and ϵ_o is that of a vacuum), R is the gas constant, T is the absolute temperature, F the Faraday constant, and C_i the molar concentration of the ith ion of valence Z in the solution.

From Eq. (10.1) it is seen that $1/\kappa$ is inversely proportional to the valence of the counterions Z and the square root of C_i. It is proportional to the square roots of T and ϵ. It is to be expected, therefore, that in solvents of high dielectric constant such as water, electrical effects will be felt much further into the solution. In low dielectric constant media such as hydrocarbons, electrical effects are very short-ranged and usually have little practical significance. Of great practical importance in all situations is the value of Z, the valence of the counterions, and C_i. For 1:1 ($M^+ X^-$) electrolytes in 0.01 M solution at room temperature, the value of $1/\kappa$ is approximately 3.0 nm; as the electrolyte concentration is raised to 0.1 and 1.0 M, the values decrease to approximately 1.0 and 0.3 nm, respectively. More significantly, as the value of Z is changed from 1 to 2 and 3, with all other factors, including C_i, held constant, the value of $1/\kappa$ falls very rapidly and the effectiveness of the electrical forces at influencing the system is rapidly lost.

The potential importance of the double layer to adsorption in charged systems can be seen from the fact that the relationship in Eq. (10.1) predicts that potential interactions between surface charges and ions in solution will be significantly affected by the conditions in the solution phase. Double-layer theory is most important in the general field of colloidal stability, in which the concentration and valence of ions present can dramatically affect the stability and utility of an electrostatically stabilized colloidal system.

10.4. THE MECHANICS OF SURFACTANT ADSORPTION

When the adsorption of a surfactant onto a solid surface is considered, there are several quantitative and qualitative points are of interest, including (1) the amount of surfactant adsorbed per unit mass or area of solid, (2) the solution surfactant concentration required to produce a given surface coverage or degree of adsorption, (3) the surfactant concentration at which surface saturation occurs, (4) the orientation of the adsorbed molecules relative to the surface and solution, and (5) the effect of adsorption on the properties of the solid relative to the rest of the system. In all cases, it is assumed that such factors as temperature and pressure are held constant.

The classical method for determining the degree of adsorption in a given system is by way of the adsorption isotherm. The basic equation describing the adsorption of one component of a binary solution onto a solid substrate is given as

$$\frac{n_0 \Delta x}{m} = n_1^s x_2 - n_2^s x_1 \tag{10.2}$$

where n_0 is the total number of moles of solution before adsorption, Δx is the change in mole fraction of component 2 with respect to 1 after adsorption, m is the mass of solid in contact with the solution, n_1^s and n_2^s are the numbers of moles of components 1 and 2 adsorbed onto the surface per gram of solid at equilibrium, and x_1 and x_2 are the mole fractions of components 1 and 2, respectively, in the liquid phase.

In the case of a dilute surfactant solution where the surfactant (component 2) is much more strongly adsorbed than the solvent (component 1), the equation simplifies to

$$n_2^s = \frac{\Delta n_2}{m} = \frac{(C_2^\circ - C_2)V}{m} = \frac{\Delta C_2 V}{m} \tag{10.3}$$

where C_2° is the molar concentration of component 2 before adsorption, C_2 is its molar concentration at adsorption equilibrium, $\Delta C_2 = C_2^{\circ} - C_2$, and V is the volume of the liquid phase in liters.

For surfactant systems, the concentration of adsorbed material can be calculated from the known amount of material present before adsorption and that present in solution after adsorption equilibrium has been reached. A wide variety of analytical methods for determining the solution concentration of surfactants are available and almost all have been used at one time or other. The utility of a specific method will depend ultimately on the exact nature of the system involved and the resources available to the investigator.

10.4.1. Adsorption and the Nature of the Adsorbent Surface

As indicated above, the nature of the solid surface involved in the adsorption process is a major factor affecting the manner and extent of surfactant adsorption. At this point it might be useful to briefly discuss some of the specific effects of solid surface characteristics on the mode of adsorption, as well as to clarify the meaning of "efficiency" and "effectiveness" as applied in the context of surfactant adsorption.

In the current context, "efficiency" is intended to define the equilibrium concentration of surfactant, C_0, in the liquid phase required to produce a given level of adsorption under defined conditions of temperature and other parameters. The "effectiveness" of adsorption refers to the amount (concentration) of surfactant actually adsorbed on the solid surface at surface saturation. The effectiveness of adsorption of a given surfactant can, of course, be determined directly from the adsorption isotherm.

When one considers the possible nature of an adsorbent surface, three principal groups readily come to mind: (1) surfaces that are essentially nonpolar and hydrophobic, such as polyethylene, polyproplyene, and Teflon; (2) those that are polar but do not possess discrete surface charges such as polyesters and such natural fibers as cotton; and (3) those that possess strongly charged surface sites. Each of these surface types is discussed, beginning with what is probably the simplest, the nonpolar, hydrophobic surface.

10.4.2. Nonpolar, Hydrophobic Surfaces

As already mentioned, adsorption of surfactants onto nonpolar surfaces is by dispersion force interactions. From aqueous solution, it is obvious that the orientation of the adsorbed molecules may be such that the hydrophobic groups are associated with the solid surface with the hydrophilic group directed toward the aqueous phase. In the early stages of adsorption it is likely that the hydrophobe will be lying (approximately) on the surface much like trains or Ls (Figure 10.11a,b). As the degree of adsorption increases, however, the molecules will gradually be oriented more perpendicular to the surface until, at saturation, an approximately close-packed assembly results (Figure 10.11c).

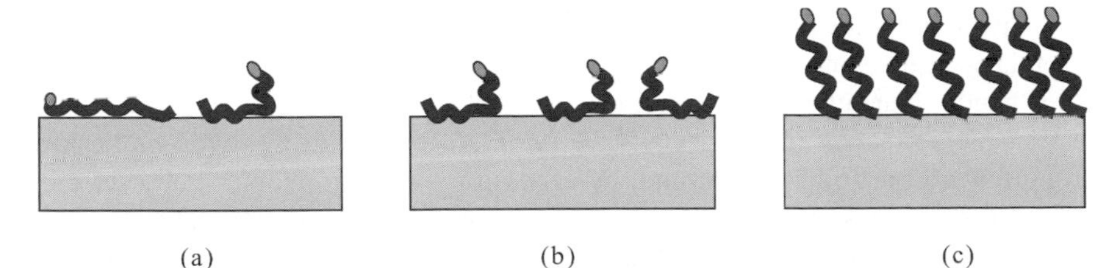

Figure 10.11. Adsorbed surfactant orientation as a function of surface coverage on a nonpolar surface: (a) low coverage—primarily trains; (b) intermediate coverage—trains and "L"s; (c) surface saturation—approximately vertical, close-packed, with near minimum area per molecule, although some tilt may be present.

It is generally found that surface saturation is attained at or near the cmc for the surfactant. In many cases the isotherm is continuous, while in others an inflection point may be found. The existence of the inflection point is usually attributed to a relatively sudden change in surfactant orientation—from train- or L-shaped to a more perpendicular arrangement.

The adsorption of surface-active agents onto nonpolar surfaces from nonaqueous solvents has been much less intensively studied than has that of aqueous systems. Generally, work has been limited to various carbon black dispersions in hydrocarbon solvents. The orientation of the adsorbed molecules in such systems appears to remain more-or-less parallel to the surface, although the exact details are found to depend greatly on the history of the carbon surface.

An important consequence of adsorption of surfactants onto a nonpolar surface is that the net character of the surface is drastically changed. If the adsorbed species is charged, the adsorbed layer imparts, to some extent, at least, the characteristics of such a surface, with all the attending strengths (e.g., increased stability in dispersed systems) and weaknesses (sensitivity to electrolyte). If the adsorbed material is nonionic, the same will generally hold true. More details on surface modification by adsorbed species are given later.

10.4.3. Polar, Uncharged Surfaces

Polar, uncharged surfaces include many of the synthetic polymeric materials such as polyesters, polyamides, and polyacrylates, as well as many natural materials such as cotton. As a result, the mechanism and extent of surfactant adsorption onto such materials have great potential technological importance. The mechanism of adsorption onto these surfaces will be more complex than that of the nonpolar case discussed above, since such factors as orientation will be determined by a balance of several forces.

The potential forces operating at a polar surface include the ever-present dispersion forces, dipolar interactions, and hydrogen bonding and other acid–base interactions. The relative balance between the dispersion forces and the uniquely polar interactions is of importance in determining the mode of surfactant adsorption. If the dispersion forces predominate, for example, adsorption will occur in a manner essentially equivalent to that for the nonpolar surfaces (Figure 10.11). If, on the other hand, polar interactions dominate, adsorption may occur in a reverse mode; that is, the surfactant molecules will be oriented more with the hydrophilic head group toward the solid surface and, by necessity, the hydrophobic group set more toward the aqueous phase (Figure 10.6b) or held more or less parallel to the solid surface. Orientation of the hydrophobic tail toward the aqueous phase can introduce its own complications and can lead to some sort of hemimicelle aggregation on the surface, at least until monolayer saturation is reached, after which a bilayer can begin to form. For now, such surface aggregation is somewhat speculative. Obviously, the net results of the two adsorption modes will be drastically different. Such differences may be of particular importance in the areas of adhesion and lubrication, where the orientation of adsorbed species could significantly affect the performance of the system.

10.4.4. Surfaces Having Discrete Electrical Charges

The final class of adsorbent surface is the most complex of the three for several reasons. From the standpoint of the nature of the surface, these materials are capable of undergoing adsorption by all the previously mentioned mechanisms. Possibly more important, however, is the fact that adsorption involving charge–charge interactions is significantly more sensitive to external conditions such as pH, the electrolyte content of the aqueous phase, and the presence of non-surface-active cosolutes than are the other mechanisms.

Materials possessing charged surfaces include almost all the inorganic oxides and salts of technological importance (silica, alumina, titania, etc.), silver halides, latex polymers containing ionic comonomers, many natural surfaces such as proteins, and cellulose. It is very important, therefore, to be able to understand the interactions of such surfaces with surfactants in order to optimize their effects in such applications as paint and pigment dispersions, papermaking, textiles, and pharmaceuticals.

Because of the large number of possible interactions in systems containing charged surfaces and ionic surfactants, it is very important to closely control all the variables in the system. As adsorption proceeds, the dominant mechanism may go from ion exchange through ion binding to dispersion or hydrophobic interactions. As a result, adsorption isotherms may be much more complex than those for the simpler systems.

The adsorption isotherms for surfactants on surfaces of opposite charge generally show three well-defined regions of adsorption in which the rates vary because of changes in the mechanism of adsorption. One interpretation of such adsorption involves three consecutive mechanisms (Figure 10.9b–d). In the early stages (region 1), adsorption occurs primarily as a result of ion exchange in which closely associated "native" counterions are displaced by surfactant molecules. During that stage the electrical characteristics (i.e., the surface charge or surface potential) of the surface may remain essentially unchanged. As adsorption continues, ion pairing of surfactant molecules with surface charges may become important (region 2), resulting in a net decrease in surface charge. Such electrical properties as the zeta potential, a measure of the surface charge density [in coulombs per square meter (C/m^2)], will tend toward zero. It is often found that in region 2 the rate of adsorption will increase significantly. The observed increase may be due to the cooperative effects of electrostatic attraction and lateral interaction among adjacent hydrophobic groups of adsorbed surfactants as the packing density increases.

As the adsorption process approaches the level of complete neutralization of the native surface charge by adsorbed surfactant, the system will go through its zero point of charge (zpc), where all the surface charges have been paired with adsorbed surfactant molecules. In that region (region 3), hydrophobic interactions between adjacent surfactant tails can predominate, often leading to the formation of aggregate structures or hemimicelles already postulated. If the hydrophobic interaction between surfactant tails is weak (because of short or bulky structures) or if electrostatic repulsion between head groups cannot be overcome (because of the presence of more than one charge of the same sign or low ionic strength), the enhanced adsorption rate of region 2 may not occur and hemimicelle formation may be absent. An additional result of the onset of dispersion-force-dominated adsorption may be the occurrence of charge reversal as adsorption proceeds. That aspect is covered in more detail later.

As mentioned above, surfaces possessing significant surface charge in aqueous solvents are especially sensitive to environmental conditions such as electrolyte content and the pH of the aqueous phase. In the presence of high electrolyte concentrations, the surface of the solid may possess such a high number of bound counterions that ion exchange is the only mechanism of adsorption available other than dispersion or hydrophobic interactions. Not only will the electrical double layer of the surface be collapsed to a few nanometers thickness; attraction between unlike charge groups on the surface and the surfactant, and repulsion between the like charges of the surfactant molecules, will be suppressed. The result will often be an almost linear adsorption isotherm, lacking any changes in slope characteristic of the mechanism changes described above.

While an increase in the electrolyte content may cause a decrease in adsorption of surfactants onto oppositely charged surfaces, it may allow an increase in adsorption of like charged molecules. That trend holds true for both the efficiency and the effectiveness of adsorption. The presence in the solution of polyvalent cations such as Ca^{2+} or Al^{3+} will generally increase the adsorption of anionic surfactants. Such ions are characteristically tightly bound to a negatively charged surface, effectively neutralizing charge repulsions. They also can serve as an efficient bridging ion by association with both the negative surface and the anionic surfactant head group (Figure 10.12).

Adsorption onto solid surfaces having weak acid or basic groups such as proteins, cellulose, and many polyacrylates can be especially sensitive to variations in solution pH. As the pH of the aqueous phase is reduced, the net charge on the solid surface will tend to become more positive. That is not to say that actual positive charges will necessarily develop; rather, ionization of the carboxylic acid

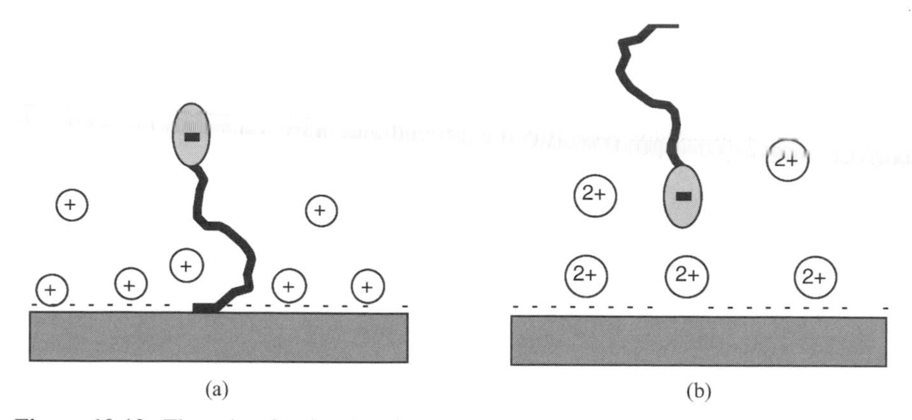

Figure 10.12. The role of polyvalent ions as potential bridging ions for the adsorption of anionic surfactants on negatively charged surfaces: (a) normal adsorption of a surfactant on a surface of the same electrical charge; (b) cation bridging leading to "inverted" adsorption.

groups will be suppressed. The net result will be that the surface may become relatively more favorable for the adsorption of surfactants of like charge (e.g., anionic surfactants onto carboxyl surfaces) and less favorable for adsorption of oppositely charged surfactants. For surfaces containing weak basic groups such as amines, the opposite would be true; that is, lowering the pH will lead to ionization of surface basic groups, increased adsorption of oppositely charged (anionic) molecules, and decreased interaction with materials of the same charge.

An increase in the temperature of an adsorbing system will usually result in a decrease in the adsorption of ionic surfactants, although the changes will be small when compared to those due to pH and electrolyte changes. Nonionic surfactants, which usually have an inverse temperature–solubility relationship in aqueous solution, will generally exhibit the opposite effect; that is, adsorption will increase as the temperature increases, often having a maximum near the cloud point of the particular surfactant.

The discussion of some of the primary effects of the solid surface character on the adsorption of surfactants admittedly has been brief and lacking in experimental detail. However, a number of excellent books and reviews cover the subject from a theoretical and experimental viewpoint in great depth.

10.5. SURFACTANT STRUCTURE AND ADSORPTION FROM SOLUTION

In this section, we turn our attention to surfactant adsorption phenomena from the viewpoint of the surfactant chemical structure, and we discuss more of the effects of external environmental factors. As usual, there is no intention to provide a complete literature review of the state of our knowledge related to surfactant structure and adsorption behavior. Rather, the discussion summarizes what is generally understood about the complex relationships among surfactant structure, solid surface characteristics, and solution environmental factors, leaving the search for more theoretical details to the interested reader. For the sake of consistency, the discussion is divided according to the surface types, as was done in the preceding section, although the order of discussion has been reversed.

10.5.1. Surfaces Possessing Strong Charge Sites

As we have seen, the primary surfactant characteristic controlling the "sense" of surfactant adsorption (i.e., the head directed outward into the liquid phase or toward the surface) onto surfaces with strong charges is the nature of the hydrophilic group of the surfactant. Obviously, for surfactants with charge opposite that of the surface, electrostatic attraction would be expected to dominate in aqueous solution, while for systems of like charge any adsorption would be expected to arise as a result of dispersion or other nonelectrostatic interactions. Within a given charge type, the exact nature of the hydrophile appears to play a minor role in adsorption. It is sometimes found that increasing the hydrated size of the hydrophile will lead to an increase in the efficiency of adsorption by the ion exchange and ion pairing mechanisms.

Once the variable of surfactant charge has been fixed, the nature of the hydrophobe becomes the major factor determining the adsorption characteristics of the system. It has generally been found that within an homologous series of surfactants, an increase in the length of the hydrophobic chain will result in an increase in the efficiency of adsorption. The usual explanation is that as the chain length increases, the free-energy gain associated with the removal of the hydrophobe from its aqueous environment and chain–chain interactions among neighboring molecules become more favorable. Some adsorption efficiencies calculated from literature data are given in Table 10.2.

When a phenyl or other aromatic group is added to the hydrophobic chain of a surfactant, it will contribute an effect equivalent to approximately 3.5 methylene groups to the free energy of adsorption. If cationic sites are present on the surface, the phenyl, and presumably other aromatics groups such as naphthalene, can be induced to interact by induced dipolar forces, resulting from the polarization of the aromatic π electrons by the positive charge. Such forces will, of course, be much weaker than direct interaction between opposite charges, but they can significantly alter the mode of orientation from a molecular standpoint.

Short alkyl branches on the hydrophobic group can have some effect on adsorption efficiency. Carbon atoms on short branches of an alkyl hydrophobe, those located between two hydrophilic groups, or those on the shorter portion of an alkyl chain with the hydrophile not substituted in the terminal position all seem to contribute an effect equal to half that of the same number of carbons in a normal-chain hydrophobe and terminally substituted hydrophile.

The efficiency of surfactant adsorption can quite often be predicted for various changes in the nature of the surfactant, in much the way adsorption at L/V interfaces was treated in Chapter 3. The effectiveness of adsorption, on the other hand, may

TABLE 10.2. Adsorption Efficiency (as $-\log C_0$) of Some Typical Surfactants on Various Substrate Types

Surfactant	Substrate	Temp. (°C)	pH (I.S.[a])	$-\log C_o$
n-$C_8H_{17}SO_4Na$	AgI	20	3 (0.001 M)	2.60
n-$C_{10}H_{23}SO_4Na$	AgI	20	3 (0.001 M)	3.89
n-$C_{12}H_{25}SO_4Na$	AgI	20	3 (0.001 M)	4.50
n-$C_{14}H_{29}SO_4Na$	AgI	20	3 (0.001 M)	5.15
n-$C_{10}H_{23}SO_4Na$	AgI	20	3 (0.0015 M)	3.40
n-$C_{12}H_{25}SO_4Na$	AgI	20	3 (0.0015 M)	4.38
n-$C_{14}H_{29}SO_4Na$	AgI	20	3 (0.0015 M)	4.78
$C_{12}H_{25}C_6H_4SO_3Na$	AgI	20	3 (0.0015 M)	4.84
n-$C_{10}H_{23}SO_4Na$	α-Al_2O_3	25	7.2 (0.002 M)	2.75
n-$C_{12}H_{25}SO_4Na$	α-Al_2O_3	25	7.2 (0.002 M)	3.55
n-$C_{14}H_{29}SO_4Na$	α-Al_2O_3	25	7.2 (0.002 M)	4.25
n-$C_{16}H_{23}SO_4Na$	α-Al_2O_3	25	7.2 (0.002 M)	5.00
n-$C_{10}H_{23}NH_3OAc$	SiO_2	25	6.7	1.75
n-$C_{12}H_{25}NH_3OAc$	SiO_2	25	6.7	2.60
n-$C_{14}H_{29}NH_3OAc$	SiO_2	25	6.7	3.45
n-$C_{16}H_{33}NH_3OAc$	SiO_2	25	6.7	4.30
n-$C_{18}H_{37}NH_3OAc$	SiO_2	25	6.7	5.15

[a] I.S. = ionic strength.

increase, decrease, or remain unchanged as a result of those same changes. The effect of an increase in the chain length of the hydrophobe, for example, will depend on the orientation of the adsorbed molecules on the surface. If adsorption is perpendicular to the substrate surface, in an approximately close-packed array, an increase in the chain length of a normal alkane will not result in any significant change in the number of moles of surfactant adsorbed per unit area of surface at saturation. Since the cross-sectional area of a perpendicularly adsorbed, straight-chain molecule does not change much with an increase in the number of units in the chain, the absence of a change in adsorption effectiveness is not a surprising result. In addition, the cross-sectional area of most adsorbed hydrophilic groups is greater than that of normal-chain hydrophobic groups, and is therefore the limiting factor in determining the number of surfactant molecules that can be adsorbed per unit area of surface.

If the mode of adsorption is less than perpendicular (L or train-shaped), or if the molecules are slightly tilted, there may be some increase in the effectiveness of adsorption as the length of the alkyl chain is increased. Again, such an effect is not surprising when one considers that greater dispersion force interaction resulting from a larger alkyl group can lead to greater lateral interactions among surfactant molecules, making possible a greater packing density for the longer chains. A similar effect may be seen when one considers the density, surface tension, cohesive energy density, and many other properties of the homologous series of n-alkanes from C_8 to C_{18} (Figure 10.13).

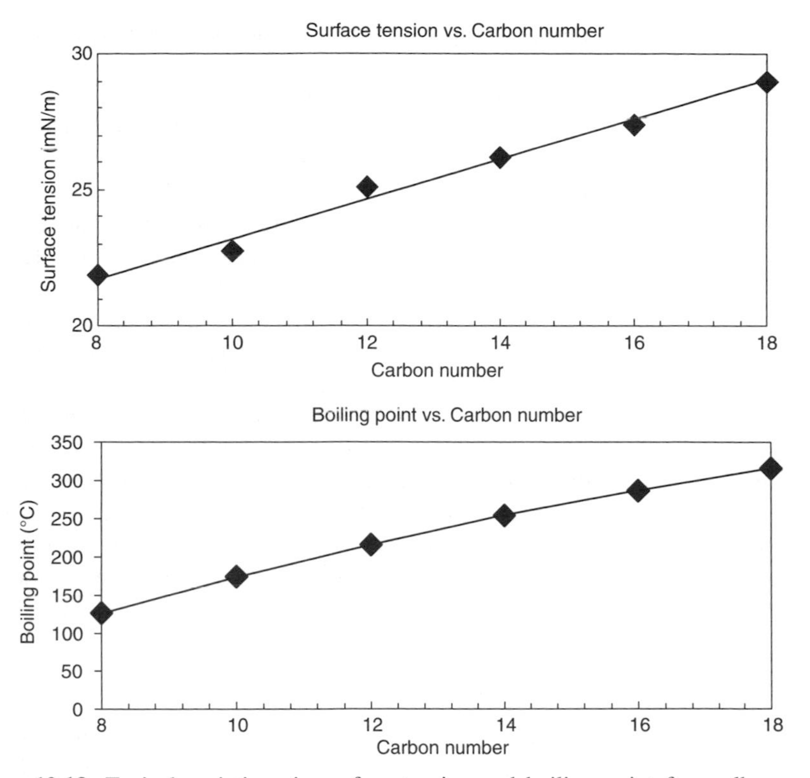

Figure 10.13. Typical variations in surface tension and boiling point for *n*-alkanes as a function of carbon number.

If the orientation of surfactant hydrophobes is parallel to the surface (e.g., as in trains), as may occur with materials that have two hydrophilic groups at either end of the molecule or contain aromatic groups that can be polarized, then the effectiveness of adsorption may be found to decrease as the chain length is increased. In a parallel orientation, a greater chain length will obviously increase the area per molecule required by the adsorbed molecule, and full surface coverage will be attained by the adsorption of fewer moles of surfactant.

While the adsorption of charged surfactants onto charged surfaces is relatively straightforward, that of nonionic species requires a bit more flexibility in interpretation. Such materials may adsorb by mechanisms significantly different from those operative in the case of ionic materials. In general, charged surfaces that also contain hydroxyl and carboxyl groups will adsorb nonionic surfactants by hydrogen bonding or by the acid–base type of interaction between the surface group and oxygen atoms in the nonionic hydrophile. Solid surfaces that contain metal oxides such as silica can interact in the opposite sense, that is, with the surface oxygen serving as the acceptor or base in the interaction.

By far the most common nonionic hydrophile is the polyoxyethylene chain. Although there has not been a great deal published concerning the adsorption of these materials at solid surfaces, it appears that the systems are very sensitive to the length of the ethylene oxide chain. Materials with relatively short POE chains tend to follow Langmuir-type adsorption isotherms in which both the efficiency and the effectiveness of adsorption decrease as the chain length is increased. Those with intermediate chains are found to adsorb by multilayer formation, while the materials containing longer chains tend not to adsorb at all.

In much the same way that changes in pH can affect the adsorption characteristics of a solid surface, so the extent and manner of adsorption can be altered by changing the nature of the surfactant molecules, especially those containing weak acid or basic groups such as carboxylic acids, nonquaternary ammonium compounds, and amphoteric molecules. In such cases, changes in solution pH may convert the surfactant from an ionic species capable of binding by ion exchange or ion binding mechanisms, to an uncharged material that can interact only through hydrogen bonding, acid–base, or dispersion forces. Solution pH changes can also alter the adsorption characteristics of nonionic surfactants containing POE or other linkages that can be protonated at low pH. At low pH, the ether linkages in POE-containing surfactant materials, in sugars, and in polyglycidols can be protonated to yield positively charged sites that will bind strongly with negative sites on the solid surface.

10.5.2. Adsorption by Uncharged, Polar Surfaces

Adsorption onto polar, uncharged surfaces occurs primarily through hydrogen bonding, acid–base, and dispersion force interactions. Any hydrogen bonding or acid–base interactions between surfactant and solid surface require that the hydrophile contain a group capable of participating in such interactions. For example, head groups that are derivatives of strong acids or bases such as sulfonic acid salts, sulfate esters, and quaternary ammonium ions would not be expected to rely greatly on such mechanisms for adsorption. Groups such as carboxylic acids, on the other hand, can interact with materials containing basic surface groups such as polyesters and polyamides. If the solid surface has –OH or –NH groups that can act as proton donors, it can interact with ether linkages such as in polyoxyethylenes. Under some circumstances, the adsorption of nonionic POE surfactants onto polyesters and polyamides exceeds that of anionic materials by a factor of > 2. Nonionic surfactants derived from straight-chain alcohols and POE are found to adsorb onto surfaces such as cotton in a close-packed monolayer with the molecules parallel to the substrate surface. It is also usually found that if the length of the POE chain is increased, both the efficiency and the effectiveness of adsorption decreases. An increase in the length of the hydrophobic chain, on the other hand, produces an enhancement in the efficiency of adsorption.

Because of the lack of charge groups in the polar materials, such factors as pH and electrolyte content would be expected to have a less pronounced effect on adsorption than in the case of charged surfaces. At extremes of pH, however, there always exists the possibility of producing charges through the protonation of –OH, –NH, or SiOH groups. In addition, the presence of high concentrations of neutral electrolytes, while having no effect in the sense of electrostatic or electrical double-layer interactions, may decrease the solubility of a surfactant and increase its interactions with the solid surface.

10.5.3 Surfactants at Nonpolar, Hydrophobic Surfaces

Because of the nature of nonpolar, hydrophobic surfaces, initial adsorption will occur almost exclusively by dispersion interactions between the surface and the hydrophobic tail of the surfactant. The orientation of the adsorption, therefore, will be with the tail on the surface and the hydrophile directed toward the solution. The efficiency and effectiveness of adsorption will be dependent largely on the size and nature of the hydrophobe; a lesser role was played by the hydrophile. Particularly important from the hydrophilic standpoint will be the extent of mutual repulsion among neighboring head groups, which may affect both the efficiency and the effectiveness of adsorption. Any condition that affects the magnitude of such electrostatic interactions (e.g., high electrolyte content) will also be expected to alter adsorption in charged surfactant systems. In the case of nonionic surfactants, where electrostatic interactions are absent and the materials normally consist of multicomponent mixtures, the role of the head group in determining the form of the adsorption isotherm is more complicated. In fact, the adsorption process for nonionic POE surfactants may act as a form of chromatography in that selected components in the mixture may be adsorbed, leaving a solution of distinctly different character.

As might be expected, the nature of the hydrophobic group (the degree of branching, unsaturation, polar substitution, the presence of aromatic groups, etc.) will play a major role in the ultimate adsorption characteristics of the system, primarily as a result of its effects on the conformation of the hydrophobic chain and its interaction with the solvent and the solid surface. Because of the relatively weak forces operative in nonpolar adsorption processes, the removal of species adsorbed in that way might be expected to be relatively easy. In fact, complete desorption in such systems usually is exceedingly difficult, and heroic measures are required to ensure complete removal of the surfactant.

10.6. SURFACTANT ADSORPTION AND THE CHARACTER OF SOLID SURFACES

When a surfactant is adsorbed onto a solid surface, its effect on the character of that surface will depend largely on the dominant mechanism of adsorption. For a highly charged surface, if adsorption is a result of ion exchange, the electrical nature of the surface will not be altered significantly, although its wetting characteristics relative to water may be altered. If, on the other hand, ion pairing becomes important, the potential at the Stern layer will decrease until it is completely neutralized. In a dispersed system stabilized by electrostatic repulsion, such a reduction in surface potential will result in a loss of stability and eventual coagulation or flocculation of the particles.

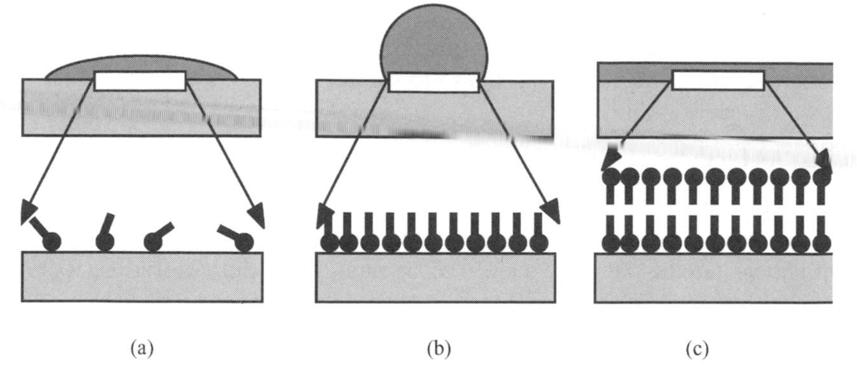

Figure 10.14. Surface charge reversal by surfactant adsorption: (a) native surface (charges omitted for clarity); (b) complete surface charge neutralization; (c) in excess of surfactant, charge reversal by bilayer adsorption.

In addition to the electrostatic consequences of specific charge–charge interactions, surfactant adsorption by ion exchange or ion pairing results in orientation of the molecules with their hydrophobic groups toward the aqueous phase; therefore the surface becomes more hydrophobic and less easily wetted by that phase. Once the solid surface has become hydrophobic, it is possible for adsorption to continue by dispersion force interactions. When that occurs, the charge on the surface will be reversed, acquiring a charge opposite in sign to that of the original surface, because the hydrophilic group will now be oriented toward the aqueous phase (Figure 10.14). In a system normally wetted by water, the adsorption process reduces the wettability of the solid surface, making its interaction with other less polar phases (e.g., air) more favorable. Industrially, the production of a hydrophobic surface by the adsorption of surfactant lies at the heart of the froth flotation process for mineral ore separation. Because different minerals have different surface charge characteristics, leading to differences in adsorption effectiveness and efficiency, it becomes possible to obtain good separation by the proper choice of surfactant type and concentration.

Charge reversal cannot, of course, occur on adsorption of nonionic surfactants. However, the character of the surface can be altered significantly with respect to its wettability by aqueous or nonpolar liquids.

The adsorption of surfactants onto a clean nonpolar surface must occur with the hydrophilic group oriented outward into the aqueous phase. Adsorption, therefore, will always result in an increase in the hydrophilic character of the surface. Such action is responsible for the generally increased dispersibility of materials such as carbon black in aqueous surfactant systems, and the stability of aqueous latex polymers in paints. The action of surfactant adsorption onto colloidal surfaces can be useful to destabilize as well as stabilize systems. It may be useful, for example, to "break" an aqueous dispersion, to isolate the dispersed material, or to facilitate the process of separating dispersed solids in the sewage treatment process, although polymers and polyvalent cation salts are most commonly employed in such instances because of their ability to "bridge" between particles and their dramatic effects on the electrical double layer, as well as their lower cost.

Although surfactant adsorption and its effect on solid surface properties often are discussed in terms of colloidal systems, the same results can be of technological importance for macrosurfaces, especially in the control of the wetting or nonwetting properties of materials in waterproofing, detergency, lubrication, the control of fluid flow through porous media (crude oil production), and corrosion control. Almost any process or product that involves the interaction of a solid phase and a liquid phase will be affected by the process of surfactant adsorption; thus the area represents a major segment of the technological application of surfactants.

10.7. WETTING AND RELATED PHENOMENA

As indicated in the preceding sections, the adsorption of surfactants at solid–liquid interfaces can play a significant role in determining the nature of the interactions between solvent and solid, and among solid surfaces, especially as related to a phenomenon such as colloidal stability. A similar role can be played by surfactants on essentially infinite surfaces related to wetting, spreading, adhesion, and lubrication. Although the basic phenomena are the same for the wetting of extended surfaces and the stabilization of colloidal particles, a number of concepts are more uniquely applied to the more extended surfaces.

While the term "wetting" may conjure up a fairly simple image of a liquid covering a surface, from a surface chemical standpoint, the situation is somewhat less clear-cut. Three classes of wetting phenomena can be defined on the basis of the physical process involved: adhesion, spreading, and immersion (Figure 10.15). The distinctions among the three may seem subtle, but they can be significant from a thermodynamic and phenomenological point of view.

"*Adhesion wetting*" refers to the situation in which a solid, previously in contact with a vapor phase, is brought into contact with a liquid phase. During the process, a specific area of solid–vapor interface, A, is replaced with an equal area of solid–liquid interface (Figure 10.15a). The free–energy change for the process is given by

$$-\Delta G = A(\sigma_{SA} + \sigma_{LA} - \sigma_{SL}) \tag{10.4}$$

where the σ's refer to the solid–air (SA), liquid–air (LA), and solid–liquid (SL) interfacial energies. The quantity in parentheses in Eq. (10.4) is known as the *thermodynamic work of adhesion*, W_a, and the equation is that of Dupré. From the equation, it is clear that any decrease in the solid–liquid interfacial energy σ_{SL} will produce an increase in the work of adhesion (and a greater energy decrease), while an increase in σ_{SA} or σ_{LA} would reduce the energy gain from the process.

Spreading applies to the situation in which a liquid (L_1) and the solid are already in contact and the liquid spreads to displace a second fluid (L_2, usually air) as illustrated in Figure 10.15b. During the spreading process, the interfacial area between the solid and L_2 is decreased by an amount A, while that between the solid and L_1

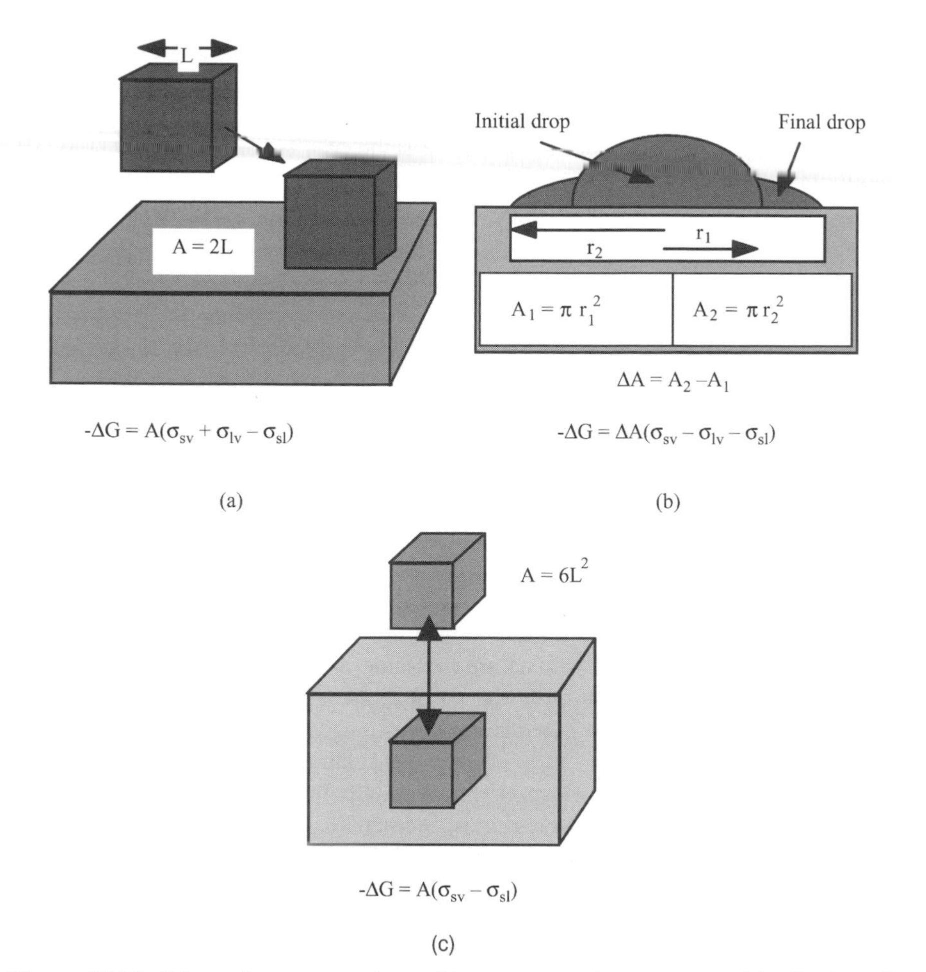

Figure 10.15. Schematic representations of important wetting processes: (a) adhesion; (b) spreading; (c) immersion.

increases by an equal amount. The interfacial area between L_1 and L_2 also increases during the process. The change in interfacial area in each case will be the same, so that the total decrease in the energy of the system will be

$$-\Delta G = A(\sigma_{SL_2} - \sigma_{SL_1} - \sigma_{1/2}) \tag{10.5}$$

where $\sigma_{1/2}$ is the interfacial tension between fluids 1 and 2. If the term in parentheses, defined as the spreading coefficient S (see Chapter 8) is positive, then L_1 will spontaneously displace L_2 and spread completely over the surface (or to the greatest extent possible). If S is negative, the spreading process as written will not proceed spontaneously.

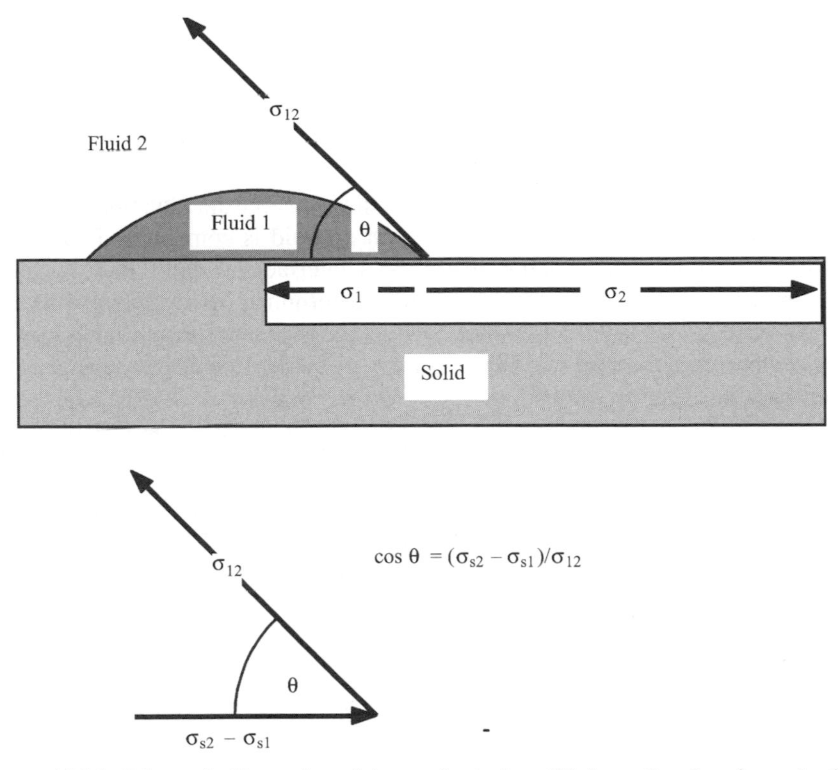

Figure 10.16. Schematic illustration of the mechanical equilibrium of surface forces leading to contact angle formation as given by Young's equation.

In a system in which a liquid is spreading over a second liquid phase, it is possible to directly measure the values for computing the value of *S*. When a solid surface is involved, the value of σ_{SL_2} is not directly available from experiment, so that an indirect route must be found to evaluate the interactions among the three component phases. The approach normally taken for such a determination is to measure the contact angle, θ, which the liquid makes with the solid of interest.

Figure 10.16 shows the general diagram for the contact angle of a liquid L_1 on a solid substrate in the presence of a second fluid L_2. At equilibrium, the contact angle measured through the liquid drop (L_1) is related to the interfacial energies between the various components through Young's equation

$$\sigma_{1/2} \cos\theta = \sigma_{SL_2} - \sigma_{SL_1} \tag{10.6}$$

or

$$\cos\theta = \frac{\sigma_{SL_2} - \sigma_{SL_1}}{\sigma_{1/2}} \tag{10.7}$$

A combination of Eq. (10.7) and the equation for the spreading coefficient S gives

$$S = \sigma_{1/2}(\cos\theta - 1) \tag{10.8}$$

It is clear from Eq. (10.8) that for $\theta > 0$, S cannot be positive or zero, and spontaneous spreading will not occur.

The third type of wetting, immersion wetting, covers the situation in which a solid substrate not previously in contact with a liquid is completely immersed in liquid L_1, completely displacing all solid–L_2 interfaces (Figure 10.15c). In this case, the free-energy change at equilibrium is determined by two factors: the component related to the solid–air interface $A\sigma_{SL_2}$ and that of the solid–liquid interface $A\sigma_{SL_1}$, where A is the total surface area of the solid. The free-energy change is then given by

$$-\Delta G = A(\sigma_{SL_2} - \sigma_{SL_1}) \tag{10.9}$$

From these relationships for wetting processes, it is clear that the interfacial energies between a solid and any contacting liquid, and the interfacial tension between the liquid and the second fluid (usually air), control the manner in which the system will ultimately perform. The ability to alter one or several of those surface energy components makes it possible to manipulate the system to attain the wetting properties desired for a given system. It is through the action of surfactants at any or all of those interfaces that such manipulation is usually achieved. We now turn our attention more specifically to the role of surfactant structure in the alteration and control of the wetting process.

10.7.1. Surfactant Manipulation of the Wetting Process

Because of its high surface tension ($\approx$72 mN/m) relative to most covalent solids (see Table 10.1), water does not spontaneously spread or wet most such materials. As pointed out earlier, for such spreading to occur, the spreading coefficient S must be positive; that is, $\sigma_{SL_2} > (\sigma_{SL_1} + \sigma_{1/2})$ [from Eq. (10.5)]. The addition of a surfactant to lower the surface tension of the aqueous phase $\sigma_{1/2}$, and possibly σ_{SL_1}, will often result in improved wetting of the solid substrate. Such will not always be the case, however. As discussed above, a surfactant may adsorb onto a solid surface with one of several molecular orientations; the predominant ones will be with the head group either pointing into the solution phase or intimately associated with a surface group on the solid. If the orientation is with the head group exposed, the value of σ_{SL_1} as well as $\sigma_{1/2}$ will be reduced and S will become more positive. If the surfactant is oriented with the hydrophobic tail exposed, σ_{SL_1} will be increased and spreading wetting will become less favored.

The penetration of water into a porous solid or fabric can also be variously affected by the lowering of $\sigma_{1/2}$ by surfactant addition. The equation for the pressure forcing the penetration of liquid into capillaries due to surface curvature is

$$\Delta P = 2\sigma_{1/2}\frac{\cos\theta}{r} = \frac{(2\sigma_{SL_2} - \sigma_{SL_1})}{r} \tag{10.10}$$

where r is the effective radius of the capillary and θ is the contact angle of the liquid at the $S/L_1/L_2$ three-phase contact line. For $\theta > 0$, the value of ΔP will depend on the relationship $\sigma_{SL_2} - \sigma_{SL_1}$, so that any change in the surface tension $\sigma_{1/2}$ that is not accompanied by a change in σ_{SL_1} will result only in an increase in cos θ. The lowering of the liquid surface tension will not change ΔP and therefore will not affect pore penetration. If, on the other hand, $\theta = 0$, Eq. (10.10) reduces to

$$\Delta P = \frac{2\sigma_{1/2}}{r} \tag{10.11}$$

and any reduction of $\sigma_{1/2}$ will reduce the pressure, leading to liquid penetration.

It should be fairly obvious from the foregoing discussions of the complexities of surfactant adsorption and wetting phenomena that clear, totally unambiguous rules relating surfactant structure to its effects on wetting are difficult to define. Each specific case must be considered carefully to ensure that the effects of the various possible interfacial interactions are incorporated. As mentioned in Chapter 9, one method for correlating surfactant structure with wetting characteristics is through the use of the hydrophile–lipophile balance (HLB). As a rule of thumb, surfactants with HLB numbers in the intermediate range of 7–9 exhibit better wetting characteristics for aqueous solutions on most solid surfaces that those with higher or lower values. As is so often necessary in the area of surfactant activity, however, due care must be taken in trying to apply even the simplest such rules.

Perhaps the most widely used test for the evaluation of the wetting power of surfactant is the Draves wetting test, in which a piece of cotton cloth with appropriate weights attached is placed on the surface of a surfactant solution and the time for complete wetting or submersion of the sample at a given surfactant concentration, temperature, electrolyte content, and other parameter(s) is determined. A standard set of conditions for the test is water of low ionic strength (< 300 ppm Ca^{2+}) at 25°C, and a surfactant concentration of 0.1% by weight. Typical wetting times for several common surfactants are given in Table 10.3. Similar tests have been developed for the wetting of powders.

From the data in the literature, it appears that optimal wetting characteristics are obtained when a normal chain hydrophobic group has a length of 12–14 carbons. Variations in structure that alter the effective chain length for most surface-active properties will also be reflected in the wetting properties. Branching in the hydrocarbon chain, for example, reduces the effect of the branched carbons to approximately two-thirds that of unbranched atoms. The optimal chain length for molecules with internally located hydrophilic groups may be increased by one or two carbons. Carbons located between the primary head group and a second polar group will usually contribute approximately half of the effect seen for similar groups in the mainchain.

It has been found that surfactants with symmetrically located internal head group substitution and ortho–substituted alkylbenzene sulfonates are better wetting agents than are their straight-chain and para-substituted analogs. The enhanced wetting activity of the nonlinear materials is generally associated with their more compact structure, allowing for faster diffusion to the solid/liquid interface and

TABLE 10.3. Wetting Times for Several Common Surfactants

		Wetting Time (sec)	
Surfactant	Concentration (wt. %)	Water	300 ppm Ca^{2+}
n-$C_{12}H_{25}SO_4^-Na^+$	0.025	>300	—
n-$C_{12}H_{25}SO_4^-Na^+$	0.05	39.9	—
n-$C_{12}H_{25}SO_4^-Na^+$	0.10	7.5	—
sec-n-$C_{14}H_{29}SO_4^-Na^+$	0.063	19.4	—
sec-n-$C_{15}H_{31}SO_4^-Na^+$	0.063	14.0	—
sec-n-$C_{16}H_{23}SO_4^-Na^+$	0.063	22	—
sec-n-$C_{17}H_{35}SO_4^-Na^+$	0.063	25	—
sec-n-$C_{18}H_{37}SO_4^-Na^+$	0.063	39	—
$C_{12}H_{25}C_6H_4SO_3^-Na^+$	0.125	6.9	—
n-$C_{10}H_{21}CH(CH_3)C_6H_4SO_3^-Na^+$	0.10	10.3	80
n-$C_{12}H_{25}CH(CH_3)C_6H_4SO_3^-Na^+$	0.10	30	>300
n-$C_{14}H_{29}CH(CH_3)C_6H_4SO_3^-Na^+$	0.10	155	>300
p,t-$C_8H_{17}C_6H_4(OCH_2CH_2)_5OH$	0.05	25	—
p,t-$C_8H_{17}C_6H_4(OCH_2CH_2)_8OH$	0.05	25	—
p,t-$C_8H_{17}C_6H_4(OCH_2CH_2)_9OH$	0.05	25	—
p,t-$C_8H_{17}C_6H_4(OCH_2CH_2)_{10}OH$	0.05	30	—
p,t-$C_8H_{17}C_6H_4(OCH_2CH_2)_{12}OH$	0.05	50	—

greater relative adsorbing efficiency. The presence of additional polar groups in the molecule (e.g., ester and amide linkages, and POE chains) usually results in a decrease in wetting power. It has been found that in a series of surfactants with the general structure

$$R(OCH_2CH_2)_nOSO_3^-Na^+$$

where R is $C_{16}H_{33}$ or $C_{18}H_{37}$ and $n = 1$–4, the Draves wetting time increased with each added OE group.

Nonionic surfactants such as POE–alcohols and fatty acids will generally pass through a minimum in wetting power, as determined by the Draves test, as the length of the POE chain is increased. For a given hydrophobic chain length, the maximum in wetting will occur for the POE chain length for which the cloud point lies just above the test temperature. As a rule, POE nonionic surfactants with an effective hydrophobic chain length of approximately 11 methylene units and POE chain length of 6–8 will exhibit optimal wetting power. In addition, POE–alcohols and thioethers are generally found to be superior to equivalent POE fatty acids.

A number of external factors that can affect the wetting power of surfactants include temperature, electrolyte content, pH, and the addition of polar organics and cosurfactants. An increase in solution temperature, for example, will generally reduce the wetting power of most ionic surfactants, presumably as a result of greater solubility and a reduced tendency for adsorption at interfaces. The optimal chain length for maximum wetting in a given surfactant series, therefore, increases with the temperature.

The addition of electrolyte will alter wetting characteristics by altering the solution properties of the surfactant. Electrolytes that cause a reduction in the cmc of a surfactant solution will normally produce improvements in wetting power. For example, surfactants with short hydrophobic tails that would show relatively poor wetting characteristics in distilled water will usually exhibit better performance in concentrated electrolyte solutions.

The addition of long-chain alcohols and nonionic cosurfactants has been reported to increase the wetting properties of many anionic surfactants. In the case of cationic surfactants, the presence of POE nonionic materials may reduce wetting power as a result of weak complex formation between the POE chain and the cationic group, and a consequent reduction of the rate of diffusion of the surfactant to the interface.

Solution pH becomes important to wetting characteristics primarily (and not surprisingly) when weakly acidic or basic groups are present in the surfactant molecule. A prime example is that of the α-sulfocarboxylic acids, which show generally better wetting behavior at low pH where the carboxyl group is not ionized.

10.7.2. Some Practical Examples of Wetting Control by Surfactants

The wetting of solid surfaces plays an important role in many technologically important processes, and an understanding of the part played by surfactants in such processes can go a long way toward solving the problem of choosing the proper material for the job. The following sections illustrate some of the principles involved by means of a brief discussion of a few of the most important processes entailing the action of surfactants at solid–liquid interfaces.

10.7.3. Detergency and Soil Removal

Detergency is unquestionably a surface and colloidal phenomenon reflecting the physicochemical behavior of matter at interfaces. Since the field is concerned principally with the removal of complex mixtures of soils and oily mixtures from equally complex solid substrates, it is not surprising that such systems do not lend themselves readily to analysis by the more fundamental theories of surface and colloid science. A rigorous treatment of the current status of detergency theory would constitute a book itself. This section summarizes some of the most important aspects of detergency and illustrates how the chemical structure of the surfactants and other components in a formulation can affect overall performance.

10.7.4. The Cleaning Process

The cleaning of a solid substrate involves the removal of unwanted foreign material from its surface. In this case the term "detergency" is restricted to systems having the following characteristics: (1) the cleaning process is carried out in a liquid medium; (2) the cleaning action is a result primarily of interfacial interactions among the soil, substrate, and solvent system; and (3) the primary process is not solubilization of the soil in the liquid phase, although that may play a minor role in the overall detergent action. Mechanical agitation and capillary action are also important aspects of the overall detergency process, especially when modern stain removal treatments are applied.

While most detergent-related applications are carried out in aqueous systems, the important nonaqueous "dry" cleaning systems also fulfill the foregoing requirements for detergency. In the following discussion, however, all references are to water as the solvent unless noted otherwise.

In detergency, as in many important technical processes, the interaction between solid substrates and dissolved or dispersed materials is of fundamental importance. Surfactants, as a class of materials that preferentially adsorb at several types of interfaces because of their amphiphilic structure, naturally play an important role in many such processes. In most adsorption processes related to detergency, it is the interaction of the hydrophobic portion of the surfactant molecule with the dispersed or dissolved soil and with the substrate that produces detergent action. Such adsorption alters the chemical, electrical, and mechanical properties of the various interfaces and depends strongly on the nature of each component. In the cleaning of textile materials with anionic surfactants, for example, the adsorption of the surfactant onto the fabric and the soil introduces electrostatic repulsive interactions that tend to reduce the adhesion between soil and fiber, lifting the soil and retarding redeposition. The process is illustrated schematically in Figure 10.17. With

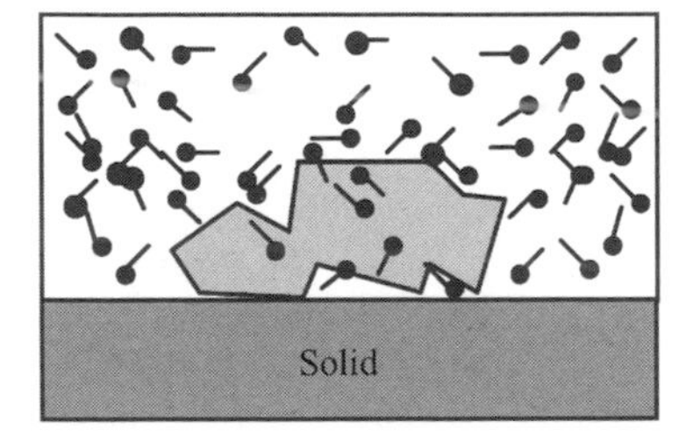

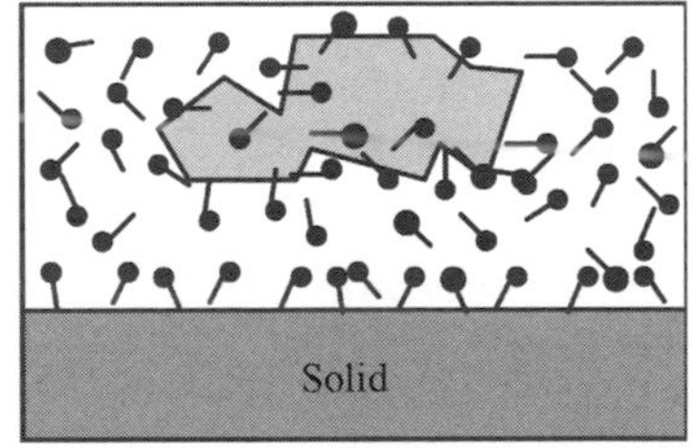

(a)

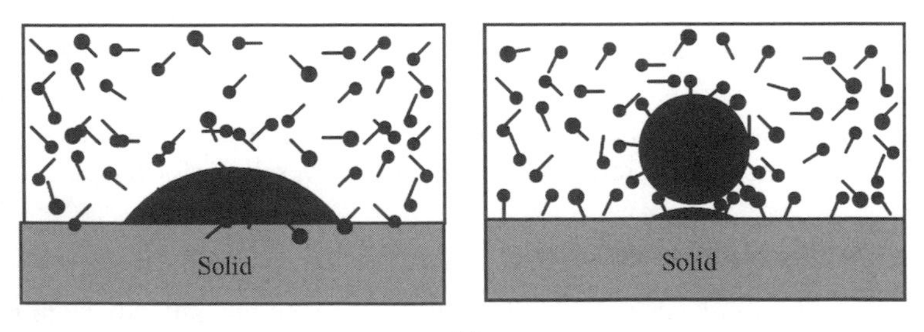

(b)

Figure 10.17. A general mechanism for solid soil removal through surfactant action: (a) solid dirt removal; (b) oily dirt removal.

nonionic surfactants, the mechanism is less clear-cut; however, steric repulsion between adsorbed surfactant layers and solubilization is of primary importance.

10.7.5. Soil Types

In general, there are two types of soil encountered in detergency situations: liquid, oily substances, and solid particulate material. Many stains on textiles such as blood, wine, mustard, catsup, and the like involve proteins, carbohydrates, and relatively high-molecular-weight pigmentlike materials that pose special problems in terms of the interfacial interactions involved. The interactions of each class of soil or stain with the solid substrate can be quite complex, and the mechanisms of soil removal may be correspondingly complex.

Solid soils may consist of various mineral compositions, carbon (soot) having a variety of surface characteristics, metal oxides and pigments, and other compounds. Liquid soils may contain skin fats (sebum), fatty acids and alcohols, vegetable and mineral oils, synthetic oils, and liquid components of creams and cosmetics. As with the solid soils, the surface chemical characteristics of the liquid soils can vary widely. It is not surprising, then, that the derivation of a comprehensive theory for detergency that can characterize every situation poses a challenge. There are some basic similarities between the two soil types, however, which allow for some generalization, while each class will also have its special requirements for efficient detergency.

The adhesion of both solid and liquid soils to solid substrates will, to a greater or lesser extent, result from dispersion and van der Waals interactions. Adsorption due to other polar forces such as acid–base interactions or hydrogen bonding is also usually of minor importance except where highly polar soils and substrates are involved. Adhesion by electrostatic interactions is generally less important for liquid soil systems, but it can become important, and in fact, determining, in the cases of some mineral and biological soils. When electrostatic forces are involved, resulting soil stain can be very difficult to remove by normal cleaning processes.

Where soil adsorption predominantly a result of dispersion and van der Waals interactions, nonpolar materials such as carbons and hydrocarbon oils can be especially difficult to remove from hydrophobic surfaces such as polyesters. More hydrophilic soils such as clays, fatty acids, and other material, on the other hand, can be more difficult to remove from hydrophilic surfaces such as cotton. Mechanical forces can also inhibit cleaning action, especially in fibrous materials with particulate soils, as a result of entrapment of the particles in the fibers. It is obvious, then, that the cleaning process can be extremely complex, and optimum results may be possible only for specifically defined systems. Like the universal solvent, the universal detergent is, in all likelihood, beyond our technological reach.

10.7.6. Solid Soil Removal

The removal of solid, particulate soils from a substrate in an aqueous cleaning bath involves the wetting of the substrate and soil by the cleaning bath followed by adsorption of surfactant and/or other components at the substrate–liquid and soil–liquid interfaces (Figure 10.17). The result is (ideally) a reduction of the energy required to separate the two phases and the creation of an electrostatic or steric barrier to retard or prevent redeposition onto the substrate.

The adhesion of small particles to a substrate may be significantly reduced by immersion in water if the interactions at the interface are favorable, as indicated by the spreading coefficient, S [Eq. (10.5)]. The presence of the water brings about the formation of an electrical double layer at each S/L interface. If both soil and substrate are negatively charged, electrostatic repulsion will reduce or eliminate adhesion forces and lead to soil removal. In addition, the presence of water may cause the substrate to swell, further reducing soil–substrate interactions. In many instances, however, the surface forces embodied in the spreading coefficient for water alone are not sufficient to bring about particle–substrate separation. The addition of surfactant can improve the situation in the ways already mentioned, but it is often found that vigorous mechanical action is necessary to make the kinetics of the process acceptable.

It is interesting to note that most commonly encountered solid soils are minerals, which usually carry a net native negative surface charge in aqueous solution. If cationic surfactants are present in the wash solution, specific adsorption through electrostatic attraction will occur, leading to a less favorable situation for soil removal. It would require the formation of a bilayer of adsorbed surfactant to attain the desired electrostatic effects for efficient removal of the soil and to prevent redeposition. For that reason, cationic surfactants are seldom encountered in normal cleaning solutions, except where their bactericidal characteristics are required. Their tendency to adsorb onto anionic substrates (especially fabrics) does have its advantages, however. The production of more hydrophobic surfaces through such adsorption lies behind the utility of cationic amphiphiles as fabric softeners. By adsorbing onto the textile fiber surfaces, such materials reduce the friction between the fabric and the contacting skin to produce the desired soft texture. The hydrophobic nature of the adsorbed surface also reduces the effects of static charge buildup, giving the material a softer, less irritating feel.

The simple act of dispersing soil particles in the cleaning bath has not been found to ensure effective cleaning. There appears to be little correlation between detergency and dispersing power for most surfactants. For example, surfactants that are excellent dispersing aids are often found to have poor detergency characteristics, and vice versa. Increased adsorption onto the soil and substrate in the cases of anionic and nonionic surfactants appears to correlate well with detergency, however, indicating that the effectiveness of a particular surfactant in the cleaning process is related to its effectiveness at separating the soil and substrate. Some of what is known about the relationships between detergency and surfactant structure is discussed below.

10.7.7. Liquid Soil Removal

Like that for solid soils, the first step in the cleaning of oily soils from a substrate is the separation of the two interfaces. Afterwards, the problem becomes that of

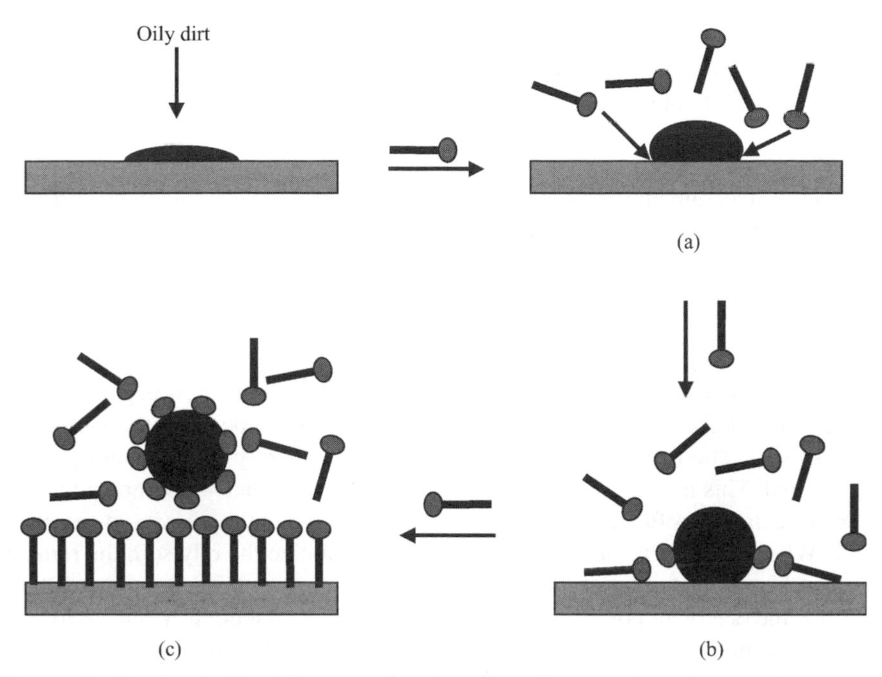

Figure 10.18. The "rollback" mechanism for oily soil removal; surfactant adsorption at O/W and S/W interfaces acts to initiate drop deformation (a), followed by "necking" of the attached drop (b), and eventual detachment (c).

keeping them apart. The primary mechanism for the soil removal process is generally felt to be the so-called "rollback" mechanism illustrated in Figure 10.18. In essence, the process involves the penetration of liquid into the soil–substrate interface by capillary action or as a result of mechanical separation. As the cleaning solution penetrates, the adsorption of surfactant at the soil–solution and solid–solution interfaces decreases the adhesive forces between the two, resulting in an increase in the soil–substrate contact angle and, eventually, in displacement of the soil by the cleaning solution.

Once the soil has been separated from the substrate, it is necessary to prevent its redeposition until it is removed in the rinsing process. There are two general mechanisms for the isolation of oily soils from the substrate: micellar solubilization and emulsification. The solubilization of oily materials in surfactant micelles is probably the most important mechanism for the removal of oily soil from substrates and follows the general tendencies outlined in Chapter 6. It has generally been observed that oily soil removal from textile surfaces becomes significant only above the cmc for nonionic surfactants, and even for some anionic materials with low cmc's. Removal efficiency reaches a maximum at several times that concentration. Since the adsorption of surfactants at interfaces involves the monomeric, rather than the micellar form, while solubilization involves only the micellar form, those results would appear to indicate that in these cases, solubilization is more important than adsorption related effects such as wetting, emulsification, and so on, in the overall cleaning process.

The degree of solubilization of oily soils depends on the chemical structure of the surfactant, its concentration in the bath, and the temperature. At low surfactant concentrations, solubilization occurs in small, roughly spherical micelles, and a relatively small amount of oil can be solubilized. At surfactant concentrations well above the cmc (10–100 times), larger micellar structures that have a greater solubilizing capacity may be encountered, or some mechanism related to microemulsion formation may take over.

In solutions of some ionic surfactants, the working concentration is seldom much above the cmc, so that solubilization may not be a significant factor in oily soil removal. For nonionic surfactants, the extent of solubilization depends on the temperature of the cleaning solution relative to the cloud point of the surfactant, since solubilization of oily materials increases significantly as the cloud point is approached. This may account in part for the observation that soil removal by nonionic surfactants is often at a maximum at temperatures in the vicinity of the cloud point. When insufficient surfactant is present to solubilize the oily soil, the remainder may be suspended by emulsification.

Since the detergent power of many surfactants cannot be directly related to their efficacy as emulsifiers, there exists some question as to the importance of emulsification as a primary soil removal mechanism and redeposition control. Certainly, for efficient solubilization to occur, the area of surfactant solution–soil interface must be maximized, which implies a reduction in the solid substrate–oil interface.

A major criticism of the emulsification mechanism is that, since most detergent class surfactants are not particularly good emulsifiers, emulsified soil droplets will be very unstable, resulting in droplet coalescence and significant redeposition. The rollback process, as well as any possible emulsification processes, will generally be aided by the addition of mechanical energy, although it is doubtful that such added energy would be sufficient to significantly overcome the inherent instability of most detergent–emulsion systems.

10.7.8. Soil Redeposition

The term "redeposition" has already been used several times. Because most cleaning processes are "batch" processes, there will always exist the possibility that soils removed from the substrate will be redeposited onto the surface as a result of a lack of colloidal stability in the dispersed soils. For oily soils removed by solubilization, the process is thermodynamically driven so that it is essentially a one-way street and redeposition will be minimal. Solid soils, on the other hand, cannot be solubilized and redeposition must be retarded by other kinetically controlled means. Emulsified oily soils, where they occur, must be handled similarly.

As pointed out earlier, one major role of surfactants at solid interfaces is to impart a degree of colloidal stability to finely divided particles in aqueous solutions. The adsorption of ionic surfactants at the soil–water and substrate–water interfaces produces an electrical double layer that retards the mutual approach of the interfaces and prevents or at least hinders redeposition. Nonionic surfactants perform the same task by the formation of a steric or entropic barrier to approach, although the efficacy of such a mechanism is probably less than the electrostatic repulsions in most aqueous systems.

10.7.9. Correlations of Surfactant Structure and Detergency

Any correlation between the detergent power of a surfactant and its chemical structure will be complicated by the existence of a wide variety of soil types that may have vastly different interactions with a given surfactant type. It is important, therefore, to specify the exact nature of the system when trying to make any general statements concerning such correlations.

For oily soils, where solubilization is the primary mechanism of soil isolation and ultimate removal, it is reasonably safe to say that any structural characteristic of the surfactant molecule that improves its solubilizing capacity as described in Chapter 6 will tend to improve its performance in such systems. Equally, if soil emulsification is important, surfactants with the proper HLB for emulsification would be expected to have advantages over those lying outside the optimum range for a given oil type. It has been reported that nonionic surfactants perform well in the process of removing and preventing the redeposition of oily soils at lower concentrations than anionic analogs, the reasoning being that their cmc is reached at lower concentrations.

As indicated earlier, the orientation of the adsorbed surfactant molecule at the solid–liquid interface will determine the physical result of the adsorption process. In detergency, the surfactant must orient with the hydrophilic head group toward the aqueous phase or the mechanisms for soil removal and prevention of redeposition will not be operative. For that reason, the detergent activity of a given surfactant in the system will also depend on the polar or ionic nature of the substrate. Both anionic and nonionic surfactants, for example, may exhibit good detergent properties on relatively nonpolar surfaces such as polyesters and nylons. On cottons and cellulose fibers, which are more hydrophilic, anionic surfactants will usually perform better than nonionic materials. It may reasonably be expected that the greater hydrophilicity of such surfaces leads to substantial polar or hydrogen bonding interactions with the POE units of the surfactant and forces its orientation to expose more of the hydrophobic tail to the aqueous phase, or causes the surfactant molecule to lie along the substrate surface (parallel) and thereby reduce the extent of adsorption. Such orientation may increase or at least not sufficiently decrease the interfacial energies at the soil–water and substrate–water interfaces and thereby retard soil removal. As already mentioned, cationic materials are generally less useful than the other types as detergents, but are especially so when the substrate has significant anionic character, leading to a reversed molecular orientation and the formation of a substantially hydrophobic surface.

Clearly, the extent and orientation of adsorption of surfactant molecules onto a solid substrate are of primary importance to their action in the detergency process. Therefore, any alterations in molecular structure that affect such adsorption characteristics should be expected also to affect detergent power. As was shown earlier in the chapter, an increase in the length of a hydrocarbon tail will result in an increase in the efficiency with which that material is adsorbed at the solid–solution interface. Similarly, modifications such as branching and internal location of the hydrophilic group will reduce that tendency. It is generally observed, therefore, that maximum detergent activity for a given carbon chain length and head group will be attained with a normal chain and terminal head group location. A number of studies have confirmed that detergent efficiency increases as the length of the hydrophobic tail is increased and as the head group is moved from internal to terminal locations. The limiting factor in all that, of course, is that the tail cannot become so long that low water solubility becomes a problem.

Although straight-chain hydrophobes with terminal head groups produce optimal detergency under ideal circumstances, the presence of electrolytes and polyvalent cations can reduce the solubility of many materials to the extent that their activity is no longer sufficient for the job. In such situations, it may be found that analogs with the hydrophilic group located internally along the chain will be superior detergents.

The nature of the surfactant head group is important for several reasons, including control of the orientation of surfactant adsorption when charged surfaces are involved and control over the sensitivity of the material to pH, electrolytes, and polyvalent ions. In the case of POE nonionic materials, increases in the length of the POE chain result in a decrease in the efficiency of adsorption on the substrate and a loss in detergent power, all other things being equal. When POE chains are inserted between the hydrophobic tail and an anionic group, such as in POE sulfonates, the material is usually found to be superior in detergent properties to the parent "normal" sulfonate. A number of other trends relating structural characteristics of surfactants to detergency have been determined, but the most generally applicable can be summarized as follows:

1. Within the limits of solubility, detergent power increases with the length of the hydrophobic chain.
2. For a given number of carbon atoms in the hydrophobic tail, maximum detergency is attained with straight rather than branched chains.
3. Terminal positioning of the head group produces superior results within a surfactant series.
4. For nonionic surfactants, optimal detergency is obtained when the cloud point of the surfactant lies just above the solution working temperature.
5. For POE nonionic surfactants, an increase in the length of the POE chain (once sufficient solubility has been attained) often results in a decrease in detergent power.
6. The insertion of a POE chain of 3–6 six units between the hydrophobic tail and the primary ionic head group generally results in superior detergent performance.
7. Characteristics of the substrate such as polarity and electrical charge that lead to adsorption via the head group will result in poor detergent action. That is

especially the case for cationic surfactants interacting with negatively charged surfaces.

10.7.10. Nonaqueous Cleaning Solutions

Although aqueous systems constitute the major portion of surfactant applications in detergency, nonaqueous or "dry" cleaning processes also constitute an important economic factor. Included in such processes are the familiar dry cleaning processes as applied to fine fabrics and articles that are adversely affected by water, as well as processes for the cleaning of various metal and ceramic parts that cannot tolerate significant exposure to moisture.

In such applications, solubility of the surfactant in the nonaqueous solvent is obviously the primary requirement. A wide variety of structures have been found useful in such applications, including nonionic materials such as POE alkylphenols, amides, and phosphate esters, sodium dialkylsulfosuccinates, alkylaryl sulfonates, and petroleum sulfonates. Although extensive experimental data are not available, it appears that the polar head group of the surfactant in such uses is of primary importance, given a hydrophobic tail sufficient to ensure the proper overall solubility.

Since the solvent system for dry cleaning is nonaqueous, the mechanisms for cleaning action will differ in sense, but not in principle, from those in aqueous detergent systems. In general, there are two processes to consider: (1) the adsorption of surfactant at the solid–soil or liquid–soil interface in such a way as to lower the solvent–soil interfacial energy and to facilitate removal and inhibit redeposition and (2) the solubilization of water-soluble soils in the interior of the reversed micelle. The first process requires that the surfactant adsorb with the hydrophobic tail oriented outward toward the nonaqueous solvent, opposite that necessary for good detergent action in water. The second step requires that water and accompanying soils have a high affinity for the polar group on the surfactant. For the prevention or retardation of redeposition, the surfactant relies on its ability to provide a steric barrier between soil and substrate, suggesting that longer hydrocarbon chains would provide better dry cleaning action. Similar mechanisms can be invoked in the application of nonaqueous surfactant systems as so-called drying mixtures with which water can be removed from metal surfaces by the solubilizing action of inverted micelles (Figure 10.19).

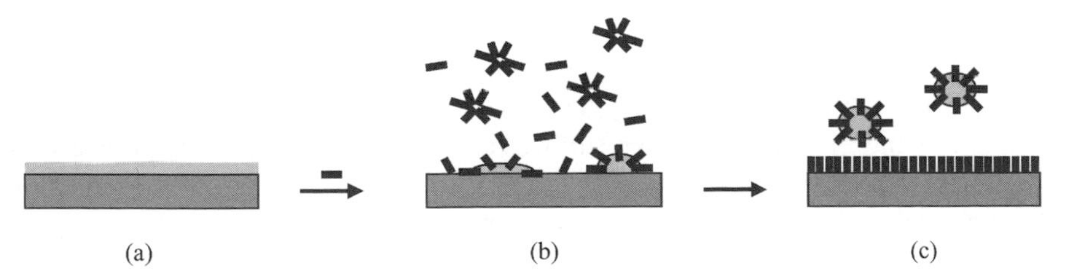

Figure 10.19. Solid surface drying by water solubilization in reversed micelles: (a) wet metal surface; (b) drying surfactant micelles begin to absorb water; (c) dry surface and solubilized waters.

10.8. ENHANCED OIL RECOVERY

Although the maximum recovery of mineral resources from known deposits and reserves has always been a goal of those in the industry, it is only since the onset of the oil crises of the 1970s that vigorous scientific efforts outside the petrochemical industry have become of importance. A review of the fundamental and applied surface chemical literature since the mid-1980s illustrates the rapid growth in the number of patents, scientific publications, books, and reviews that address the problems of resource recovery, and possible solutions. Although an in-depth discussion of the present state of knowledge in the field of secondary and tertiary oil recovery is not possible here, it is useful to introduce the major mechanisms for attaining petroleum recovery enhancement that are related to surfactants.

Initial or primary oil recovery is accomplished primarily by use of the inherent energy of the oil reservoir—that is, the pressure of the gases and volatile hydrocarbons trapped under high pressures and temperatures in the rock formation. For shallow wells, mechanical pumping may be used. Additional recovery may be accomplished by the injection of water or steam into the rock to maintain a high pressure in the system and force additional oil to the surface through production wells. The use of such mechanisms can normally result in the recovery of about 40% of the potential oil in the formation. Beyond that point, more drastic (and more expensive) measures must be employed. Such measures may involve the use of surfactants and polymers for the alteration of the interfacial and rheological properties of the oil deposit and the fluids injected to facilitate movement of the crude toward production wells.

There are four primary mechanisms by which enhanced oil recovery can be attained through the use of surface active additives: (1) the production of very low ($<10^{-3}$ mN/m) interfacial tension between the oil and the aqueous flooding solution, (2) the spontaneous emulsification or microemulsification of the trapped oil, (3) the reduction of the interfacial rheological properties at the oil–aqueous solution interface, and (4) control of the wettability of the rock pores to optimize oil displacement.

A major role of low interfacial tensions in enhanced oil recovery is to facilitate the removal of small oil drops entrapped in pores that have been wetted by water. The interfacial tension between water and a common crude oil is about 30 mN/m, and typical pore sizes may be in the range of 10 mm, so that the capillary pressure, as calculated with Eq. (10.10), retarding flow of the trapped oil will be greater than the normal pressure gradients encountered in oil formations. In discussions of liquid flow through capillary systems, it is common to refer to the "capillary number," which is the ratio of the surface tension of the liquid to its viscosity. A small capillary number implies easier flow through the system, so that a reduction in interfacial tension should produce an improvement in oil extraction.

"Spontaneous emulsification," as the term implies, refers to the formation of small drops of oil in the aqueous flooding solution in the absence of mechanical agitation. Assuming pore sizes of approximately 10 mm, it is clear that the formation of drops of 1–2 mm diameter or less should facilitate the flow of oil in the system. Since mechanical agitation in an oil reservoir is obviously impractical, the emulsification process must require a minimum of energy input. This energy is usually supplied by the diffusion of water-soluble components from the oil phase to the aqueous phase, resulting in the creation of turbulence at the interface and emulsification.

In Chapters 8 and 9, it was noted that one mechanism for the stabilization of emulsions, foams, and other fluid interfaces was the presence of a viscous or elastic interfacial layer. The extraction of oil from porous rock formations with varying pore sizes requires the expansion and contraction of oil–water interfaces. The presence of a highly viscous interfacial layer could greatly inhibit such action. Because oil deposits naturally contain surface-active components that adsorb at the O/W interface, such elastic films are commonly encountered. To counteract their effect, it is necessary to displace the elastic interfacial film with one possessing more favorable interfacial rheological properties.

One of the major areas of research into enhanced oil recovery has been modification of the wetting characteristics of oil-bearing rocks by the addition of surface-active materials. In this case, the goal is to enhance the extent and rate of wetting of the rock by aqueous solutions so that the petroleum deposits can be more efficiently displaced by water pumped into the formation. It may also be that wetting phenomena could be used to assist in mechanically breaking or fracturing the rock formation to produce larger pores and, as a result, facilitate oil removal. Unfortunately, the very nature of oil-bearing rock formations means that there may be great variations in the surface characteristics of the pore surfaces, wide differences in wettability, and significant differences in the requirements for the appropriate wetting agent.

In addition to the four mechanisms for surfactant action in enhanced oil recovery cited above, work continues on the use of polymeric additives to control the rheological properties of the aqueous flooding solutions. In the use of such processes in conjunction with surfactant additives, it is important to consider potential interactions between the polymer and the surfactant. As more surface-active chemicals are employed in the recovery process, control of the formation and breaking of emulsions and microemulsions produced in the recovery process can become important as a postproduction problem.

A new area of research related to oil production is the use of biosurfactants. The use of such materials is so far limited by cost and availability, but work is currently being carried out on the in situ production of such materials by selected microorganisms. Such underground production, while conceptually interesting, carries with it a number of significant problems, not the least of which is finding organisms that can survive and flourish under the harsh environmental conditions found in oil deposits. Nevertheless, hardy organisms are being discovered and/or developed every day, so the not-too-distant future may see significant developments in that area.

In view of the foregoing observations, and in consideration of the sometimes harsh chemical environments present in oil reservoirs, a number of points must be considered in the selection of suitable oil recovery surfactants. Some of those factors are (1) the production of a low O/W interfacial tension; (2) the compatibility of the surfactant with other additives such as polymers; (3) the long-term chemical stability of the surfactant under the conditions encountered in the oil-bearing rock (temperature, pressure, etc.); (4) the activity of the surfactant under the conditions of use, including the salinity or electrolyte content of the aqueous phase; (5) the solubility characteristics of the surfactant in the oil and water phases, including mesophase formation, cloud points, and Krafft temperature; and, of course, (6) economics.

In principle, the physical concepts discussed in earlier chapters and above in this chapter for the adsorption and activity of surfactants at L/L and S/L interfaces will apply equally well to crude oil–water interfaces and porous rock deposits. Unfortunately, the reality of the situation is such that the best laboratory models of oil-bearing rock formations only qualitatively reproduce what is found thousands of feet below the surface. As a result, the general principles that work so well for emulsification and detergency fall short of answering many of the questions that arise in an actual petroleum recovery situation. In addition, because reservoir conditions and crude oil characteristics differ greatly among Texas, Saudi Arabian, and North Sea fields, components effective in one area may perform less well in others. As a result, a great deal remains to be learned before we can take full advantage of the potential for surfactants to increase the availability of petroleum resources to fuel our technological development.

10.9. SUSPENSIONS AND DISPERSIONS

The suspension or dispersion of solid particles in liquid media is an immensely important technological process related to many of the major chemical areas, including foods, pharmaceuticals, paints and inks, cosmetics, and agricultural products. The ability to prepare suspensions of the proper particle size, and to maintain the stability of such dispersions for extended periods of time, quite often involves the use of one or more surfactants. The role of the surfactant may be related to the preparation process or to the long-term stability of the system or both. In any case, the proper choice of surfactant will be important to the ultimate success of the process.

It is usually considered that there are two basic mechanisms for the preparation of solid suspension in liquid media—by *condensation*, in which the particles are built up from basic molecular units (emulsion polymerization, crystallization, etc.), and *dispersion*, in which small particles are formed in the suspending liquid by breaking up or grinding larger solid units. In each case, the presence of a surfactant can have a significant effect on the characteristics of the final product.

In condensation processes, the surfactant may be important in all stages of the process from nucleation through particle growth to ultimate stabilization. The exact role played will depend on the details of the system under consideration. In emulsion polymerization, for example, nucleation may occur in monomer-swollen micelles, so that the size and number of micelles (i.e., the nature and concentration of the surfactant) will ultimately determine the number of particles formed and their final size. In other polymerization systems, nucleation may occur from small, dissolved oligomers, in which case the micelles present solubilize unreacted monomer and function as reservoirs to feed the growing polymer particles. The surfactant will also play a major role in the stability of the system. In suspension polymerization, where nucleation and particle growth definitely occur in an emulsified monomer droplet, the final particle size will depend on the size of the initial monomer drops and therefore the characteristics of the surfactant.

In dispersion processes, a new solid–liquid interface is formed, leading to an increase in the potential energy of the system. One role of the surfactant in such processes is to reduce the interfacial energy at the S/L interface, facilitating the formation of new interface and retarding the aggregation of already formed particles. In porous solids, the surfactant may assist in the dispersion process by improving the wetting of the channels by the liquid, thereby accelerating breakup. Additional roles related to the weakening of solid structure due to adsorption at crystal defects have also been suggested.

The role of surfactants in stabilizing solid suspensions is, again, one of great academic and technological importance. Because of the vast literature available concerning the fundamental and practical aspects of the subject, its pursuit will be left to the interested reader. Suffice it to say that the nature of the surfactant to be used (its adsorption properties, electrical charge characteristics, rheological properties in solution, etc.) should always be considered early in preliminary formulation processes.

PROBLEMS

10.1. If a solid has a high surface energy, would one expect that to result in an increase or decrease in solubility with decreasing particle size? Why?

10.2. How many nearest neighbors are there for a sphere in the surface of a hexagonal closepacked array when the sphere is (a) part of a terrace; (b) part of a monotonic step; (c) adjacent to a kink in a step; (d) isolated atop a terrace?

10.3. What is the particle size of a colloidal silica if 25% of the silicon atoms are on the surface? What is the approximate surface area per gram? Assume a density of 2.3 g/cm^3.

10.4. A fresh mica surface is prepared under three sets of conditions—in air, under argon, and in a vacuum—and the surface energy determined. Will the surface energies determined be equal? If not, rank them in order of increasing value and give your reason(s) for the order chosen.

10.5. A polymeric material is being evaluated for use in prosthetic devices. Initial in vitro tests showed the material to cause no apparent problems of blood compatibility. Long-term animal tests, however, resulted in the formation of dangerous blood clots in the region of the implant. Suggest an explanation for the observed results.

10.6. In a situation of competitive adsorption of two polymeric materials, A and B, from solution, A will be more readily adsorbed than B if (a) the molecular weight of A is greater than that of B; (b) A is more soluble than B; (c) the molecular weight of A is smaller that of than B; (d) all of the above; (e) none of the above.

10.7. A spreading monolayer of camphor can be used to propel a toy boat through the water. The motion is produced by the effect of a monolayer of camphor on the water surface tension at the rear of the boat. Is the propelling effect a result of (a) a permanent increase in σ; (b) a permanent decrease in σ; (c) a transient increase in σ; (d) a transient decrease in σ. Will the effect continue as long as there is camphor available, or will it reach some point at which movement will cease?

10.8. A compound is found to adsorb onto a glass surface in such a way that the resulting adsorbed layer may be either hydrophilic or hydrophobic, depending on the concentration of adsorbate, time of adsorption, and temperature. In all probability, the process(es) involved is (are) (a) monolayer adsorption; (b) random multilayer adsorption; (c) oriented multilayer adsorption; (d) all of these; (e) none of these.

10.9. An air bubble 2×10^{-6} m in diameter is attached to a hydrophobic surface. What is the expected contact angle, θ, given the following data: $\sigma_{LV} =$ 72.5 mN/m, $\sigma_{LS} = 45$ mN/m, and $\sigma_{SV} = 22$ mN/m.

10.10. The surface and interfacial tensions for a series of liquids are given in the table below. On the basis of that information, predict whether n–octanol will spread at the water–mercury interface. Will hexane? If the alcohol spreads at the water–mercury interface, what molecular orientation do you predict for the alcohol?

Interface	σ(mN/m)	Interface	σ(mN/m)
Water–air	72	Mercury–water	375
n-Octanol–air	28	Mercury–octyl alcohol	348
Hexane–air	18	Mercury–hexane	378
Mercury–air	476	Water–octyl alcohol	9
Water–hexane	50		

10.11. The surface tensions of sodium and mercury at 100°C were found to be 220 and 460 mN/cm, respectively, and their contact angles on quartz were measures as 66° and 143°, respectively. Calculate a value for the surface tension (energy) of the quartz sample.

10.12. The contact angle is proportional to $(\sigma_{SV} - \sigma_{SL})$; therefore addition of a surfactant that adsorbs at the SL interface should decrease σ_{SL}, increase the quantity in parentheses, and reduce θ. However, in flotation systems such addition increases θ. Explain what is incorrect or misleading about the opening statement.

BIBLIOGRAPHY

GENERAL READINGS

Adamson, A. W; Gast, A. P., *Physical Chemistry of Surfaces*, 6th ed., Wiley-Interscience; New York, 1997.

Evans, D. F.; Wennerstrom, H., *The Colloidal Domain*, 2nd ed., Wiley-VCH, New York, 1999.

Hiemenz, P. C.; Rajagopalan, R., *Principles of Colloid and Surface Chemistry*, 3rd ed., (revised and expanded), Marcel Dekker, New York, 1997.

Israelachvili, J., *Intermolecular and Surface Forces: With Applications to Colloidal and Biological Systems*, 2nd ed., Academic Press, San Diego, CA, 1992.

Rosen, M. J., *Surfactants and Interfacial Phenomena*, 3rd ed., Wiley, Hoboken, NJ, 2004.

Tanford, C., *The Hydrophobic Effect. Formation of Micelles and Biological Membranes*, 2nd ed., J. Wiley, New York, 1980.

CHAPTER BIBLIOGRAPHIES

Chapter 1—An Overview of Surfactant Science and Technology

Attwood, A. T.; Florence, A. T., *Surfactant Systems, Their Chemistry, Pharmacy, and Biology*, Chapman and Hall, London, 1983.

D. T.; Ginn, M. E.; Ahah, D. J., Eds., *Surfactants in Chemical/Process Engineering*, Surfactant Science Series Vol. 28, Marcel Dekker, New York, 1988.

Rosen, M. J., *Surfactants in Emerging Technology*, Surfactant Science Series Vol. 26, Marcel Dekker, New York, 1987.

Schwuger, M. J., *Detergents and the Environment*, Surfactant Science Series Vol. 65, Marcel Dekker, New York, 1996.

Swisher, R. D., *Surfactant Biodegradation*, Surfactant Science Series Vol. 18, Marcel Dekker New York, 1986.

Chapter 2—The Organic Chemistry of Surfactants

Balzer, D.; Luders, H., eds., *Nonionic Surfactants: Alkyl Polyglucosides*, Surfactant Science Series Vol. 91, Marcel Dekker, New York, 2000.

Surfactant Science and Technology, Third Edition by Drew Myers

Esumi, K.; Ueno, M., eds., *Structure-Performance Relationships in Surfactants*, 2nd ed., Surfactant Science Series Vol. 112, Marcel Dekker, New York, 2003.

Hill, R. M., ed., *Silicone Surfactants*, Surfactant Science Series Vol. 86, Marcel Dekker, New York, 1999.

Holmberg, K., ed., *Novel Surfactants: Preparation, Applications, and Biodegradability*, 2nd ed., Surfactant Science Series Vol. 114, Marcel Dekker, New York, 2003.

Kissa, E., *Fluorinated Surfactants and Repellents*, 2 ed., Surfactant Science Series Vol. 97, Marcell Dekker, New York, 2001.

Kosaric, N., *Biosurfactants: Production, Properties, and Applications*, Surfactant Science Series Vol. 48, Marcel Dekker, New York, 1983.

Lomax, E., *Amphoteric Surfactants*, Surfactant Science Series Vol. 59, Marcel Dekker, New York, 1996.

Nace, V. M., ed., *Nonionic Surfactants: Polyoxyalkylene Block Copolymers*, Surfactant Science Series Vol. 60, Marcel Dekker, New York, 1996.

Richmond, J. M., *Cationic Surfactants: Organic Chemistry*, Surfactant Science Series Vol. 34, Marcel Dekker, New York, 1990.

Rieger, M. M.; Rhein, L. D., Eds., *Surfactants in Cosmetics*, 2nd ed., Marcel Dekker, New York, 1997.

Rosen, M.; J., *Surfactants and Interfacial Phenomena*, 3rd ed., Wiley, Hoboken, NJ, 2004.

Stache, H., *Anionic Surfactants: Organic Chemistry*, Surfactant Science Series Vol. 56, Marcel Dekker, New York, 1995.

Chapter 3—Fluid Surfaces and Interfaces

Adamson, A. W; Gast, A. P. *Physical Chemistry of Surfaces*, 6th ed., Wiley-Interscience, New York, 1997.

Evans, D. F.; Wennerstrom, H., *The Colloidal Domain*, 2nd ed., Wiley-VCH, New York, 1999.

Hartland, S., ed., *Surface and Interfacial Tension*, Surfactant Science Series Vol. 119, Marcel Dekker, New York, 2004.

Hiemenz, P. C.; Rajagopalan, R., *Principles of Colloid and Surface Chemistry*, 3rd ed., (revised and expanded), Marcel Dekker, New York, 1997.

Myers, D. Y., *Surfactant Science and Technology*, 2nd ed., Wiley-VCH, New York, 1992.

Schick, M. J., ed., *Nonionic Surfactants: Physical Chemistry*, Surfactant Science Series Vol. 23, Marcel Dekker, New York, 1987.

Chapter 4—Surfactants in Solution: Monolayers and Micelles

Abe, M; Scamehorn, J. F., eds., *Mixed Surfactant Systems*, Surfactant Science Series Vol. 124, Marcel Dekker, New York, 2004.

Eicke, H. F.; Parfitt, G. D., *Interfacial Phenomena in Apolar Media*, Surfactant Science Series Vol. 21, Marcel Dekker, New York, 1986.

Israelachvili, J., *Intermolecular and Surface Forces: With Applications to Colloidal and Biological Systems*, 2nd ed., Academic Press, San Diego, CA, 1992.

Mittal, K. L.; Shah, D. O., eds., *Adsorption and Aggregation of Surfactants in Solution*, Surfactant Science Series Vol. 109, Marcel Dekker, New York, 2002.

Rieger, M. M.; Rhein, L. D., eds., *Surfactants in Cosmetics*, 2nd ed., Marcel Dekker, New York, 1997.

Rosen, M. J., *Surfactants and Interfacial Phenomena*, 3rd ed., Wiley, Hoboken, NJ, 2004.

Shaw, D. O., *Micelles, Microemulsions, and Monolayers: Science and Technology*, Marcel Dekker, New York, 1998.

Shinoda, K.; Nakagawa, T.; Tamamushi, B.; Isemura, T., *Colloidal Surfactants, Some Physicochemical Properties*, Academic Press, New York, 1963.

Chapter 5—Higher-Level Surfactant Aggregate Structures: Liquid Crystals, Continuous Biphases, and Microemulsions

Bourrel, M.; Schechter, R. S., *Microemulsions and Related Systems: Formulation, Solvency, and Physical properties*, Surfactant Science Series Vol. 30. Marcel Dekker, New York, 1988.

Fendler, J. H., *Membrane Mimetic Chemistry*; Wiley-Interscience, New York, 1982.

Friberg, S. E.; Lindman, B., *Organized Solutions: Surfactants in Science and Technology*, Surfactant Science Series Vol. 44, Marcel Dekker, New York, 1992.

Garti, N.; Sato, K., *Crystallization and Polymorphism of Fats and Fatty Acids*, Surfactant Science Series Vol. 31, Marcel Dekker, New York, 1988.

Garti, N., ed., *Thermal Behavior of Dispersed Systems*, Surfactant Science Series Vol. 93, Marcel Dekker, New York, 2000.

Israelachvili, J., *Intermolecular and Surface Forces: With Applications to Colloidal and Biological Systems*, 2nd ed., Academic Press, San Diego, CA, 1992.

Laughlin, R. G., in Brown, G. H., ed., *Advances in Liquid Crystals*, Vol. 3, Academic Press, New York, 1978.

Morrow, N., *Interfacial Phenomena in Petroleum Recovery*, Surfactant Science Series Vol. 36, Marcel Dekker, New York, 1990.

Prince, L. M., *Microemulsions: Theory and Practice*, Academic Press, New York, 1977.

Rosoff, M., *Vesicles*, Surfactant Science Series Vol. 62, Marcel Dekker, New York, 1996.

Solans, C.; Kunieda, H., eds., *Industrial Applications of Microemulsions*, Surfactant Science Series Vol. 66, Marcel Dekker, New York, 1996.

Chapter 6—Solubilization and Micellar and Phase Transfer Catalysis

Christian, S. D.; Scamehorn, J. F., *Solubilization in Surfactant Aggregates*, Surfactant Science Series Vol. 55, Marcel Dekker, New York, 1995.

Dehmlow, E. V.; Dehmlow, S. S., *Phase Transfer Catalysis*, Weinheim, New York; VCH, New York, 1993.

Jones, R.; Jones, R. A., *Quaternary Ammonium Salts: Their Use in Phase-Transfer Catalysed Reactions*, Academic Press, New York, 2000.

Scamehorn, J. F.; Harwell, J. H., *Surfactant -Based Separation Processes*; Surfactant Science Series Vol. 33, Marcel Dekker, New York, 1989.

Starks, C. M.; Liotta, C. L.; Halpern, M., *Phase-Transfer Catalysis: Fundamentals, Applications, and Industrial Perspectives*, Chapman and Hall, New York, 1994.

Texter, J., ed., *Reactions and Synthesis in Surfactant Systems*, Surfactant Science Series Vol. 100, Marcel Dekker, New York, 2001.

Chapter 7—Polymeric Surfactants and Surfactant–Polymer Interactions

Bailey, F. E.; Koleske, J., *Alkylene Oxides and Their Polymers*, Surfactant Science Series Vol. 35, Marcel Dekker, New York, 1990.

Bender M., ed., *Interfacial Phenomena in Biological Systems*, Surfactant Science Series Vol. 39, Marcel Dekker, New York, 1991.

Goodwin, J. W., *Colloids and Interfaces with Surfactants and Polymers—an Introduction*, Wiley, New York, 2004.

Holmberg, K.; Jonsson, B.; Kronberg, B.; Lindman, B., *Surfactants and Polymers in Aqueous Solution*, 2nd ed., J. Wiley, New York, 2002.

Kwak, J. C. T., ed., *Polymer-Surfactant Systems*, Surfactant Science Series Vol. 77, Marcel Dekker, New York, 1998.

Magdassi, S., *Surface Activity of Proteins: Chemical and Physiochemical Modifications*, Marcel Dekker, New York, 1996.

Malmsten, M., ed., *Biopolymers at Interfaces*, Surfactant Science Series Vol. 110, Marcel Dekker, New York, 2003.

Nace, V. M., ed., *Nonionic Surfactants: Polyoxyalkylene Block Copolymers*, Surfactant Science Series Vol. 60, Marcel Dekker, New York, 1996.

Nnanna, I.; Xia, J., eds; *Protein-Based Surfactants: Synthesis, Physicochemical Properties, and Applications*, Surfactant Science Series Vol. 101, Marcel Dekker, New York, 2001.

Piirma, I., *Polymeric Surfactants*, Surfactant Science Series Vol. 42, Marcel Dekker, New York, 1992.

Radeva, T., ed., *Physical Chemistry of Polyelectrolytes*, Surfactant Science Series Vol. 99, Marcel Dekker, New York, 2001.

Chapter 8—Foams and Liquid Aerosols

Berkman, S.; Egloff, G., *Emulsions and Foams*, Reinhold Publishing, New York, 1961.

Bikerman, J. J., *Foams*, Springer-Verlag, New York, 1973.

Garrett, P. R., *Defoaming*, Surfactant Science Series Vol. 45, Marcel Dekker, New York, 1992.

Morrison, I. D.; Ross, S "*Colloidal Dispersions: Suspensions, Emulsions, and Foams*." John Wiley and Sons, Inc., New York, 2002.

Nguyen, A. V.; Schulze, H. J. "*Colloidal Science of Flotation*". Surfactant Science Series Vol. 118, Marcel Dekker, Inc, New York, 2003.

Prud'homme, R.; Khan, S.-A., eds., *Foams: Theory, Measurements, Applications*, Surfactant Science Series Vol. 57, Marcel Dekker, New York, 1995.

Chapter 9—Emulsions

Becher, P., *Emulsions: Theory and Practice*, 3rd ed., American Chemical Society, New York, 2001.

Becher, P., ed., *Encyclopedia of Emulsion Technology*, Vols. 1–4, Marcel Dekker, New York, 1983–1987.

Evans, D. F.; Wennerstrom, H., *The Colloidal Domain.*, 2nd ed., Wiley-VCH, New York, 1999.

Sjoblom, J., *Emulsions and Emulsion Stability*, Surfactant Science Series Vol. 61, Marcel Dekker, New York, 1996.

Chapter 10—Solid Surfaces and Dispersions

Berg, J. C., *Wettability*, Surfactant Science Series Vol. 49, Marcel Dekker, New York, 1993.

Cutler, W. G.; Kissa, E., *Detergency*, Surfactant Science Series Vol. 20, Marcel Dekker, New York, 1986.

Evans, D. F.; Wennerstrom, H., *The Colloidal Domain.*, Wiley—VCH, New York, 1994.

Index

Surfactant Science and Technology, Third Edition by Drew Myers